Advances in 21st Century Human Settlements

Indexed by SCOPUS

This Series focuses on the entire spectrum of human settlements – from rural to urban, in different regions of the world, with questions such as: What factors cause and guide the process of change in human settlements from rural to urban in character, from hamlets and villages to towns, cities and megacities? Is this process different across time and space, how and why? Is there a future for rural life? Is it possible or not to have industrial development in rural settlements, and how? Why does 'urban shrinkage' occur? Are the rural areas urbanizing or is that urban areas are undergoing 'ruralisation' (in form of underserviced slums)? What are the challenges faced by 'mega urban regions', and how they can be/are being addressed? What drives economic dynamism in human settlements? Is the urban-based economic growth paradigm the only answer to the quest for sustainable development, or is there an urgent need to balance between economic growth on one hand and ecosystem restoration and conservation on the other – for the future sustainability of human habitats? How and what new technology is helping to achieve sustainable development in human settlements? What sort of changes in the current planning, management and governance of human settlements are needed to face the changing environment including the climate and increasing disaster risks? What is the uniqueness of the new 'socio-cultural spaces' that emerge in human settlements, and how they change over time? As rural settlements become urban, are the new 'urban spaces' resulting in the loss of rural life and 'socio-cultural spaces'? What is leading the preservation of rural 'socio-cultural spaces' within the urbanizing world, and how? What is the emerging nature of the rural-urban interface, and what factors influence it? What are the emerging perspectives that help understand the human-environment-culture complex through the study of human settlements and the related ecosystems, and how do they transform our understanding of cultural landscapes and 'waterscapes' in the 21st Century? What else is and/or likely to be new vis-à-vis human settlements – now and in the future? The Series, therefore, welcomes contributions with fresh cognitive perspectives to understand the new and emerging realities of the 21st Century human settlements. Such perspectives will include a multidisciplinary analysis, constituting of the demographic, spatio-economic, environmental, technological, and planning, management and governance lenses.

If you are interested in submitting a proposal for this series, please contact the Series Editor, or the Publishing Editor:

Bharat Dahiya (bharatdahiya@gmail.com) or
Loyola D'Silva (loyola.dsilva@springer.com)

More information about this series at http://www.springer.com/series/13196

Giuseppe T. Cirella
Editor

Human Settlements

Urbanization, Smart Sector Development, and Future Outlook

Springer

Editor
Giuseppe T. Cirella
Faculty of Economics
University of Gdansk
Sopot, Poland

ISSN 2198-2546 ISSN 2198-2554 (electronic)
Advances in 21st Century Human Settlements
ISBN 978-981-16-4030-8 ISBN 978-981-16-4031-5 (eBook)
https://doi.org/10.1007/978-981-16-4031-5

This Springer imprint is published by the registered company Springer Nature Singapore Pte Ltd.
The registered company address is: 152 Beach Road, #21-01/04 Gateway East, Singapore 189721, Singapore

ai miei

Series Editor's Foreword

Human Settlements: Urbanization, Smart Sector Development, and Future Outlook, edited by Prof. Dr. Giuseppe T. Cirella, is a welcome addition to the Scopus-indexed Springer book series, *Advances in 21st Century Human Settlements*. The book spans innovative findings in urban development, energy efficiency, agriculture security, and policy development. It includes a number of important case studies and highlights novel research in the sciences and social sciences. Professor Cirella and the accompanied co-authors neatly stitch these multifaceted ideas into a first-rate compendium that offers insight into human settlements research and associated alternatives.

My first face-to-face meeting with Prof. Cirella was in Bangkok, on 28 December 2019. Just before the onset of the COVID-19 outbreak, he noted he had completed a month-long visiting professorship in China and had arrived in Thailand thereafter. We met over a Siamese lunch and spoke about the book manuscript he submitted earlier in the year to the same book series. He struck me as a thorough gentleman and a knowledgeable professor who is dedicated to his craft. After lunch, he left to catch a bus that would take him to southern Thailand for the new year and then back to Europe where he resides. In 2020, Prof. Cirella completed what he calls the first part of a series of edited books, namely, *Sustainable Human–Nature Relations: Environmental Scholarship, Economic Evaluation, Urban Strategies* [1]. Key findings from this book focused on sustainability thinking by considering how and from where contemporary schools of thought emerged. A more purist three pillar approach to sustainability was applied by examining the human-nature relationship and the alternatives for a cleaner, safer society. The current book is the second part.

Seeking the answers to the enquiries of why and how we live where we live is the founding premise to bettering human settlements and establishing a cohesive interface with the natural world. Understanding the spectrum of different challenges human settlements face is fundamental to our well-being. This is interconnected with the three pillars of sustainability, i.e., the environmental, social, and economic, and the five critical dimensions—also known as the five Ps of sustainable development, i.e., people, planet, prosperity, peace, and partnership [2]. These significant concepts formulate the United Nations Sustainable Development Goals (SDGs) and veer towards a world in harmony with nature. The New Urban Agenda [3], adopted at the United Nations Conference on Housing and Sustainable Urban Development (popularly known as Habitat III) in Quito, Ecuador, on 20 October 2016, "presents a paradigm shift based on the science of cities and lays out standards and principles for the planning, construction, development, management, and improvement of urban areas along its five main pillars of implementation: national urban policies, urban legislation and regulations, urban planning and design, local economy and municipal finance, and local implementation" [4: p. x]. This global agenda also works as an "accelerator" of SDGs, especially that of SDG 11: Make cities and human settlements inclusive, safe, resilient, and sustainable [4, 5].

As the world's population approaches the eight billion mark—the majority which will live in urban areas [6], our ability to adapt to these humanmade environments has come under increasingly close scrutiny. Cities and urban centers often have dynamic economies that build on higher levels of factor productivity. They also feature growing levels of poverty and intra-urban inequality. Cities and urban centers tend to be overcrowded and polluted that results in poor livability [7]. Evolving human settlement patterns, either from historical settlements or newly developed ones, have transformed traditional egalitarian societies into economic classes; this has also led to the widescale urbanization of rural landscapes. The demographic transition and migratory transformation from rural to urban makes it difficult to imagine a world without the city. They make up the centers of political power and act as hubs of modern culture, i.e., the urban system that formulates the skeleton of society.

Against this background, the current book presents 13 well-written contributions that are divided into three parts. Focusing on *Urbanization*, the first part of the book, 'Urban Development Needs', includes four chapters that discuss: urban challenges and future development related to human settlements; adaptive knowledge sharing in turbulent times vis-à-vis urban disaster risk and knowledge management; the disaster risk of human settlements; and research and development within public transport systems. As such, urban settlements encompass a broad range of functional traits—e.g., cities from developed versus developing countries. Nonetheless, understanding the relationships between cities and the surrounding countryside can be measured against external forces. Such forces are inherent to every urban center that influences adjacent regions with its spatio-economic dominance. Agricultural producers, for instance, from the urban region sell many of their products at city markets, and customers from smaller, regional towns come to the main urban center for specialized goods and services. Central to urbanization and asserted by several contributions in this book, issues such as adaptive knowledge sharing, urban disaster risks, and transportation, are looked at in this context.

Smart sector development is divided into two perspectives that focus on energy efficiency and agriculture security. The five chapters featuring in the second part, 'Perspectives in Energy Efficiency and Agriculture Security', examine: energy transition in maritime transport and related solutions and costs; sustainability and renewable energy education for children of the next generation in Israel; fostering sustainable development in regard to green energy policy in the European Union and the United States; shelterbelt planning in agriculture and its application in Bulgaria; and synchronizing agricultural trade regulations with a case study from Subang regency in Indonesia.

Future outlook is explored in the third part, 'Policy Development and Future Scenarios.' It includes five chapters that discuss: environmental safety in the SDGs using a public survey; SDGs, conflict and fragility as an Anglophone crisis in Cameroon; the Pearl's social and environmental failures and related development challenges; and changing dynamics with COVID-19 and its future outlook. This is specific to urban populations and how they affect the development patterns and decision to establish certain infrastructure for a business and sense of belonging. The examination of policy development challenges alongside several forward-thinking future scenarios and case research pieces helps to join together the type of sustainability the SDGs are founded on. The make-up of urban settlement patterns is fast becoming the human settlement formula of present-day. Considering the COVID-19 pandemic, city arrangements, i.e., via circulatory systems, have been forced to rethink commuting patterns, delineate infectious zones (e.g., city lockdowns), and evaluate well-being on a mass-scale.

All in all, this edited volume brings fresh insights on human settlements in relation to three themes: (i) the urban development needs, (ii) energy efficiency and agriculture security, and (ii) future outlook. The various essays in the book present and discuss case studies based on primary and secondary data. The book will be of interest to students and researchers of human settlements and sustainability from around the world.

Bharat Dahiya
Director, Research Center for
Sustainable Development and Innovation
School of Global Studies
Thammasat University
Bangkok, Thailand

References

1. Cirella GT (2020) Sustainable human–nature relations. Springer, Singapore
2. United Nations (2015) Transforming our world: the 2030 agenda for sustainable development. In: Sustainable Development Knowledge Platform https://sustainabledevelopment.un.org/post2015/transformingourworld/publication. Accessed 27 Jun 2021
3. United Nations (2017) The New Urban Agenda: A/RES/71/256. Habitat III and United Nations, Quito
4. UN-Habitat (2020) The new urban agenda illustrated. UN-Habitat, Nairobi
5. United Nations (2021) The 17 Goals. Department of Economic and Social Affairs, Sustainable Development, New York
6. PRB (2020) 2020 world population data sheet. In: Population Reference Bureau https://interactives.prb.org/2020-wpds/. Accessed 26 Jun 2021
7. Economist Intelligence Unit (2021) The global liveability index 2021. https://www.eiu.com/n/campaigns/global-liveability-index-2021. Accessed 27 Jun 2021

Preface

The answers to the questions of why and how people live where they live as well as how they maintain and integrate with one another are fundamental human settlement issues rooted in history and culture. The study of human settlements has deep historical heritage interlinking with the access to resources, the fortification from adversaries, and the mythos of great cities and civilizations. Cities have played a central role in redefining the interface between human beings and nature. They have revolutionized the human experience by taming natural surroundings and building environments that are extremely human-friendly. Often, as a result, this interrelationship is reduced to living exclusively within the confines of cities without experiencing wilderness or the untamed. Studies into human settlements obviously intertwine a multiplicity of fields from the sciences, social sciences, and humanities. The human-nature relationship is at the core of creating a healthy environment, regardless of the philosophical and custodial arguments of which one controls the other. This book is divided into three parts: it looks at urban development trends, explores perspectives in energy efficiency and agriculture security, and considers policy development and future scenarios in human-nature relations. It is a compendium of multidisciplinary work that establishes scientific merit by challenging directions of modernity and offering reference to alternatives.

The first part begins with four contributions that examine urban development trends by focusing on city challenges, adaptive knowledge sharing, disaster risk, and public transport systems. These contributions interconnect the human interface via urbanization, ramifications of settlement location, and how to maintain and integrate settlements, accordingly. Human settlements are comprehensive. They are shaped by human ecology and the relationship between humans and their interaction with their environment. "Human Settlements: Urban Challenges and Future Development" explores urban morphology and landscape ecology as a pretense to a broader examination of the vast scholarship of why humans settle where they settle—with the focus on cities. Urbanization is explored in conjunction with urban environmental pollution and well-being. The chapter elucidates the importance of climate with reference to energy efficiency and food security, and compares cities from the global North versus the global South. In light of the COVID-19 pandemic, informal settlements are also considered. The next two chapters consider catastrophic events and disaster risk management in three high-risk prone areas. "Adaptive Knowledge Sharing in Turbulent Times: Urban Disaster Risk and Knowledge Management" uses adaptive knowledge sharing to show how urban disaster risk reduction and management can effectively mitigate risk and strengthen community disaster resilience. Case research from Typhoon Ketsana that hit Marikina City, Metro Manila, the Philippines in 2009 is looked at in terms of complexity, i.e., channels, behavior and locality, and multiplicity, with reference to social networks and strategies. The following four linkages are identified: disaster prevention and hazard mitigation, disaster preparedness, disaster response, and recovery and rehabilitation. "Understanding the Disaster Risk of Human Settlements: Case Research" examines risk occurrences and vulnerability by analyzing the socio-political circumstances following two other catastrophic natural disasters: the Aceh tsunami in 2004 and the Haiti earthquake in 2010. In this chapter, the pressure and release model is adopted as a framework to discern, i.e., from the events, if the progression of vulnerability and safety in both countries has improved. An examination of the measures implemented to mitigate the impact of disasters and trigger policy-makers to better consider such events is stressed. "Research and Development Within Public Transport Systems" circles back to the cityscape and examines the importance of developing a complete solution to public transport systems, so as to make them accessible, affordable, available, and acceptable. Case research from Vietnam and Ukraine is investigated in an effort to recommend comprehensive public transport strategy and structure.

The second part shifts the focus of human settlements towards two basic survival-oriented sectors: energy and agriculture. Three contributions are dedicated to energy efficiency (i.e., energy transition in maritime transport, renewable energy education, and green energy policy), and two are focused on agriculture security (i.e., shelterbelt planning and synchronizing agricultural trade regulations). "Energy Transition in Maritime Transport: Solutions and Costs" is a technical examination into the technologies and schematics of converting maritime transport into a cleaner and more energy efficient sector. It considers modern human settlements' reliance on shipping and maritime-related services as key provisions of modern life. Technical research examines how to reduce the carbon footprint of the entire transshipment service by utilizing renewable energy sources from different propulsion engines in ships to the use of on-shore power. "Sustainability and Renewable Energy Education: Children of the Next Generation" is an education study from the Holon Institute of Technology in Israel that integrates a practicum for its students to teach elementary pupils about energy consumption, reliance on fossil fuels, and the difference between non-renewable and finite energy. Important pedagogical instruction includes how to turn waste into a resource and what energy conversion and renewable energy mean. "Fostering Sustainable Development: Green Energy Policy in the European Union and the United States" expands upon the concept of sustainable development by looking at climate, environmental pollution, green energy policy, and historical policy developments. Two key parties are considered: the European Union and the United States. The chapter is pertinent to the current global movement to create green deal policy. "Shelterbelt Planning in Agriculture: Application from Bulgaria" is the first of two chapters that focalizes on agricultural security. This chapter explicitly looks at the use of wind protection belts in agriculture throughout Bulgaria. Complimentary climate research is conducted to explore environmental trends and create guidelines for shelterbelts, green system-based structures, and optimal species composition. The agriculture security measures mentioned will assist in protecting agricultural landscapes and aid in piecing together best practices. "Synchronizing Agricultural Trade Regulations: Case Study from Subang Regency" examines government agricultural policy in Indonesia with a specific emphasis on product trading and lack of fairness, i.e., via the exploitation of farmers. Legal arguments are introduced and elucidate possible reasons for the unfairness, e.g., inadequate laws and regulations and an overburdened bureaucracy. Pineapple is used as the example cash crop in an attempt—i.e., to synchronize national, provincial, and city level difficulties—to better develop agriculture security via price stabilization, growth stimulus, and legal protection.

The third part expands the policy development and future outlook of human settlements by considering the United Nations Sustainable Development Goals (SDGs) and reflects on future scenarios. "Environmental Safety in the Sustainable Development Goals: Public Survey" breaks down SDGs in terms of three aspects of environmental safety: energy preference, air pollution, and climate change—via direct and indirect relations. Empirical research from Hungary is then conducted and correlated in terms of the SDGs' environmental safety measures, and policy recommendations are presented. "Sustainable Development Goals, Conflict, and Fragility: Anglophone Crisis in Cameroon" evaluates SGDs in terms of instability, conflict, and fragility. A case study of the Anglophone crisis in Cameroon is looked at in terms of deep-rooted effects of colonialism and a call for international cooperation. Key questions considered include: how do SDGs and their targets address the challenge of Africa's historical instability and fragility? How does long-standing instability and fragility affect the achievability of the goals? With the 2030 SDGs deadline fast-approaching, this chapter digs deeply into the significance of human cooperation and global settlement disparities. "The Pearl's Social and Environmental Failures: Development Challenges" identifies case research from The Gulf region's past construction and communicates those lessons in future development to aid the process of successful innovation. This chapter identifies the social and environmental failures of Qatar's Pearl Island, in order to analyze how MIA Park and Hamad Port overcome those challenges. Wealthy, rapidly developing Middle Eastern countries create desert cities with budgets exceeding billions of dollars to create unique landscapes few places in the world could match. Implemented policy initiatives indicate improvements from the previous decade's short-term thinking. "Changing Dynamics with COVID-19: Future Outlook" concludes with a look at human activities and the speed of excessive development as problematic to human health and sustainability. Discourse into the COVID-19 pandemic is assessed with broad-ranging effects from the economy and agriculture to tourism and transportation. Specifically, climate change is reflected on in terms of agricultural areas and shrinkage in production patterns. Moreover, population growth pressure in terms of resource usage is connected to the accessibility of food and land use. Arguments are constructed to create sensitive areas where human habitation as well as human and economic activities, as well as human and economic activities, should be limited.

The inspiration of this book dates back to the Second International Conference on Sustainability, Human Geography, and Environment. It is complementary to the book, *Sustainable Human-Nature Relations: Environmental Scholarship, Economic Evaluation, Urban Strategies*, from the same book series, *Advances in 21st Century Human Settlements*, published in 2020. As an extension to the three topics of that book, this book integrates urbanization patterns and development trends, smart sector development in energy and agriculture, and stimulates a future outlook for a bigger picture of sustainability and the general betterment of human settlements.

Sopot, Poland

Giuseppe T. Cirella
Faculty of Economics, University of Gdansk

Contents

Editor and Contributors

About the Editor

Prof. Dr. Giuseppe T. Cirella Professor of Human Geography, works at the Faculty of Economics, University of Gdansk, Sopot, Poland. He specializes in development and environmental social science, human geography, and sustainability. His interdisciplinary background also includes socio-political research throughout Eastern Europe, Africa, and China. After completing a Doctor of Philosophy (Ph.D.) at Griffith University, Australia within the Centre for Infrastructure Engineering and Management that, developing a sustainability-based index, he founded the Polo Centre of Sustainability (www.polocentre.org) in Italy. Notably, he has held professorships and scientific positions at Saint Petersburg State University, Saint Petersburg (Russia), Inner Mongolia University of Science and Technology, Baotou (China), Life University, Sihanoukville (Cambodia), and Free University of Bozen, Bozen (Italy). In his early career, he worked with the Canadian International Development Agency in Indonesia as well as with Radarsat International in Brazil.

Contributors

Solomon T. Abebe Polo Centre of Sustainability, Imperia, Italy

Tran N. Anh Faculty of Economics, University of Gdansk, Sopot, Poland

Ifaa Banner Holon Institute of Technology, Holon, Israel

Giuseppe T. Cirella Faculty of Economics, University of Gdansk, Sopot, Poland

Ernest Czermański Faculty of Economics, University of Gdansk, Sopot, Poland

Winifredo Dagli Department of Science Communication, College of Development Communication, University of the Philippines Los Baños, Laguna, Philippines

Yulia Einav Faculty of Engineering, Holon Institute of Technology, Holon, Israel

Rini H. Ekawati School of Business and Management, Bandung Institute of Technology, Bandung, Indonesia

Andrea Farkas National University of Public Service, Budapest, Hungary

Hen Friman Faculty of Engineering, Holon Institute of Technology, Holon, Israel

Herry Hudrasyah School of Business and Management, Bandung Institute of Technology, Bandung, Indonesia

Cengiz Kahraman Faculty of Engineering, Istanbul University-Cerrahpasa, Istanbul, Turkey

Ella Kozemko Faculty of Economics, University of Gdansk, Sopot, Poland

Nurrani Kusumawati School of Business and Management, Bandung Institute of Technology, Bandung, Indonesia

Bryan Joel S. Mariano Department of Geography, College of Social Sciences and Philosophy, University of the Philippines Diliman, Quezon, Philippines

Janos Mika Eszterhazy Karoly University, Eger, Hungary

Ikrom Mustofa Piarea Environment and Technology, Bogor, Indonesia;
Water System and Global Change Group, Wageningen University and Research, Wageningen, the Netherlands

Samuel Mwangi Institute of Political Science, Tübingen University, Tübingen, Germany

Christian Orobello Faculty of Economics, University of Gdansk, Sopot, Poland

Mustika S. Purwanegara School of Business and Management, Bandung Institute of Technology, Bandung, Indonesia

Michael J. Rosciszewski-Dodgson Faculty of Science and Engineering, University of Liverpool, Liverpool, UK

Alessio Russo School of Arts, University of Gloucestershire, Cheltenham, UK

Veselin M. Shahanov Landscape Architecture Department, University of Forestry, Sofia, Bulgaria

Yafa Sitbon Holon Institute of Technology, Holon, Israel

Chelsea R. Spring Faculty of Economics, University of Gdansk, Sopot, Poland

Katerina Streltsova Faculty of Economics, University of Gdansk, Sopot, Poland

Abbreviations

CAS	Complex adaptive systems
CFR	Code of Federal Regulations
CZM	Coastal zone management
DRR	Disarmament, demobilization, and reintegration
DRRM	Disaster risk reduction and management
EEA	European Environment Agency
EPA	United States Environmental Protection Agency
ETS	Emissions trading system
EU ETS	European Union emissions trading system
FAO	Food and Agriculture Organization of the United Nations
GAM	Free Aceh Movement
GDP	Gross domestic product
GHG	Greenhouse gas
GSDP	General Secretariat for Development Planning
HIT	Holon Institute of Technology
ICOR	Incremental capital output ratio
ICT	Information and communication technology
IDPs	Internally displaced persons
IMO	International Maritime Organization
IPCC	Intergovernmental Panel on Climate Change
KM	Knowledge management
NAPs	National Allocation Plans
NASA	National Aeronautics and Space Administration
NGO	Non-Governmental Organization
OECD	Organisation for Economic Co-operation and Development
PAR	Pressure and release
RGGI	Regional Greenhouse Gas Initiative
RPJMD	Regional Medium-Term Development Plan
RPJMN	National Medium-Term Development Plan
SDGs	United Nations Sustainable Development Goals
SNA	Social network analysis
TEU	Twenty-foot equivalent unit

UIP	Unemployment insurance programs
UNDP	United Nations Develop Programme
UNFCC	United Nations Framework Convention on Climate Change
UNHCR	United Nations High Commissioner for Refugees
WHO	World Health Organization

Urban Development Trends

Human Settlements: Urban Challenges and Future Development

Giuseppe T. Cirella, Samuel Mwangi, Katerina Streltsova, Solomon T. Abebe, and Alessio Russo

Abstract Human settlements are comprehensive, i.e., shaped by human ecology and the relationship between humans as a social being and biological organisms and their interaction with their environments. This chapter explores urban morphology and landscape ecology as a pretext to a wider examination of the vast scholarship of why humans settle where they settle—with the focus on cities. The movement away from rural to urban is considered in conjunction with urban energy use, agriculture and food security, and sustainability. Maladaptation to climate change is considered in the context to urban environmental pollution, human health and well-being, and quality of life. Cities have a unique opportunity to advance policies that ensure the energy supply and food production are reliable, affordable, and environmentally sustainable. In terms of energy research, direct effects on people, communities, and countries in terms of economic growth, health, safety, the environment, education, and employment are investigated. Agricultural data is presented from a global perspective with specific land use and land cover specificities. Food security, food health, and food production are interfaced with regional populations and agricultural land use. An overview of cities from the Global North versus the Global South is assessed in terms developmental parameters—including city-to-city climate action. These city variances, specific to developed and developing countries, indicate megacities in the North have relatively high affluent and stable populations while those in the South have rapid expanding and overcrowded ones. Case-specific research into the effects of the COVID-19 pandemic on informal settlements is looked at in terms of direct and indirect impacts. The complexity of these issues signposts different

G. T. Cirella (✉) · K. Streltsova
Faculty of Economics, University of Gdansk, Sopot, Poland
e-mail: gt.cirella@ug.edu.pl

S. Mwangi
Institute of Political Science, Tübingen University, Tübingen, Germany

S. T. Abebe
Polo Centre of Sustainability, Imperia, Italy

A. Russo
School of Arts, University of Gloucestershire, Cheltenham, UK
e-mail: arusso@glos.ac.uk

G. T. Cirella (ed.), *Human Settlements*, Advances in 21st Century Human Settlements,
https://doi.org/10.1007/978-981-16-4031-5_1

types of human settlements and conditions and veers toward piecing together the urban challenges and future development of the twenty-first century.

Keywords Human ecology · Urban morphology · Green infrastructure · Urban energy · Climate change · Agriculture and food security · Cities · Informal settlements · COVID-19 support

1 Human Ecology: Urban Morphology and Landscape Ecology

Human ecology expands on the relationship between humans as social beings and biological organisms and their interaction with their environment. It has become the epicenter of international debate, revolving around the sustainable development agenda, in every corner of the world. In the field of sustainable urbanism, urban morphology and landscape ecology have become crucial in analyzing perspectives of future of human settlements [1, 2]. This is because cities play a crucial role in the rise of greenhouse gas (GHG) emissions and urban pollution. Cities account for a small percentage of the earth's surface, yet they host the highest proportion of the world's population. In 2018, it was estimated that 53% of the world's population lived in urban settlements [3]. Moreover, cities consume about 75% of the world's energy and emit about 60% of the world's CO_2 [4–6]. The urban population is projected to increase above 60% by 2030 [7]. This creates an urgent need to examine sustainable energy consumption, low-carbon emission, and climate adaptation. The vast literature on urban morphology and landscape ecology enable us to understand human ecology in the urban setting, including architectural archetypes, energy systems, and inhabitant behavior [8]. As a result, there is a growing scholarship on knowledge management perspectives in terms of urbanization and human ecology where cities have become information hubs [9–12]. As such, urban morphology is a key approach to study human ecology. It analyzes the formative and transformative process of an urban area and draws from various disciplines, including urban geography, planning and archeology, anthropology, urbanism, and architectural history [13–15]. Urban morphological concepts enable us to understand how current urban environments, as habitats, were formed and how urban phenomena such as energy use, land value, urban agriculture, urban microclimate, and mobility exist. As well, it brings together the interrelationship and complexity of various segments of urban settlements and their relationship with non-urban areas. Urban morphology is a method of identifying, structuring, and investigating sets of relationships of various multidimensional complexities ultimately to inform policy-makers on strategic urban planning. This is notwithstanding that urbanization is a sociopolitical act, i.e., different urban processes are driven by economic and political concerns [15, 16]. Scholars have studied urban morphology in different ways. The four broad approaches are typo-morphological, configurational, historico-geographical, and spatial analytical. Each urban form uses different tools and methods to examine human ecology. These approaches differ in explaining levels of complexity and compositional hierarchy in the structure of the urban open and built form. Typo-morphological analysis emerged from urban design analysis in Europe and North America and is mainly confined to historical urban forms. Mainly relevant to urban surveying and planning, e.g., studies showing harmonious development of European towns and cities [17], it has been used to study urban areas globally [18]. Other investigative research has used configurational urban morphology, commonly using space syntax methods (i.e., to analyze urban configurations) such as spatial categories and structures, aerial differentiation, and other urban structural sub-systems, to better understand these areas. Such an analysis is essential in understanding the relationship between urban structural configuration and social capital [19]. The historico-geographical approach is concerned with the physical forms of cities, the agents, and processes that shape those forms over time [14]. The spatial analysis of urban morphology relies on analytical cartography and visual communication tools to illustrate the formation and transformation of urban forms [20, 21]. These human ecology approaches each have their own strengths and challenges in the assembly of urban morphology research. To have a complete understanding of urbanization, scholars combine different urban morphology forms to the study of human ecology [18].

Landscape ecology is concerned with the interrelationship between spatial patterns and ecological processes. It borrows from economics, sociology, geography, earth sciences, and computer application to study urban settlements, i.e., open and built landscapes [22]. It also deals with the generation and dynamics of constellations in ecosystems and their relationship with urban structures, communities, and ecosystem processes [23, 24]. Scholars have used landscape ecology to examine the structure and composition of heterogeneity of landscapes and their effect on people, the environment, and climate change [25, 26]. The future of human ecology examines our understanding of climate change and biodiversity in urban settings—interlinked by urban morphology and landscape ecology. Some of the new concepts like urban green spaces are becoming more mainstream and are gaining traction in how cities are designed [27, 28]. Other concepts such as smart cities incorporate Internet-based development, e.g., digital infrastructure, renewable energy, data management systems, cloud computing, and the Internet of things [19, 20, 29]. In other cases, peri-urban areas in developing countries undergo physical transformations to become cities [30] in which urbanization is often linked to deforestation, flooding, and desertification [31, 32]. The future of human ecology underpins the sustainable urbanization process, including clean energy consumption and decarbonization in cities. There is a vast scholarship of climate change adaptation, where cities spearhead climate action [33, 34], where emerging threats of detrimental risks of adaptation are not met—the so-called maladaptation dilemma [35] and business-as-usual mentality [36]. In this chapter, a breakdown on why humans settle where they settle—with a focus on why people settle in cities—is examined. The movement away from rural to urban is looked at in conjunction with urban energy, agriculture, a North–South overview, and case research into the effects of the COVID-19 pandemic on informal settlements.

2 Green Cities, Urban Agriculture, and Sustainable Energy

The health and well-being of people are becoming increasingly dependent on the quality of urban settlements [28, 37, 38]. Over the last 30 years, urban settlements have experienced dramatic growth [39]. Also, food, energy, raw materials, consumer products, and economic production are all connected to global networks, and these long-distance transactions produce substantial GHG emissions [39]. Thus, having access to clean, accessible, and reliable energy has become the pillar of prosperity and economic development [40]. As a result, cities have a unique opportunity to advance policies that ensure the energy supply and food production are reliable, affordable, and environmentally sustainable [39]. Urban design and landscape architecture are some of the more important tools for creating sustainable urban settlements. There is a need to consider design approaches and strategies that would work at a city scale such as a green infrastructure-led design approach that "creates healthier more socially cohesive and biodiverse urban environments and a connected city ecosystem for people and wildlife that also builds in resilience measures against climate change in the form of storm, flood, heat, drought, and pollution protection" [41]. In particular, green infrastructure can improve cities' adaptive capacity by preparing for and responding to shocks and systemic changes brought on by severe weather and natural disasters [42]. Ecosystem services, i.e., provided by green infrastructure, can contribute to a more energy-efficient and less carbon-intensive urban metabolism [43].

Green infrastructure can help to mitigate the negative effects of the energy sector, by (1) reducing energy consumption, (2) supplying bioenergy, and (3) capturing and storing carbon [44]. Green roofs can contribute to decreasing building energy as well as mitigating the urban heat island in cities [45]. According to Tsoka et al. [46], the shading effect of dense trees can result in energy savings of 54% in a dense urban area in Thessaloniki, Greece. A critical review for evidence-based urban greening in North America, Ko [47] found, in contrast to buildings without trees, buildings with trees used 2.3–90% less cooling energy and 1–20% less heating energy due to windbreak effects. A study conducted by Nowak et al. [48] in the USA, found that trees and forests in urban and community areas annually reduce electricity use by about 38.8 million MWh (i.e., USD 4.7 billion), heating use by 246 million MMBtus (i.e., USD 3.1 billion), and an average reduction in national residential energy use due to trees is 7.2%. If implemented throughout an urban watershed, green infrastructure strategies such as low impact development in the USA or surface water management systems and sustainable drainage systems in the UK can have significant energy cost savings to municipal water pollution control facilities [49]. Moreover, green infrastructure implementation in urban regeneration projects can have a positive impact on the economic value of target buildings as well as their larger contexts of open spaces, housing, and public facilities [50]. The ecocity of Augustenborg, Sweden, is an excellent example that incorporates blue and green infrastructure to address issues such as flooding, new renewable energy sources, sustainable construction, recycling systems, and sustainable transportation [44, 51]. Augustenborg's design had several tangible benefits, including: increased biodiversity by 50% (i.e., by creating natural habitats), reduced emissions from the neighborhood by more than 20% (i.e., due to energy efficiency protocols), and increased renewable energy usage (i.e., accounting for 80–85% of it is used in the heating systems city-wide) [51]. Green infrastructure not only reduces energy consumption but can provide food security in cities, e.g., via edible green infrastructure (i.e., urban agriculture, allotment gardens, edible commons, and edible green roofs) [27, 28, 52, 53]. An illustrative difference between gray and green cities is shown in Fig. 1.

Several studies have highlighted that edible green infrastructure and urban agriculture can be very productive, bolstering environmental and social benefits over intensive farming since crops are typically grown with little chemical inputs, travel shorter distances, and are intended for local use [54]. According to Russo and Cirella [27, 28, 55], edible green infrastructure can also regenerate sounder urban settlements. Urban woody biomass, for instance, from pruning residues can be used as bioenergy [56, 57]. The direct burning of pruning residues for electricity generation can be advantageous not only because it saves fossil fuels and creates new economic opportunities, but it also results in low CO_2 power generation [58]. Winzer et al. [59] calculated that the clearance wood from fruit trees could generate an energy potential of 191,000 MJ * ha^{-1}. Clinton et al. [60] estimated urban ecosystem services provided by urban vegetation—globally—could result in annual food production of 100–180 million tons, energy savings ranging from 14 to 15 billion kilowatt-hours, nitrogen sequestration between 100,000 and 170,000 tons, and avoided stormwater

Fig. 1 Gray city (e.g., fossil fuel society, no sustainable transportation, and no recycling) versus green city (e.g., green infrastructure, 10-min walk to a park, sustainable energy, urban agriculture, sustainable transportation, and recycling). Vector files designed by Macrovector and Freepik

runoff between 45 and 57 billion m^3 annually. It is believed that a sustainable planning and management vision that promotes integrated green space, a multimodal transportation system, sustainable food production, energy efficiency, and mixed-use growth should guide the design and planning of urban settlements [61]. Future research should investigate the implementation of futuristic algae-powered buildings for the production of bioenergy and biofuel in urban environments [62]. To explore this factor, the need to better understand human settlements in relation to energy is looked at followed by an examination of land use in terms of agriculture and food security.

3 Human Settlements in Relation to Energy Needs

The relationship of energy with human development is identified in many studies [63, 64] directly affecting people, communities, and countries in terms of economic growth, health, safety, the environment, education, and employment [65–67]. However, a study by Wu et al. [68] using panel data claimed that in 105 countries, this relationship is only short term. In contrast, research from Wang et al. [69] illustrated renewable energy consumption did not improve the human development process; as such, fossil fuels (i.e., oil, coal, and gas) still lead the global energy supply [70] accounting for around 500 EJ annually [71]. The global installed capacity is much less (i.e., 50% in China, Brazil, Canada, and the USA) from the total capacity potential of 3721 GW [72]. Bioenergy, which is generated from biological sources, has a large potential (i.e., 3500 EJ annually) [73]; however, the production of biofuels is comparatively low [74]. The emergence of wind as a source of energy has taken a superior lead in renewable sources [75]. Another important source of energy is direct solar energy; the World Energy Council [72] shows that the total energy from solar radiation is more than 7500 times the world's annual energy consumption.

The share of global primary renewable energy could rise from 11% in 2019 [76] to 63% in 2050 [77]. The IRENA [78] project clearly manifests that about 50% of renewable energy will need to be from wind, solar, and biofuel energy sources with 24%, 15%, and 10% share, respectively, in 2050 (Fig. 2). Estimations using energy demand models show that the amount of primary energy from biomass—if supplied cost effectively—is approximately 50–250 EJ per year while the global primary energy use is predicted to be approximately 600–1040 EJ per year by 2050 [72, 79]. This confirmation, at least in principle, indicates the biomass potential and demand could increase to one-third of the global energy demand [80]. The motivation of using renewable energy is steadily being accrued in many countries via fossil fuel price hiking [81], clean energy subsidies, technological advancements, and policy targets in-line with the United Nations Sustainable Development Goals (SDGs) [82, 83]. For instance, the European Union has recently revised its 2030 target from 27 to 32% set back in 2014 [84]. The Government of India set a renewable energy target of 175 GW by 2022, which includes 60 GW from wind and 100 GW from solar energy [85, 86]. About 11% of the total energy demand and 17% of all electricity

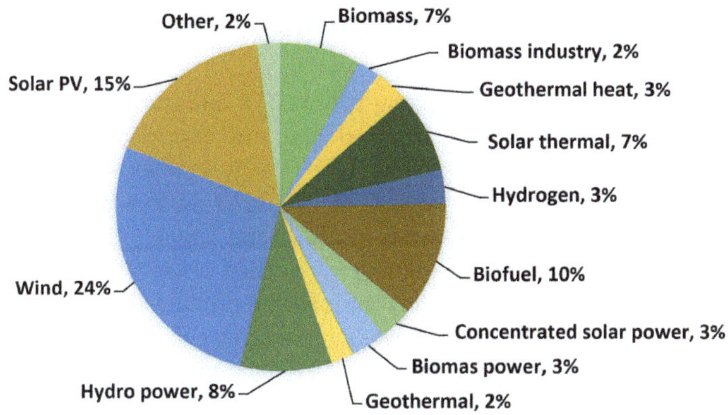

Fig. 2 Projected renewables in total world energy consumption, 2050 [72, 77–79]

generation in the USA are supplied from renewable energy resources according to the latest data from its Energy Information Administration [87]. Likewise, China has also set targets to reduce its carbon emissions per unit of gross domestic product (GDP) by 65% by 2030 in-line with its 2005 levels—playing a pivotal role in its energy grid. China's target for non-fossil fuel share in total energy demand is 20% by 2030 [88]. Russia, one of the largest fossil fuel resources in the world [77], has implemented energy trials of in excess of 5 GW from wind and solar energy since 2013, with projections of exceeding its 2024 target of 5.9 GW [77, 78, 89].

A transition away from fossil fuels to low-carbon solutions will play an essential role, as energy-related polluting emissions represent two-thirds of all GHGs [90, 91]; however, experience has shown that energy transitions take time, typically half a century from the first market uptake to the majority of market share [92]. Therefore, business opportunities, energy transition benefits, and self-determination of individuals will still need to be at the core of such change [93, 94]. Moreover, problems in the energy sector go beyond traditional government research and development, as it will require appropriate policy incentives and long-term perspectives—both currently lacking [78]. The potential of this energy transition is not yet fully appreciated by many policy-makers and analysts. Yet, there will be a critical threshold if many of the SDGs are to be met by 2030. From the top-down, human settlements in relation to energy must be forward-looking with the prospect of just and fair growth, that being, additional investments by 2050 should support an increase in global GDP, jobs, and environmental benefits.

4 Agriculture and Human Settlements

Over the last three hundred years, spatial patterns of land use and land cover have changed significantly [95]—specifically following the expansion of human settlements and economic development [96–101]. Croplands and pastures have increased by 279 million ha (i.e., 16.7%) between 1985 and 2013—a trend that dates back to the 1950s [102–104]. Agricultural lands establish the largest biome on the planet [105] with typological make up, including (1) a third of the global ice-free landmass [106], (2) providing food and other agricultural products for the rapidly rising human population (e.g., the increase in cereal production per capita from 0.29 to 0.39 tons per person between 1961 and 2014 [107]), (3) the major livelihood for 40% of the world's population, and (4) contributing to about 30% of GDP in low-income countries [108]. Lowder et al. [109] showed that 69% of the world's farmlands exist in Southeast Asia, South Asia, and Sub-Saharan Africa, with 30% of their produce coming from holdings less than 2 ha in size [110]. Despite all these, 800 million people remain undernourished [107] and approximately 2 billion suffer from micronutrient deficiencies [111, 112]. The main reason for the human nutrition gap is global dependency on a very few crops for energy, i.e., 84% of calories is generated from just 17 crops [113]. This is demonstrated by the dominance of white rice in the diet of Southeast Asian and South Asian regions, which experience micronutrient deficiency in prevalence of about 30% [114]. Moreover, in some regions such as in Sub-Saharan Africa, noticeable drops in micronutrient density in diets have been observed in recent decades—moving away from fruits, nuts, and pulses toward calorie-dense, but nutrient-poor foods (i.e., maize, rice, wheat, and vegetable oils) [114]. Furthermore, there is also an indication of declining trends in the nutritional quality of crops for some items detected in the USA [115]. The world produces 22% less fruit and vegetables than required to meet the World Health Organization recommendation to consume per day to achieve a healthy diet [116].

A common manifestation of climate change is the strengthening of hazardous climate events that effect agriculture such as floods, droughts, and irregular heat–cold fluctuations. "Climate change is a major issue for agricultural sustainability, and changes in farming practices will be necessary both to reduce emissions and to adapt to a changing climate and to new social expectations" [117]. Currently, agricultural lands are being degraded at an annual depletion nutrient rate of 10 million ha [118] in which clean phosphorous reserves are predicted to run out in only 20–50 years [119] (i.e., used for waterlogging and salinity control of irrigated areas [120]). Spatial diversity of cropping has also declined as large amounts of farmland grow monocultures in which the doubling and tripling of annual crops are degrading soils—globally [121]. Traditional varieties—vital for maintaining biodiversity—have been reduced due to industrial agricultural transition [122] even though farmers in traditional agroecosystems often maintain high varietal and species diversity on their farms as well as across communities and regions. This is much more prominent for staple rather than non-staple crops [123]. Moreover, price instability of agricultural food commodities has been escalating in the last decade [107, 124]. Farmers producing for the global market are particularly vulnerable because they are facing increasingly unpredictable market trends while the cost of agricultural inputs is increasing following the rise in the price of oil [125, 126]. Simultaneously, recent phenomena like fast urban expansion [38, 127] and land grabbing [128–130] are consuming agricultural land. In light of a continuously increasing world population, approaches and solutions to conquer these crises are immediately needed. Agricultural science and practice are asked to provide solutions to both alleviate the effects of climate change and increase adaptation of cropping and farming systems [52, 117]. According to the Millennium Ecosystem Assessment [131], the unsustainable production of food, feed, fiber, and fuel strongly degrades global ecosystems and the services those systems provided for human existence, including the provision of pure water, recycling of organic matter and nutrients, and adaptation to climate and weather events [27, 52]. Such degradation has not been hindered or overturned hitherto in spite of the fact that sustainability has become the focus of agricultural policy [118, 132]. A shift toward sustainable agricultural production demands the adoption of more system-oriented strategies that include farm-derived inputs and productivity based on ecological processes and functions [133]. However, studies show, for instance, overall yield gaps of organic farming for all crops are estimated to be 25% based on 316 comparisons [134] and 20% based on 362 comparisons [135].

The world's population is projected to be approximately 10 billion by 2050 [136] with greater per capita consumption of meat, refined fats, refined sugars, alcohols, and oils [137]. One of the two future projections of crop production for 2050 is a 60% growth in aggregate production (i.e., USD-weighted) from a 2005 to 2007 baseline and 100–110% increase in caloric demand. These studies have resulted in a doubling of food production requirement by 2050 [137, 138]. Research by Tomlinson [139], however, challenges this by utilizing the initial Food and Agriculture Organization of the United Nations (FAO) estimate of a 70% increase by 2050 as not a normative estimate but rather a projection of the most likely future. Moreover, the indicated FAO estimate is not of production or caloric input but of the USD-weighted aggregate production—exclusive of fruit and vegetables. Another recent study also commented on the doubling narrative by ignoring baselines [126], indicating only a 25–70% increase is needed between 2014 and 2050. Until recently, the prime focus of agricultural science was on supply-side solutions meeting the sustainable food security challenge. However, recent research has indicated the essential and massive advantage of demand-side solutions [137, 138, 140]. For example, Erb et al. [141] explored 500 different future scenarios for feeding the world in 2050 including the exclusion of further deforestation and found feasible biophysical options in nearly two-thirds of their scenarios—all requiring cropland intensification. Cassidy et al. [140] estimated that shifting the current combination of crops away from biofuels and animal feed would itself increase global caloric inputs by 70%. This study also calculated the approximate equivalent of all yield gains met in maize, wheat, and rice during the period 1965–2009. Notably, less-extreme shifts toward decreasing meat consumption, waste, and the demand for non-food agricultural products could greatly decrease the environmental impacts of the food system [138]. Generally speaking, increasing population, market infrastructure, and climate change are major driving forces that are transforming the agricultural industry. New management options and methods of production are required for sustainable agriculture [142]. Discovering best management options in terms of climate degradation is a key factor under consideration in agronomic research to sustain future crop productivity [143]. The use of crop simulation, i.e., as a decision-making tool, could make up an important and viable research alternative to the betterment of sustainable agriculture, the relationship between human settlements and arable land, and the future of agriculture relative to food projections [144]. This type of focused approach is paramount to the different ways human settlements are built and dictate, to large extent, the way people live. In terms of the urban construct, an elucidative look at cities in the north and the south will help piece together some perspective on the geography of feeding the world as well as pinpointing varying reasons for their urban morphology.

5 Cities: The North Versus the South

In the twenty-first century, cities have become the loci of the future for human settlement. An emerging primacy of cities, as places of international action, plays key parts of society's structure in the era of globalization [145–149]. They are also key agents of sustainable development [150, 151]; however, some cities are located in the Global North while others are in the South. Due to differences in the country's level of development and urban morphological processes, cities in the two worlds have generalizable characteristics—either differences or similarities—which to some extent define their relations. There are different ways to look at cities in the North and the South, including the analysis of land area (i.e., open and built environment) and demography, energy consumption, pollution, urban settlement (i.e., structure density of cities), and physical characteristics such as urban spatial structure which define the state of urbanization. The country's development level is critical, such that cities in the Global North and the Global South could have the same size but be defined by different socioeconomic rationales [152] (Fig. 3). Demographic transition in cities has shifted the locus of urban population from the Global North to the Global South [148, 153]. In this regard, the urban demographic transition is defined as "the historical period in which the population growth in cities structurally changes the settlement of territories" [154]. Also, cities in the Global South have a higher density than those in the North. Urban density is measured by determining the level of compactness, i.e., land use diversity, natural environment preservation, and efficient public transport facilities. The peripheral squatter settlement (i.e., city slums) in developing countries has the highest compactness [155]. Even though cities in the Global North and the Global South experience relatively similar inequalities in urban green spaces [155], cities' compactness in the South is generated by less strict land use planning rather than an absolute lack of it. However, some phenomena such as water and air pollution are not directly attributed to higher densities but rather because of weak environmental regulation and enforcement.

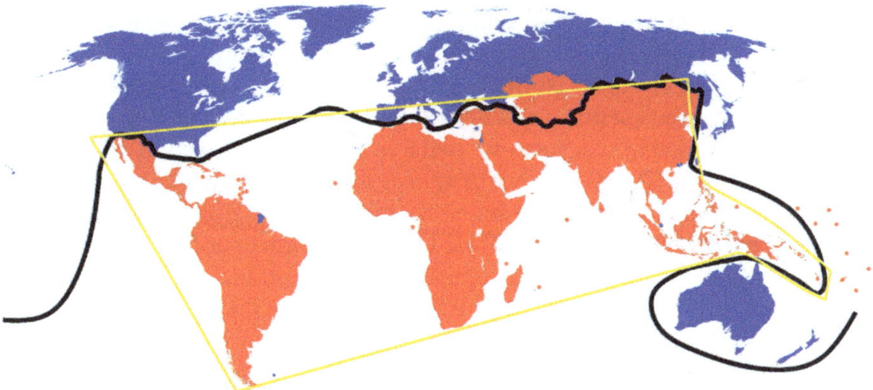

Fig. 3 World map of the Global North (blue) and Global South (red) overlaid on the Brandt Line (black line) [156] that divides developed and developing countries from forty years ago, and Ganguly and Mobley's South outline (yellow line) [157], adapted from Wikipedia Commons [158]

Most megacities are located in the Global North, while large cities in the Global South follow the Global North's spatial development trajectory. But again, megacities in the North have relatively high affluent and stable populations while those in the South are expanding more rapidly and become overcrowded. The urban landscape between the Global North and Global South is different. In developed countries, cities are built more outwardly (i.e., extensive) and up (i.e., have tall buildings) than in developing countries [152, 153]. The former has more developed transport systems compared to the South. For instance, special group-friendly features in cities are some characteristics of modern cities. Special groups like the disabled or the elderly require extra care within cities [159]. Age-friendly features are extensive within cities in developed countries than in developing ones. These urban structures correlate with the higher older population and urbanization designs in the North [160]. As such, smart city technologies are common and increasingly becoming a norm in most cities in the North. The technology is recently diffusing to the South, though, there are still inadequate studies on smart cities initiatives in developing countries [161]. Even before making an effort to become smart cities, most urban policies in the South aim to make cities functional by maintaining the provision of essential public goods such as public transport and sewerage. In addition, urban governance aims to regulate migration from rural to already the crowded urban centers [162].

Contemporarily, cities have become a major source of pollution through increased GHG emissions. As a response, and in-line with SDGs, cities are engaged in climate adaptation planning, which include a combination of social, structural, institutional, and technological measures to adapt [163–165]. While cities in the South are less industrialized and generate far lower levels of emissions than those in the North, climate adaptation planning is a top priority agenda in both. In the North, adaptation planning is well established [166, 167], especially where cities share sustainability-oriented knowledge and experience through city-to-city networks

[168–172]. However, cities in the South are in their early stages of adaptation planning and tend to learn best practices from their North counterparts [167]. Owing to the advanced city networks in the North and the homogeneity of city features and urban challenges, climate adaptation planning diffuses easily between cities and countries. Still, there lacks established South-to-South city networks as well as between the North and the South. There is a need for more research into the relationship between cities in the North and the South and whether climate adaptation and mitigation planning in cities in the South are diffused from the North or are locally generated. Researchers have also noted the risks of maladaptation in climate action, where policies and practices by some actors might either fail to meet their objectives or might increase the vulnerability of other groups or sectors in the future needs to be considered [173–176]. Maladaptation is defined as "an action taken ostensibly to avoid or reduce vulnerability to climate change that impacts adversely on, or increases the vulnerability of other systems, sectors, or social groups" [177]. One of the inadequately researched areas in the role of cities in climate action is how common forms of maladaptation in the North differ from those in the South. It remains unclear whether maladaptation in the North has a real or potential impact on the South and vice versa.

In the North, cities play a crucial role in climate action [33] with emerging roles of city diplomacy [178–180] and foreign policy of cities [150]. There are few studies on the differentiated role of cities in foreign policy and city diplomacy in climate action, first in the South and between the North and the South. As well, there is a need for more examination of relations between cities in the North and the South and how they promote climate adaptation and mitigation while addressing the already known maladaptation. The concept of the sustainable city underpins economic, environmental, and social sustainability. As such, a sustainable city should not be a goal but rather a principle of efficient provision of livelihoods based on social equity and justice [181]. Most city dwellers in some of the fastest-growing cities in the Global South do not have access to basic amenities such as clean and reliable energy. The challenge of urban energy access is rampant to the low-income population, especially in Africa and Asia. Hence, the old carbon-intensive development model used in the Global North is not reasonable in the Global South [181]. The goal-based sustainable city discourse popular in the North does not adequately capture problems of cities in the South and, therefore, might be inappropriate and misleading in the development of urban cities in the South [182, 183]. Based on the differential impact of climate action in cities between the North and the South, the idea of justice in climate adaptation has increasingly become popular [184–187]. By integrating environmental regulations in the urban development–climate adaptation nexus, researchers and policy-makers are able to generate a balanced (i.e., beyond neoliberal) comparison of sustainable cities in the two worlds [148]. Anguelovski et al. [164] recommend the need for a win-win climate adaptation solution with balanced costs and benefits for both the North and the South. In the South, a key concern is whether the efforts toward climate adaptation adequately prioritize the needs of the most vulnerable and marginalized cohorts or leave them in a worsened state. Notwithstanding this marginalization, the world—united—faces the COVID-19 pandemic in which human settlements, i.e., from the Global North and the Global South, must work to alleviate the outbreak that is central to cities (i.e., as hubs of transmission). Informal settlements, outside of the mainstream, are a distressing example of how urban centers have fallen short, especially during the pandemic, and where the North and the South can come to terms with bettering the human condition and our relationship with one another.

6 Informal Settlements: Impact of the COVID-19 Pandemic

The COVID-19 pandemic has changed the way people and societies function, globally and across sectors. Some circumstances have been replicated in informal settlements, advocating a type of renewal-to-revitalization concept. This idea encompasses three main value systems: human-centered, planetary health, and transdisciplinary (i.e., where the general act of human settlements exactly recognizes the relationship between people and nature alongside urban systems) [188]. Informal settlements have turned into epicenters for the pandemic [189, 190]. For an estimated one billion people, who represent the majority of the urban population in many low- and middle-income countries, these settlements are their neighborhoods. As such, informal settlements are facing considerable challenges and limitations due to the COVID-19 crisis [191]. A direct impact on communities living in informal settlements that impose city lockdowns—intended to prevent the spread of the disease—further creates impoverishment [191]. Two key measures to prevent the disease's spread are social distancing and increased hygiene. However, von Seidlein et al. [192] elucidate that both the lack of access to basic needs and overcrowding in slum areas, especially throughout cities in the Global South, have worsened. Regardless of the well-known fact that demographic groups with specific age or health conditions are more vulnerable to the virus, there is significantly less recognition of effect of inequality in access to measurements on a population's exposure to COVID-19 or its capacity to respond, which could augment existing vulnerabilities and create new ones [193]. Undoubtedly, public health officials recognize that most assailable groups are people living in informal habitats and contingent to informal livelihoods [194]. COVID-19 has seriously affected informal communities by virtue of the poor medical care services and undeveloped infrastructure; in addition, there is the lack of resources to develop appropriate living conditions by local governments that has further compounded the problem. As such, there are several issues on how COVID-19 affects informal settlements in terms of state intervention. To date, restrictions have made income harder to generate (i.e., by laying off employees or closing down) as well as rendering low-income households more vulnerable to infection, food shortages, and lack of digital services (e.g., limitations to online education and Internet access). One of the main goals to understanding the impact of the pandemic on these settlements is to uncover the appropriate strategies and tactics that should be used to minimize negative influences. Utilizing an exploratory qualitative stance, one may ask how such communities are affected in the near(est) future and what might be the impeding results caused by the pandemic? To answer these questions, global human settlement patterns and processes alongside social needs and health concerns would be where to start.

Almost every corner of the world has been influenced by the pandemic. From its outset with no pharmaceutical solution, governments were forced to implement special rules—i.e., regulations—and public policies in order to prevent the spread of the virus and control social behavior. COVID-19 has a significant direct and indirect impact on people in poverty, living in informal settlements—i.e., slums, shanty towns, and favela communities [195]. Many studies address the dramatic evidence of social and economic effects of government regulations as well as their huge negative effect on informal settlements. Such communities often develop strong social networks to share social and economic resources and fight for the access to private and public facilities; however, research indicates that possibilities for collaboration and social interaction have been significantly reduced since the pandemic [196]. Hence, those living in slum areas are at a structural disadvantage to overcome challenges and deal with the health crisis since governments are often not capable of implementing needed health targets. For instance, households in informal settlements have been affected by lack of space (i.e., to practice social distancing), overburdened infrastructure, lack of savings, loss of income, shortage of food, hunger and diseases, anxiety and depression, and poor access to education. Moreover, many people in informal settlements are not formally employed and depend on the informal sector to support their families. As such, informal sectors hardly generate extra income during hard times like the COVID-19 lockdowns [197]. These groups are paying a high toll since their employment loss is significantly larger than the loss reported in the general public. Data shows that labor participation has decreased by 30% which represents around a 50% drop in comparison with pre-pandemic levels as well as a drop in more than 40% in labor force participation [198]. Special financial support plans and unemployment insurance programs (UIP) were implemented by a number of governments. The participation of informal communities in UIP increased to 17% in September last year. Similar to that, to date, participation in financial pro bono support programs has increased from 33 to 37% during the pandemic [196]. However, more than half of the people who applied for support have not received any help. Regrettably, one of the limitations to apply for government support is usually central to the informally employed, leaving a large group of informal dwellers ineligible. The lack of any regular employment places positions informal dwellers in a very peculiar and unfortunate circumstance. In particular, large unemployment hikes have affected migrating populations, (i.e., people who travel from slum areas to urban centers for work), since more than 40% have now been fired during the pandemic and, as a result, have returned home wage less and stressed. These informal communities struggle to progress beyond a dead-end situation and are often limited to a life of crime and violence [199]. Moreover, an increase in domestic abuse places women and girls at much higher risk with obstruct access to protective services where medical care services are already limited. Shortage of food among an already underprivileged people has escalated the various infections and deficiency diseases (i.e., especially in children), notwithstanding the risks these populations face going out in the community and returning home infected—infecting their household. Economic stress has led to depression with shame among many informal communities when one loses their employment since security is a major health challenge [200, 201].

The absence of governmental attention is confirmed by the creation, location, and operation of quarantine centers. Quarantine and care centers are in part an acknowledgment of the compressed conditions of life in slums. Unfortunately, these are usually centers allocated quite far from the settlements themselves. Some quarantine hubs are initiated in high pollution zones and in unused slum buildings with inappropriate living conditions and infrastructure. For informal dwellers, environmental costs can originate from inadequate or no provision for piped water, paved roads, sanitation, and high levels of street garbage [202]. Accordingly, around 25% of the world's population live in informal settlements, with 213 million residents added since 1990 [197]. Such an increase places a heavy demand on land and natural resources which lead to harmful effects on the environment. As a result, the possibility of contracting new infections in an overburdened slum area could easily enable accompanying illnesses among the community [197]. In terms of access to information, guidelines given by police services, health authorities, and the government regarding the pandemic are limited due to the fact that majority of informal settlements cannot afford either a television or Internet access. Respectfully, a number of non-governmental organizations (NGOs) have started to provide brochures written in the local language to raise awareness; however, more than half of informal residents around the world are illiterate [197]. Even though informal settlements are considered to be temporary, central governments should consider placing a high priority in strengthening the infrastructure in these areas as a strategy to curtail the impact of pandemics like COVID-19 [190]. In doing so, there is a need for both governments and all types of human settlements to form partnerships as that can ensure that the virus will be countered effectively.

The COVID-19 pandemic is a wake-up call for city authorities to rethink their engagement with the people living in informal settlements [203]. The impact of COVID-19 will be most devastating in poor and densely populated urban areas, especially for the people living in slums worldwide as well as for refugees, internally displaced people, and migrants. For the one billion people who live in informal settlements, they face extremely tough conditions—even at the best of times—where many residents do not have access to sanitation or on-site water and face the constant threat of forced eviction and overcrowding. Along with that, vulnerable populations are also experiencing excess all-causing mortality related to disruptions in healthcare services [194], critical to aiding and preventing the spread of COVID-19 via disinfecting, physical distancing, and quarantine for the infected. However, these essentials are almost impossible to follow unless governments assist these communities in resolving their problems by establishing immediate measures. NGOs together with the authorities have to react quickly and provide economic support and social protection for those living in this state [204]. There is a need for central governments to establish strong working policies to support vulnerable communities living in informal settlements, such that policies could act as guarantors of care for these communities during pandemics [205]. It is necessary to prioritize the building of permanent households, communication initiatives, and pandemic awareness programs to mitigate impact in informal communities. Mobile medical centers to support testing, diagnostics, and early treatment need to be deployed. It is obvious that in most slums health clinics are not able to provide needed care for a large number of people; hence, distance, cost, and mistreatment by medics in those areas must be prioritized [194]. Moreover, organized food banks should be set in order to provide food with nutritional value to these communities. Hunger crises, especially during the lockdown of informal settlements, make it imperative that NGOs and central authorities create food provision services for these vulnerable communities [197]. There are grounds for an interdisciplinary approach in helping human settlements overcome these challenges caused by the global health crisis. Public health authorities have to be compelled to roll out strategic communication campaigns that focus on pandemic awareness to alleviate struggles caused by COVID-19. As such, the complexity of differing human settlements amounts to a broad vision of human ecology. As this chapter has pointed out, urban morphology in relation to the different typologies of infrastructure, energy usage, and agriculture—compounded by geographic location—sets the stage for urban challenges and future development in the twenty-first century.

References

1. Chokhachian A, Perini K, Giulini S, Auer T (2020) Urban performance and density: generative study on interdependencies of urban form and environmental measures. Sustain Cities Soc 53:101952. https://doi.org/10.1016/j.scs.2019.101952
2. D'Acci L (2019) On urban morphology and mathematics. In: Modeling and simulation in science, engineering and technology. Springer, Zurich, pp 1–18
3. World Bank (2021) Urban population (% of total population). https://data.worldbank.org/indicator/SP.URB.TOTL.IN.ZS. Accessed 14 Mar 2021
4. UN-Habitat (2014) Energy: the challenge. In: United Nations humanities settlements program. https://unhabitat.org/topic/energy. Accessed 15 Mar 2021
5. Privitera R, Palermo V, Martinico F et al (2018) Towards lower carbon cities: urban morphology contribution in climate change adaptation strategies. Eur Plan Stud 26:812–837. https://doi.org/10.1080/09654313.2018.1426735
6. Shi Z, Fonseca JA, Schlueter A (2017) A review of simulation-based urban form generation and optimization for energy-driven urban design. Build Environ 121:119–129. https://doi.org/10.1016/j.buildenv.2017.05.006
7. United Nations (2017) The World's cities in 2018: data booklet. In: United Nations department of economics and social affairs. https://www.un.org/en/events/citiesday/assets/pdf/the_worlds_cities_in_2018_data_booklet.pdf. Accessed 15 Mar 2021
8. Salat S (2009) Energy loads, CO_2 emissions and building stocks: Morphologies, typologies, energy systems and behaviour. Build Res Inf 37:598–609. https://doi.org/10.1080/09613210903162126
9. Evans G (2009) Creative cities, creative spaces and urban policy. Urban Stud 46:1003–1040. https://doi.org/10.1177/0042098009103853
10. Yigitcanlar T, O'Connor K, Westerman C (2008) The making of knowledge cities: Melbourne's knowledge-based urban development experience. Cities 25:63–72. https://doi.org/10.1016/j.cities.2008.01.001

11. Yigitcanlar T, Velibeyoglu K, Martinez-Fernandez C (2008) Rising knowledge cities: the role of urban knowledge precincts. J Knowl Manag 12:8–20. https://doi.org/10.1108/13673270810902902
12. Israilidis J, Odusanya K, Mazhar MU (2021) Exploring knowledge management perspectives in smart city research: a review and future research agenda. Int J Inf Manage 56:101989. https://doi.org/10.1016/j.ijinfomgt.2019.07.015
13. Hall T, Barrett H (2018) Urban Geography, 5th edn. Routledge, New York
14. Oliveira V (2019) An historico-geographical theory of urban form. J Urban Int Res Placemaking Urban Sustain 12:412–432. https://doi.org/10.1080/17549175.2019.1626266
15. Kropf K (2018) The handbook of urban morphology. Wiley, West Sussex
16. Kropf K (2013) Urbanism, politics and language: the role of urban morphology. Riv Geogr Ital 69:117–132
17. Trache H (2001) Promoting urban design in development plans: Typo-morphological approaches in Montreuil, France. Urban Des Int 6:157–172. https://doi.org/10.1057/palgrave.udi.9000052
18. Zhang Y, Li X (2020) What new insights can the combination of the historico-geographical and configurational approaches to urban morphology offer? Dublin Hist Rec 88:84–96
19. Rashid M (2019) Space syntax: a network-based configurational approach to studying urban morphology. In: Modeling and simulation in science, engineering and technology. Springer, Zurich, pp 199–251
20. Boeing G (2021) Spatial information and the legibility of urban form: Big data in urban morphology. Int J Inf Manage 56:102013. https://doi.org/10.1016/j.ijinfomgt.2019.09.009f
21. Zaki SA, Azid NS, Shahidan MF et al (2020) Analysis of urban morphological effect on the microclimate of the urban residential area of Kampung Baru in Kuala Lumpur using a geospatial approach. Sustainability 12:7301. https://doi.org/10.3390/su12187301
22. Gergel SE, Turner MG (2017) Learning landscape ecology. Springer, New York
23. Ingegnoli V (2013) Landscape ecology: a widening foundation. Springer, Berlin
24. Naveh Z, Lieberman AS (2013) Landscape ecology: theory and application. Springer, New York
25. Forman RT (2014) Urban ecology: science of cities. Cambridge University Press, Cambridge
26. Azhdari A, Soltani A, Alidadi M (2018) Urban morphology and landscape structure effect on land surface temperature: evidence from Shiraz, a semi-arid city. Sustain Cities Soc 41:853–864. https://doi.org/10.1016/j.scs.2018.06.034
27. Russo A, Cirella GT (2019) Edible urbanism 5.0. Palgrave Commun 5:1–9. https://doi.org/10.1057/s41599-019-0377-8
28. Russo A, Cirella GT (2018) Modern compact cities: how much greenery do we need? Int J Environ Res Public Health 15:2180. https://doi.org/10.3390/ijerph15102180
29. Deakin M, Reid A (2018) Smart cities: under-gridding the sustainability of city-districts as energy efficient-low carbon zones. J Clean Prod 173:39–48. https://doi.org/10.1016/j.jclepro.2016.12.054
30. Cobbinah PB, Gaisie E, Owusu-Amponsah L (2015) Peri-urban morphology and indigenous livelihoods in Ghana. Habitat Int 50:120–129. https://doi.org/10.1016/j.habitatint.2015.08.002
31. Cirella GT, Iyalomhe F, Jensen A, Akiyode O (2018) Exploring community of practice in Uganda's public sector: environmental impact assessment case study. Sustainability 10:2502. https://doi.org/10.3390/su10072502
32. Cirella GT, Iyalomhe FO (2018) Flooding conceptual review: sustainability-focalized best practices in Nigeria. Appl Sci 8:1558. https://doi.org/10.3390/app8091558
33. Watts M (2017) Commentary: cities spearhead climate action. Nat Clim Chang 7:537–538. https://doi.org/10.1038/nclimate3358
34. Araos M, Berrang L, Ford JD et al (2016) Climate change adaptation planning in large cities: a systematic global assessment. Environ Sci Policy 66:375–382. https://doi.org/10.1016/j.envsci.2016.06.009

35. Magnan AK, Schipper ELF, Burkett M et al (2016) Addressing the risk of maladaptation to climate change. Wiley Interdiscip Rev Clim Chang 7:646–665. https://doi.org/10.1002/wcc.409
36. Pozzer A, Zimmermann P, Doering UM et al (2012) Effects of business-as-usual anthropogenic emissions on air quality. Atmos Chem Phys 12:6915–6937. https://doi.org/10.5194/acp-12-6915-2012
37. Vujcic M, Tomicevic-Dubljevic J, Zivojinovic I, Toskovic O (2019) Connection between urban green areas and visitors' physical and mental well-being. Urban For Urban Green 40:299–307. https://doi.org/10.1016/j.ufug.2018.01.028
38. Russo A, Cirella GT (2020) Urban sustainability: integrating ecology in city design and planning. In: Cirella GT (ed) Sustainable human-nature relations: environmental scholarship, economic evaluation, urban strategies. Springer, Singapore, pp 187–204
39. UN-Habitat (2009) Sustainable urban energy planning: a handbook for cities and towns in developing countries. UN-Habitat and UNEP, Geneva
40. Chu S, Majumdar A (2012) Opportunities and challenges for a sustainable energy future. Nature 488:294–303. https://doi.org/10.1038/nature11475
41. ARUP (2014) Cities alive: rethinking green infrastructure. Work Pap 161
42. Carter JG, Cavan G, Connelly A et al (2015) Climate change and the city: building capacity for urban adaptation. Prog Plann 95:1–66. https://doi.org/10.1016/j.progress.2013.08.001
43. Perrotti D, Stremke S (2020) Can urban metabolism models advance green infrastructure planning? Insights from ecosystem services research. Environ Plan B Urban Anal City Sci 47:678–694. https://doi.org/10.1177/2399808318797131
44. European Commission (2014) Green Infrastructure in the energy sector
45. Susca T (2019) Green roofs to reduce building energy use? A review on key structural factors of green roofs and their effects on urban climate. Build Environ 162:106273. https://doi.org/10.1016/j.buildenv.2019.106273
46. Tsoka S, Leduc T, Rodler A (2021) Assessing the effects of urban street trees on building cooling energy needs: the role of foliage density and planting pattern. Sustain Cities Soc 65:102633. https://doi.org/10.1016/j.scs.2020.102633
47. Ko Y (2018) Trees and vegetation for residential energy conservation: a critical review for evidence-based urban greening in North America. Urban For Urban Green 34:318–335. https://doi.org/10.1016/j.ufug.2018.07.021
48. Nowak DJ, Appleton N, Ellis A, Greenfield E (2017) Residential building energy conservation and avoided power plant emissions by urban and community trees in the United States. Urban For Urban Green 21:158–165. https://doi.org/10.1016/j.ufug.2016.12.004
49. Spatari S, Yu Z, Montalto FA (2011) Life cycle implications of urban green infrastructure. Environ Pollut 159:2174–2179. https://doi.org/10.1016/j.envpol.2011.01.015
50. Hsu K-W, Chao J-C (2020) Economic valuation of green infrastructure investments in urban renewal: the case of the Station District in Taichung Taiwan. Environments 7:56. https://doi.org/10.3390/environments7080056
51. BSHF (2014) Eco-city Augustenborg. Winner, World Habitat Award. Work Pap, Sweden, pp 1–43
52. Russo A, Escobedo FJ, Cirella GT, Zerbe S (2017) Edible green infrastructure: an approach and review of provisioning ecosystem services and disservices in urban environments. Agric Ecosyst Environ 242:53–66. https://doi.org/10.1016/j.agee.2017.03.026
53. Sardeshpande M, Rupprecht C, Russo A (2021) Edible urban commons for resilient neighbourhoods in light of the pandemic. Cities 109:103031. https://doi.org/10.1016/j.cities.2020.103031
54. Nicholls E, Ely A, Birkin L et al (2020) The contribution of small-scale food production in urban areas to the sustainable development goals: a review and case study. Sustain Sci 15:1585–1599. https://doi.org/10.1007/s11625-020-00792-z
55. Russo A, Cirella GT (2020) Edible green infrastructure for urban regeneration and food security: case studies from the Campania region. Agriculture 10:358. https://doi.org/10.3390/agriculture10080358

56. Nowak DJ, Greenfield EJ, Ash RM (2019) Annual biomass loss and potential value of urban tree waste in the United States. Urban For Urban Green 46:126469. https://doi.org/10.1016/j.ufug.2019.126469
57. Russo A, Escobedo FJ, Timilsina N et al (2014) Assessing urban tree carbon storage and sequestration in Bolzano, Italy. Int J Biodivers Sci Ecosyst Serv Manag 10:54–70. https://doi.org/10.1080/21513732.2013.873822
58. Sagani A, Hagidimitriou M, Dedoussis V (2019) Perennial tree pruning biomass waste exploitation for electricity generation: the perspective of Greece. Sustain Energy Technol Assessments 31:77–85. https://doi.org/10.1016/j.seta.2018.11.001
59. Winzer F, Kraska T, Elsenberger C et al (2017) Biomass from fruit trees for combined energy and food production. Biomass Bioenerg 107:279–286. https://doi.org/10.1016/j.biombioe.2017.10.027
60. Clinton N, Stuhlmacher M, Miles A et al (2018) A global geospatial ecosystem services estimate of urban agriculture. Earth's Futur 6:40–60. https://doi.org/10.1002/2017EF000536
61. ASLA (2021) Sustainable urban development. American Society of Landscape Architects. https://www.asla.org/sustainableurbandevelopment.aspx
62. Chew KW, Khoo KS, Foo HT et al (2021) Algae utilization and its role in the development of green cities. Chemosphere 268:129322. https://doi.org/10.1016/j.chemosphere.2020.129322
63. Hendry DF, Juselius K, Hendry D, Juselius K (2000) Explaining cointegration analysis: part 1. Energy J 21:1–42
64. Sanchez-Loor DA, Zambrano-Monserrate MA (2015) International journal of energy economics and policy causality analysis between electricity consumption, real gross domestic product, foreign direct investment, human development and remittances in Colombia, Ecuador and Mexico. Int J Energy Econ Policy 5:746–753
65. UNDP (2018) Statistical update 2018, human development reports. United Nations Development Program. http://hdr.undp.org/en/content/human-development-indices-indicators-2018-statistical-update. Accessed 23 July 2020
66. UNDP (2018) Industrialization with a human face. United Nations Development Programme, Addis Ababa, Ethiopia
67. UNDP (2019) Human development index. United Nations Development Programme. http://www.hdr.undp.org/en/2019-report. Accessed 17 July 2019
68. Wu Q, Maslyuk S, Clulow V (2010) Energy consumption transition and human development. Monash University, Melbourne
69. Wang Z, Danish ZB, Wang B (2018) Renewable energy consumption, economic growth and human development index in Pakistan: evidence form simultaneous equation model. J Clean Prod 184:1081–1090. https://doi.org/10.1016/j.jclepro.2018.02.260
70. Owusu PA, Asumadu-Sarkodie S (2016) A review of renewable energy sources, sustainability issues and climate change mitigation. Cogent Eng 3:1167990. https://doi.org/10.1080/23311916.2016.1167990
71. IEA (2008) Key world energy statistics 2020—analysis. In: IEA OECD. https://www.iea.org/reports/key-world-energy-statistics-2020. Accessed 15 Mar 2021
72. World Energy Council (2013) World energy resources 2013 survey: summary. World Energy Council, London
73. Hoogwijk M, Faaij A, Eickhout B et al (2005) Potential of biomass energy out to 2100, for four IPCC SRES land-use scenarios. Biomass Bioenergy 29:225–257. https://doi.org/10.1016/j.biombioe.2005.05.002
74. Ajanovic A (2011) Biofuels versus food production: does biofuels production increase food prices? Energy 36:2070–2076. https://doi.org/10.1016/j.energy.2010.05.019
75. Manwell JF, McGowan JG, Rogeres AL (2010) Wind energy explained: theory, design and application, 2nd edn. Wiley, West Sussex
76. Ritchie H, Roser M (2020) Renewable energy. In: Our world data. https://ourworldindata.org/renewable-energy. Accessed 17 Mar 2021
77. IRENA (2017) Renewable energy prospects for the Russian Federation. IRENA, Abu Dhabi
78. IRENA (2019) Innovation landscape for a renewable-powered future. IRENA, Abu Dhabi

79. World Energy Council (2001) Living in one world. In: Chapter 5. Concerns about sustainable history sustain. http://www.worldenergy.org/wec-geis/publications/reports/liow/the_concerns/sustainability.asp. Accessed 13 Dec 2017
80. UNEP (2007) Global environment outlook 4. United Nations Environment Programme, Nairobi
81. Abbasi T, Premalatha M, Abbasi SA (2011) The return to renewables: will it help in global warming control? Renew Sustain Energy Rev 15:891–894. https://doi.org/10.1016/j.rser.2010.09.048
82. Allen C, Metternicht G, Wiedmann T (2016) National pathways to the sustainable development goals (SDGs): a comparative review of scenario modelling tools. Environ Sci Policy 66:199–207. https://doi.org/10.1016/j.envsci.2016.09.008
83. Nilsson M, Griggs D, Visbeck M (2016) Policy: map the interactions between sustainable development goals. Nature 534:320–322. https://doi.org/10.1038/534320a
84. European Commission (2020) Renewable energy. In: Energy. https://ec.europa.eu/energy/topics/renewable-energy_en. Accessed 15 Mar 2021
85. MNRE (2015) Tentative state-wise break-up of renewable power target to be achieved by the year 2022. Ministry of New Renewable Energy, Government of India. https://policy.asiapacificenergy.org/node/3652. Accessed 15 Mar 2021
86. Kota S, Bayne SB, Nimmagadda S (2015) Offshore wind energy: a comparative analysis of UK, USA and India. Renew Sustain Energy Rev 41:685–694. https://doi.org/10.1016/j.rser.2014.08.080
87. EIA (2018) Renewable energy explained. U.S. Energy Information Administration. https://www.eia.gov/energyexplained/renewable-sources/. Accessed 15 Mar 2021
88. NDRC (2016) China's national climate change program. National assessment report on climate change. http://www.china.org.cn/english/environment/213624.htm
89. Power Technology (2018) Is Russia finally ready to embrace renewable energy? In: Power technology. https://www.power-technology.com/features/russia-renewable-energy/. Accessed 15 Mar 2021
90. IPCC (2013) AR5 climate change 2013: the physical science basis—IPCC. Working Group I of the Intergovernmental Panel on Climate Change, Geneva
91. IPCC (2017) AR6 climate change 2021: the physical science basis—IPCC. Intergovernmental Panel on Climate Change, Geneva
92. Sovacool BK (2016) How long will it take? Conceptualizing the temporal dynamics of energy transitions. Energy Res Soc Sci 13:202–215. https://doi.org/10.1016/j.erss.2015.12.020
93. Grayson M (2017) Energy transitions. Nature 551:S133. https://doi.org/10.1038/d41586-017-07507-y
94. Mey F, Diesendorf M (2018) Who owns an energy transition? Strategic action fields and community wind energy in Denmark. Energy Res Soc Sci 35:108–117. https://doi.org/10.1016/j.erss.2017.10.044
95. Turner BL, Clark WC, Kates RW et al (1990) The earth as transformed by human action: global and regional changes in the biosphere over the past 300 years edited by. Cambridge University Press, New York
96. Grigg DB (1987) The industrial revolution and land transformation. In: Wolman MG, Fournier FGA (eds) Land transformation in agriculture. Wiley, Chichester
97. Lewis MW, McNeill JR (2000) Something new under the Sun: an environmental history of the twentieth-century world. Geogr Rev 90:149. https://doi.org/10.2307/216186
98. Richards JF (1990) Land transformation. In: Turner BL, Clark WC, Kates RW et al (eds) The earth as transformed by human action: global and regional changes in the biosphere over the past 300 years. Cambridge University Press, New York
99. Kaereem MA (2021) Strategic planning for agricultural development. National Institute of Agricultural Extention Management
100. Griskevicius V, Cantú SM, Van Vugt M (2012) The evolutionary bases for sustainable behavior: Implications for marketing, policy, and social entrepreneurship. J Public Policy Mark 31:115–128

101. Testa F, Russo MV, Cornwell TB et al (2018) Social sustainability as buying local: effects of soft policy, meso-level actors, and social influences on purchase intentions. J Public Policy Mark 37:152–166. https://doi.org/10.1509/jppm.16.215
102. Bennett AJ, Bending GD, Chandler D et al (2012) Meeting the demand for crop production: the challenge of yield decline in crops grown in short rotations. Biol Rev 87:52–71. https://doi.org/10.1111/j.1469-185X.2011.00184.x
103. Capper JL, Cady RA, Bauman DE (2009) The environmental impact of dairy production: 1944 compared with 2007. J Anim Sci 87:2160–2167. https://doi.org/10.2527/jas.2009-1781
104. Carlson KM, Gerber JS, Mueller ND et al (2017) Greenhouse gas emissions intensity of global croplands. Nat Clim Chang 7:63–68. https://doi.org/10.1038/nclimate3158
105. Ellis EC, Ramankutty N (2008) Putting people in the map: anthropogenic biomes of the world. Front Ecol Environ 6:439–447. https://doi.org/10.1890/070062
106. Ramankutty N, Evan AT, Monfreda C, Foley JA (2008) Farming the planet: 1. Geographic distribution of global agricultural lands in the year 2000. Global Biogeochem Cycles 22:GB1003. https://doi.org/10.1029/2007GB002952
107. FAO (2015) The state of food insecurity 2015. The Food and Agricultural Organisation of United Nations. http://www.fao.org/publications/sofi/2015/en/. Accessed 15 Mar 2021
108. World Bank (2019) World Bank open data. In: World Bank. https://data.worldbank.org/. Accessed 27 July 2019
109. Lowder SK, Skoet J, Raney T (2016) The number, size, and distribution of farms, smallholder farms, and family farms worldwide. World Dev 87:16–29. https://doi.org/10.1016/j.worlddev.2015.10.041
110. Herrero M, Thornton PK, Power B et al (2017) Farming and the geography of nutrient production for human use: a transdisciplinary analysis. Lancet Planet Heal 1:e33–e42. https://doi.org/10.1016/S2542-5196(17)30007-4
111. Tulchinsky TH (2010) Micronutrient deficiency conditions: global health issues. Public Health Rev 32:243–255. https://doi.org/10.1007/BF03391600
112. FAO, IFAD, UNICEF et al (2018) The state of food security and nutrition in the world 2018. Building climate resilience for food security and nutrition. Rome
113. West PC, Gerber JS, Engstrom PM et al (2014) Leverage points for improving global food security and the environment. Science (80-) 345:325–328. https://doi.org/10.1126/science.1246067
114. Beal T, Massiot E, Arsenault JE et al (2017) Global trends in dietary micronutrient supplies and estimated prevalence of inadequate intakes. PLoS One 12:e0175554. https://doi.org/10.1371/journal.pone.0175554
115. Davis DR, Epp MD, Riordan HD, Davis DR (2004) Changes in USDA food composition data for 43 garden crops, 1950 to 1999. J Am Coll Nutr 23:669–682. https://doi.org/10.1080/07315724.2004.10719409
116. Siegel KR, Ali MK, Srinivasiah A et al (2014) Do we produce enough fruits and vegetables to meet global health need? PLoS One 9:e104059. https://doi.org/10.1371/journal.pone.0104059
117. Fleming A, Vanclay F (2010) Farmer responses to climate change and sustainable agriculture. A review. Agron Sustain Dev 30:11–19. https://doi.org/10.1051/agro/2009028
118. Pimentel D (2006) Soil erosion: a food and environmental threat. Environ Dev Sustain 8:119–137. https://doi.org/10.1007/s10668-005-1262-8
119. Figueres C, Schellnhuber HJ, Whiteman G et al (2017) Three years to safeguard our climate. Nature 546:593–595. https://doi.org/10.1038/546593a
120. Lal R (2009) Soils and world food security. Soil Tillage Res 102:1–4. https://doi.org/10.1016/j.still.2008.08.001
121. Cassman KG, Dobermann A, Walters DT, Yang H (2003) Meeting cereal demand while protecting natural resources and improving environmental quality. Annu Rev Environ Resour 28:315–358. https://doi.org/10.1146/annurev.energy.28.040202.122858
122. Pereira HM, Navarro LM, Martins IS (2012) Global biodiversity change: the bad, the good, and the unknown. Annu Rev Environ Resour 37:25–50. https://doi.org/10.1146/annurev-environ-042911-093511

123. Jarvis DI, Brown AHD, Pham HC et al (2008) A global perspective of the richness and evenness of traditional crop-variety diversity maintained by farming communities. Proc Natl Acad Sci USA 105:5326–5331. https://doi.org/10.1073/pnas.0800607105
124. Akram QF (2009) Commodity prices, interest rates and the dollar. Energy Econ 31:838–851. https://doi.org/10.1016/j.eneco.2009.05.016
125. Mitchel D (2008) A note on rising food prices. World Bank, Washington, DC
126. Hunter MC, Smith RG, Schipanski ME et al (2017) Agriculture in 2050: recalibrating targets for sustainable intensification. Bioscience 67:386–391. https://doi.org/10.1093/biosci/bix010
127. Cohen B, Muñoz P (2016) Sharing cities and sustainable consumption and production: towards an integrated framework. J Clean Prod 134:87–97. https://doi.org/10.1016/j.jclepro.2015.07.133
128. Gemeda BS, Abebe BG, Paczoski A et al (2019) What motivates speculators to speculate? Entropy 22:59. https://doi.org/10.3390/e22010059
129. de Schutter O (2011) How not to think of land-grabbing: three critiques of large-scale investments in farmland. J Peasant Stud 38:249–279. https://doi.org/10.1080/03066150.2011.559008
130. Gemeda BS, Abebe BG, Cirella GT (2020) How efficient is urban land speculation? In: Cirella GT (ed) Advances in 21st century human settlements. Springer, Singapore, pp 101–121
131. MEA (2005) Ecosystems and human well-being: wetlands and water synthesis. Millennium Ecosystem Assessment, Washington, DC
132. Pimentel D, Harvey C, Resosudarmo P et al (1995) Environmental and economic costs of soil erosion and conservation benefits. Science (80-)267:1117–1123. https://doi.org/10.1126/science.267.5201.1117
133. Garnett T, Godfray CJ (2012) Sustainable intensification in agriculture. Navigating a course through competing food system priorities. Food Climate Research Network and the Oxford Martin, Oxford
134. Seufert V, Ramankutty N, Foley JA (2012) Comparing the yields of organic and conventional agriculture. Nature 485:229–232. https://doi.org/10.1038/nature11069
135. De Ponti T, Rijk B, Van Ittersum MK (2012) The crop yield gap between organic and conventional agriculture. Agric Syst 108:1–9. https://doi.org/10.1016/j.agsy.2011.12.004
136. United Nations (2017) World population prospects: the 2017 revision. United Nations Department of Economics and Social Affairs. https://www.un.org/development/desa/publications/world-population-prospects-the-2017-revision.html. Accessed 15 Mar 2021
137. Tilman D, Clark M (2014) Global diets link environmental sustainability and human health. Nature 515:518–522. https://doi.org/10.1038/nature13959
138. Foley JA, Ramankutty N, Brauman KA et al (2011) Solutions for a cultivated planet. Nature 478:337–342. https://doi.org/10.1038/nature10452
139. Tomlinson I (2013) Doubling food production to feed the 9 billion: a critical perspective on a key discourse of food security in the UK. J Rural Stud 29:81–90. https://doi.org/10.1016/j.jrurstud.2011.09.001
140. Cassidy ES, West PC, Gerber JS, Foley JA (2013) Redefining agricultural yields: from tonnes to people nourished per hectare. Environ Res Lett 8:034015. https://doi.org/10.1088/1748-9326/8/3/034015
141. Erb KH, Lauk C, Kastner T et al (2016) Exploring the biophysical option space for feeding the world without deforestation. Nat Commun 7:1–9. https://doi.org/10.1038/ncomms11382
142. Doré T, Clermont-Dauphin C, Crozat Y et al (2008) Methodological progress in on-farm regional agronomic diagnosis. A review. Agron Sustain Dev 28:151–161. https://doi.org/10.1051/agro:2007031
143. Xiong W, Holman I, Conway D et al (2008) A crop model cross calibration for use in regional climate impacts studies. Ecol Modell 213:365–380. https://doi.org/10.1016/j.ecolmodel.2008.01.005
144. Amanullah M, Kailasam C, Safiullah A et al (2009) Crop simulation growth model in Cassava. Res J Agri Biol Sci 31:838–851

145. Csomós G (2017) Cities as command and control centres of the world economy: an empirical analysis, 2006–2015. Bull Geogr 38:7–26. https://doi.org/10.1515/bog-2017-0031
146. Herrschel T, Newman P (2017) Cities as international actors: urban and regional governance beyond the nation state. Palgrave Macmillan, London
147. Malasenkova AA, Lavrov IR (2020) Global cities: Sydney. J Gov Polit 1:1–11
148. Parnell S, Robinson J (2012) (Re)theorizing cities from the global south: looking beyond neoliberalism. Urban Geogr 33:593–617. https://doi.org/10.2747/0272-3638.33.4.593
149. Toly NJ (2008) Transnational municipal networks in climate politics: from global governance to global politics. Globalizations 5:341–356. https://doi.org/10.1080/14747730802252479
150. Kosovac A, Acuto M, Jones TL (2020) Acknowledging urbanization: a survey of the role of cities in UN frameworks. Glob Policy 11:293–304. https://doi.org/10.1111/1758-5899.12783
151. De Guimarães JCF, Severo EA, Felix Júnior LA et al (2020) Governance and quality of life in smart cities: towards sustainable development goals. J Clean Prod 253:119926. https://doi.org/10.1016/j.jclepro.2019.119926
152. Jedwab R, Loungani P, Yezer A (2021) Comparing cities in developed and developing countries: population, land area, building height and crowding. Reg Sci Urban Econ 86:103609. https://doi.org/10.1016/j.regsciurbeco.2020.103609
153. Jedwab R, Loungani MP, Yezer A (2019) How should we measure city size theory and evidence within and across rich and poor countries. IMF, Geneva
154. Dumont G-F (2018) Urban demographic transition. Urban Dev Issues 56:13–25. https://doi.org/10.2478/udi-2018-0009
155. Rigolon A, Browning M, Lee K, Shin S (2018) Access to urban green space in cities of the global south: a systematic literature review. Urban Sci 2:67. https://doi.org/10.3390/urbansci2030067
156. Brandt W (1980) North-south: a programme for survival: report of the independent commission on international development issues. Independent Commission on International Development Issues, London
157. Ganguly D, Mobley C (2021) Global south humanities lab. Academy of Global Humanities and Critical Theory. https://aghct.org/research-and-opportunities/611. Accessed 24 Mar 2021
158. RGS (2021) A 60 second guide to … the global north/south divide. Global Learning Program. moz-extension://b25df011-8780-4be7-9be7-6714e4992816/enhanced-reader.html?openApp&pdf=https%3A%2F%2Fwww.rgs.org%2FCMSPages%2FGetFile.aspx%3Fnodeguid%3D9c1ce781-9117-4741-af0a-a6a8b75f32b4%26lang%3Den-GB. Accessed 23 Mar 2021
159. Cirella GT, Bąk M, Kozlak A et al (2019) Transport innovations for elderly people. Res Transp Bus Manag 30:100381. https://doi.org/10.1016/j.rtbm.2019.100381
160. Plouffe L, Kalache A (2010) Towards global age-friendly cities: determining urban features that promote active aging. J Urban Heal 87:733–739. https://doi.org/10.1007/s11524-010-9466-0
161. Vu K, Hartley K (2018) Promoting smart cities in developing countries: policy insights from Vietnam. Telecomm Policy 42:845–859. https://doi.org/10.1016/j.telpol.2017.10.005
162. Duranton G (2008) Viewpoint: from cities to productivity and growth in developing countries. Can J Econ Can d'économique 41:689–736. https://doi.org/10.1111/j.1540-5982.2008.00482.x
163. Ford JD, Berrang-Ford L, Biesbroek R et al (2015) Adaptation tracking for a post-2015 climate agreement. Nat Clim Chang 5:967–969. https://doi.org/10.1038/nclimate2744
164. Anguelovski I, Shi L, Chu E et al (2016) Equity impacts of urban land use planning for climate adaptation. J Plan Educ Res 36:333–348. https://doi.org/10.1177/0739456X16645166
165. Ford JD, Berrang-Ford L (2016) The 4Cs of adaptation tracking: consistency, comparability, comprehensiveness, coherency. Mitig Adapt Strateg Glob Chang 21:839–859. https://doi.org/10.1007/s11027-014-9627-7
166. Olazabal M, Ruiz De Gopegui M, Tompkins EL et al (2019) A cross-scale worldwide analysis of coastal adaptation planning. Environ Res Lett 14:124056. https://doi.org/10.1088/1748-9326/ab5532

167. Le TDN (2020) Climate change adaptation in coastal cities of developing countries: characterizing types of vulnerability and adaptation options. Mitig Adapt Strateg Glob Chang 25:739–761. https://doi.org/10.1007/s11027-019-09888-z
168. Campbell T (2013) Beyond smart cities: how cities network, learn and innovate. Taylor and Francis, New York
169. Keiner M, Kim A (2007) Transnational city networks for sustainability. Eur Plan Stud 15:1369–1395. https://doi.org/10.1080/09654310701550843
170. Mocca E (2017) City networks for sustainability in Europe: an urban-level analysis. J Urban Aff 39:691–710. https://doi.org/10.1080/07352166.2017.1282769
171. Tosun J, Leopold L (2019) Aligning climate governance with urban water management: insights from transnational city networks. Water 11:701. https://doi.org/10.3390/w11040701
172. Haupt W, Chelleri L, van Herk S, Zevenbergen C (2020) City-to-city learning within climate city networks: definition, significance, and challenges from a global perspective. Int J Urban Sustain Dev 12:143–159. https://doi.org/10.1080/19463138.2019.1691007
173. Bicknell J, Dodman D, Satterthwaite D (2009) Adapting cities to climate change: understanding and addressing the development challenges. Routledge, London
174. Guodaar L, Asante F, Eshun G et al (2020) How do climate change adaptation strategies result in unintended maladaptive outcomes? Perspectives of tomato farmers. Int J Veg Sci 26:15–31. https://doi.org/10.1080/19315260.2019.1573393
175. Juhola S, Glaas E, Linnér BO, Neset TS (2016) Redefining maladaptation. Environ Sci Policy 55:135–140. https://doi.org/10.1016/j.envsci.2015.09.014
176. Chung Y, Noh H, Honda Y et al (2017) Temporal changes in mortality related to extreme temperatures for 15 cities in northeast Asia: adaptation to heat and maladaptation to cold. Am J Epidemiol 185:907–913. https://doi.org/10.1093/aje/kww199
177. Barnett J, O'Neill S (2010) Maladaptation. Glob Environ Chang 20:211–213. https://doi.org/10.1016/j.gloenvcha.2009.11.004
178. Burksiene V, Dvorak J, Burbulytė-Tsiskarishvili G (2020) City diplomacy in young democracies: the case of the baltics. In: Amiri S, Sevin E (eds) City diplomacy. Springer International Publishing, Cham, pp 305–330
179. Chan DK, Hong (2016) City diplomacy and "glocal" governance: Revitalizing cosmopolitan democracy. Innovation 29:134–160. https://doi.org/10.1080/13511610.2016.1157684
180. Kuşku-Sönmez E (2014) Regional cooperation in the black sea basin: what role for city diplomacy? Southeast Eur Black Sea Stud 14:489–507. https://doi.org/10.1080/14683857.2014.967944
181. Westphal MI, Martin S, Zhou L, Satterthwaite D (2017) Powering cities in the global south: how energy access for all benefits the economy and the environment. World Resource Institute. www.citiesforall.org. Accessed 15 Mar 2021
182. Rana MMP (2009) Sustainable city in the global north and south: goal or principle? Manag Environ Qual An Int J 20:506–521. https://doi.org/10.1108/14777830910981195
183. Rana MMP (2011) Urbanization and sustainability: challenges and strategies for sustainable urban development in Bangladesh. Environ Dev Sustain 13:237–256. https://doi.org/10.1007/s10668-010-9258-4
184. Chu E, Anguelovski I, Roberts D (2017) Climate adaptation as strategic urbanism: assessing opportunities and uncertainties for equity and inclusive development in cities. Cities 60:378–387. https://doi.org/10.1016/j.cities.2016.10.016
185. Chu E, Michael K (2019) Recognition in urban climate justice: marginality and exclusion of migrants in Indian cities. Environ Urban 31:139–156. https://doi.org/10.1177/0956247818814449
186. Porter L, Rickards L, Verlie B et al (2020) Climate justice in a climate changed world. Plan Theory Pract 21:293–321. https://doi.org/10.1080/14649357.2020.1748959
187. Jenkins K (2018) Setting energy justice apart from the crowd: lessons from environmental and climate justice. Energy Res Soc Sci 39:117–121. https://doi.org/10.1016/j.erss.2017.11.015
188. French M, Ramirez-Lovering D, Sinharoy SS et al (2020) Informal settlements in a COVID-19 world: moving beyond upgrading and envisioning revitalisation. Cities Heal 1–4. https://doi.org/10.1080/23748834.2020.1812331

189. Hakovirta M, Denuwara N (2020) How COVID-19 redefines the concept of sustainability. Sustainability 12:3727. https://doi.org/10.3390/su12093727
190. Bhide A (2020) Informal settlements, the emerging response to COVID and the imperative of transforming the narrative. J Soc Econ Dev 1–10. https://doi.org/10.1007/s40847-020-00119-9
191. Lau LS, Samari G, Moresky RT et al (2020) COVID-19 in humanitarian settings and lessons learned from past epidemics. Nat Med 26:647–648. https://doi.org/10.1038/s41591-020-0851-2
192. von Seidlein L, Alabaster G, Deen J, Knudsen J (2021) Crowding has consequences: prevention and management of COVID-19 in informal urban settlements. Build Environ 188:107472. https://doi.org/10.1016/j.buildenv.2020.107472
193. Gupte J, Mitlin D (2020) COVID-19: what is not being addressed. Environ Urban 095624782096396. https://doi.org/10.1177/0956247820963961
194. Corburn J, Vlahov D, Mberu B et al (2020) Slum health: arresting COVID-19 and improving well-being in urban informal settlements. J Urban Heal 97:348–357. https://doi.org/10.1007/s11524-020-00438-6
195. Gupta S, Nguyen TD, Rojas FL et al (2020) Tracking public and private responses to the COVID-19 epidemic: evidence from state and local government actions. NBER Working Paper
196. Gil D, Domínguez P, Undurraga EA, Valenzuela E (2021) The socioeconomic impact of COVID-19 in urban informal settlements. medRxiv 2021.01.16.21249935. https://doi.org/10.1101/2021.01.16.21249935
197. Nyashanu M, Simbanegavi P, Gibson L (2020) Exploring the impact of COVID-19 pandemic lockdown on informal settlements in Tshwane Gauteng Province, South Africa. Glob Public Health 15:1443–1453. https://doi.org/10.1080/17441692.2020.1805787
198. Zhang W, Wang K, Yin L et al (2020) Mental health and psychosocial problems of medical health workers during the COVID-19 epidemic in China. Psychother Psychosom 89:242–250. https://doi.org/10.1159/000507639
199. Winter SC, Obara LM, McMahon S (2020) Intimate partner violence: a key correlate of women's physical and mental health in informal settlements in Nairobi, Kenya. PLoS One 15:e0230894. https://doi.org/10.1371/journal.pone.0230894
200. Harris J, Cook T, Gibbs L, et al (2018) Searching for the impact of participation in health and health research: challenges and methods. Biomed Res Int 9427452. https://doi.org/10.1155/2018/9427452
201. Jennings Mabery M, Gibbs-Scharf L, Bara D (2013) Communities of practice foster collaboration across public health. J Knowl Manag 17:226–236. https://doi.org/10.1108/13673271311315187
202. Adekola PO, Iyalomhe FO, Paczoski A et al (2021) Public perception and awareness of waste management from Benin City. Sci Rep 11:306. https://doi.org/10.1038/s41598-020-79688-y
203. Van Belle S, Affun-Adegbulu C, Soors W et al (2020) COVID-19 and informal settlements: an urgent call to rethink urban governance. Int J Equity Health 19:81. https://doi.org/10.1186/s12939-020-01198-0
204. Buckley RM (2020) Targeting the world's slums as fat tails in the distribution of COVID-19 cases. J Urban Heal 97:358–364. https://doi.org/10.1007/s11524-020-00450-w
205. Leach M, Scoones I, Stirling A (2010) Governing epidemics in an age of complexity: narratives, politics and pathways to sustainability. Glob Environ Chang 20:369–377. https://doi.org/10.1016/j.gloenvcha.2009.11.008

Adaptive Knowledge Sharing in Turbulent Times: Urban Disaster Risk and Knowledge Management

Bryan Joel S. Mariano, Winifredo Dagli, and Giuseppe T. Cirella

Abstract Knowledge management in development discourse has been largely shaped by the emergence and complexity of modern times. The increasing number and intensity of catastrophes as a result of a changing climate render a new social order with high complexity, uncertainty, and ambiguity. Designing knowledge sharing strategies can be informed by complexity theory and its application to organizational environments in enabling a framework for strengthening institutional adaptive capacity. This chapter investigates adaptive knowledge sharing and interlinks urban disaster risk reduction and management in the context of case research from Typhoon Ketsana that hit Marikina City, Metro Manila, the Philippines in 2009. An in-depth examination of knowledge sharing in terms of complexity—channels, behavior and locality, and multiplicity—is looked at with reference to social networks and strategies to effectively mitigate risk and strengthen community disaster resilience. Using social network analysis, four linkages are identified: disaster prevention and hazard mitigation, disaster preparedness, disaster response, and recovery and rehabilitation. Categorical findings on adaptive knowledge sharing are presented.

Keywords Knowledge sharing · Complexity · Disaster management · Social network analysis · The Philippines

B. J. S. Mariano (✉)
Department of Geography, College of Social Sciences and Philosophy, University of the Philippines Diliman, Quezon, Philippines
e-mail: bsmariano@up.edu.ph

W. Dagli
Department of Science Communication, College of Development Communication, University of the Philippines Los Baños, Laguna, Philippines
e-mail: wbdagli1@up.edu.ph

G. T. Cirella
Faculty of Economics, University of Gdansk, Sopot, Poland
e-mail: gt.cirella@ug.edu.pl

G. T. Cirella (ed.), *Human Settlements*, Advances in 21st Century Human Settlements,
https://doi.org/10.1007/978-981-16-4031-5_2

1 Introduction

The scourge of disasters which brought unparalleled scale of damages demand new strategies that would effectively reduce risks and uncertainties in complex organizational environments. In today's knowledge-based economies, knowledge is said to be a key driver of sustainable development [1]. From this conjecture, decision-makers and various development organizations realize the potential of knowledge management (KM) in addressing development problems. There is a placement of the creation, transfer, and management of knowledge at the core of their activities [2]. KM in the context of development is viewed as the optimization of identification, sharing, and use of knowledge among local communities and development workers [3]. One of the core processes in KM is knowledge sharing, which covers social interaction and communication that leads to learning. Knowledge sharing is defined as a process of mutual exchange of knowledge to enable action and new knowledge [4]. The main goal is either to create new knowledge through the combination of existing ones or become better in utilizing existing ones [5]. Knowledge sharing is a complex process that involves the interaction of several agents within a complex system [6].

Knowledge that emerges in complex environments is due to unpredictable, dynamic, and nonlinear interaction of multiple agents; however, knowledge sharing systems are often designed on the basis of who, what, where, why, and how one will get certain knowledge in a given context [7]. This mechanistic view is being challenged by more decentralized and adaptive ways in which actors and institutions in increasingly complex organizational environments are able to reconnect and reorganize in the face of enduring stress or abrupt shocks [8]. The age of complexity poses a huge challenge to many institutions in thinking of new ways to understand the complex, interrelated, and ever-changing world. Scientists, particularly physicists, evolutionary biologists, and ecologists, described complex systems as self-organizing aggregates of individuals whose dynamic interactions create large-scale patterns, which in turn influence individuals' actions [9, 10]. Adapting to the complexity of the environment brought by disasters is undoubtedly one of the biggest challenges in the development sector. Disaster risk reduction and management (DRRM) involves cross-scale and multilevel coordination among various agents from different governing bodies (i.e., local and national government, local and international non-governmental organizations, civil society groups, the private sector, and grassroots communities). It needs to have a holistic approach and include stakeholders at all levels to effectively mitigate risks and enable resiliency to perturbations brought by disasters. This makes the DRRM environment in an urban setting a complex and knowledge-intensive endeavor. This chapter highlights how adaptive knowledge sharing can be fostered in a complex organizational landscape by analyzing the dynamics of knowledge sharing within the broader context of disaster risk reduction and urban development in Marikina City, the Philippines.

2 Knowledge Sharing in the Age of Complexity

The world's economies refer to knowledge as one of the most significant assets [11]. There is in fact a growing consensus among development organizations to look at development problems differently through the knowledge perspective. According to Talisayon and Suministrado [3], managing and sharing knowledge in the context of development is viewed as the optimization of identification, sharing, and use of knowledge among local communities and development workers. Moreover, the World Bank Institute [12] considers knowledge as something that can contribute to development because it acts as an enabler of institutions and governance through policymaking, as a driver of competitiveness and productivity, and as a facilitator of welfare and human security, among others. This gives a clear picture of how knowledge is positioned at the core of development issues—including DRRM. An alternative framing of the place of knowledge in global development is that the dominant forms of knowledge and knowledge production have become the means to sustain an undesirable status quo where inequality is deeply entrenched both globally and within countries. The ways in which scientific knowledge and technology have been appropriated to serve the interests of the powerful and privileged have undoubtedly contributed to the sustainability challenges that development institutions are trying to address [13].

Knowledge sharing as a research arena has gained much attention from researchers in various fields due to potential benefits in the learning of individuals, groups, and within and across organizations. Several scholars similarly argue that there are four generations of KM [14–16]. The first generation of KM is referred to as the rationalist approach, wherein the purpose is knowledge transfer and is primarily information and communication technology (ICT)-driven. It views knowledge as something that can be transferred in a linear fashion which often neglects the local context of development. The study of knowledge sharing is rooted from technology transfer and innovation literature, wherein the focus is on storing and reusing codified knowledge with the aid of ICTs [17, 18]. This view has been challenged because it does not fully recognize the inherent collaboration and social interaction needed to effectively share knowledge. From this, the perspective has shifted to a more people-centric approach which focuses on the emergence of knowledge from context-specific social interactions [15, 19]. The second generation is focused on organizational learning and how knowledge can be used to increase efficiency and effectiveness of development processes. Meanwhile, the third generation gives importance to knowledge sharing across organizational boundaries through communities of practice. Lastly, the fourth generation is referred to as the post-rationalist approach, which is geared toward the flow and emergence of knowledge from social construction of individual practices in relation to its context [15, 16]. This involves situated learning, wherein the emergence and construction of knowledge in local context-specific practices are embedded within specific socio-spatial and institutional contexts [15]. The practice-based approach considers the specific context in which local knowledge is developed as well as the multiple knowledge that agents hold [19]. Building upon the four generation, a fifth generation of KM for development has been proposed and identified as development knowledge ecology which highlights the interrelatedness of the elements in a system [16].

It is noteworthy, however, that these generations of KM are often interrelated, borrowing and navigating key features from one another. For the purpose of this chapter, discussions will focus on the fifth generation of KM as proposed by Cummings et al. [16] where the positioning of adaptive knowledge sharing is examined. One of the basic features of this paradigm is the recognition of complexity and emergence. Complex adaptive systems (CASs) are considered complex because of the nonlinear interactions among agents within the system [10]. Although scholars have not agreed on a universal definition for a complex system, there are several basic features of a complex system that they have similarly described. Complex systems are made up of agents that are said to be inherently adaptive [20]. These adaptive agents interact with each other that engender co-evolution which, in turn, contribute in shaping the entirety of CAS [21–23]. Janus and Paulo [24] suggest that complexity theory can provide new ways of improving aid systems, which institutionally encompasses mechanisms for disaster risk management. In complex systems, emergence occurs in the edge of chaos where several agents interact, self-organize, and create new kinds of order [25]. Emergence points out how the unpredictable behavior of systems emerge based on the interconnectedness and interaction of agents with its environment [20]. Outcomes of micro-level interactions are largely uncertain and unpredictable, although longitudinal analysis may point to early warning indicators of a looming system transition [26]. Because of this, over-controlling approaches such as top-down techno-managerial fixes and risk management approaches based on known or calculated risks can be counterproductive. Instead, more flexible and contingent responses are better designed in dealing with uncertainties or areas of decision-making where probabilities and potential outcomes are not known [27]. Moreover, there should be minimum rules to allow space for innovation and maximize the system's adaptive capability [20]. The concepts of interconnectedness and nature of interaction or knowledge sharing, which includes the locality and multiplicity of knowledge, are integral in the study of emergence in complex systems.

Sellnow et al. [28] argued that natural disasters have interactions with human structures and processes in highly complex and unpredictable manner. Management of disaster risk is inherently complex and dynamic. It entangles adaptive knowledge sharing for effective decision-making and coordination among multiple agents and organizations across various levels and locations. Likewise, looking at complexity theory through the concept of feedback can give a new perspective in viewing disasters. While disasters destabilize the natural order of human systems [29] as well as social and ecological systems [30], it also gives space for the emergence of new order. This emergence opens up an opportunity for organizations to realize its maximum adaptive capacity in such complex environments and transform themselves into a more desired state [30]. Emergence points out how the overall properties of a complex system evolve from interconnectedness and interaction of the system with its environment as well as in the local context how agent's interaction must be considered highly important [20, 31]. This makes the whole system different from the sum of its parts.

2.1 Knowledge Sharing Channels, Knowledge Sharing Behavior and Locality, and Multiplicity of Knowledge

Included in the nature of knowledge sharing are the channels, also called medium or tools, being used when sharing knowledge. These are crucial in the knowledge sharing process because they serve as platforms to share complex tacit knowledge, translate tacit knowledge to more explicit forms, and combine explicit knowledge forms. These channels include documents (e.g., reports, manuals, proceedings, meeting minutes, and databases), face-to-face communication (i.e., formal such as meetings, trainings, and lectures vs. non-formal such as coffee break talks), and computer-mediated communication (e.g., ICTs, social networking sites such as Facebook, Twitter, YouTube, Skype, Intranet, databases, mobile phone usage, and e-mail). The channels that are available and are being used by agents in the complex system are highly important. This shows how the system adapts to the requirements of the ever-changing environment. For instance, because of geographical boundaries, it might not be possible to use face-to-face communication, so video conference and online communities of practice might be the viable alternative.

One of the research areas within the knowledge sharing literature that have been studied by several scholars is organizational culture. According to Rivera-Vazquez et al. [32], knowledge sharing processes are influenced by the culture of organization. As such, a culture of knowledge sharing and face-to-face communication is highly important and encouraged [33]. To be effective in sharing knowledge, organizations must identify and overcome cultural barriers. In a study conducted by Oye et al. [34], culture is considered as one of the critical factors that affect knowledge sharing in the workplace. According to Zin Aris [35], drawing upon the organizational culture, there are three major factors that affect the knowledge sharing behavior of members in an organization. These include individual factors which include individual attitude, organizational factors which include management system, incentive system, and the organizational culture itself, and lastly, the technological factor. Other researchers articulated similar typologies of factors affecting knowledge sharing behavior [35]. These studies have their focus on the antecedents of knowledge sharing behavior such as motivation [34, 36] and trust [37–39]. In a survey conducted by Al-Alawi et al. [40] in several private and public organizations in Bahrain, it was indicated that trust, communication, information systems, rewards, and organization structure in organizations play an important role in breaking down the obstacles to knowledge sharing. Some similarities to the findings of Al-Alawi et al. [40] are present in the results of another survey conducted by Amayah [41] where different factors affected knowledge sharing i.e., enablers were social interaction, rewards, and organizational support, while barriers were lack of both courage to share and empathy. Meanwhile, the phrase that "knowledge is power" may lead to knowledge hoarding of individuals [42]. Ipe [42] postulated that power and status in the organization determine the agents' motivation for knowledge sharing and direction of knowledge flow.

There is a realization on the potential of local knowledge in enhancing development effectiveness in the issues of climate change [43], disaster management [44], and agriculture [45]. Beckford and Barker [45] argued that local knowledge is highly important in coping with risk and uncertainty in an ever-changing world. The knowledge from local people is a significant asset when responding to emergencies like disasters and crises. The United Nations International Strategy for disaster reduction's [46] publication "Indigenous Knowledge for Disaster Risk Reduction: Good Practices and Lessons Learned from Experiences in the Asia–Pacific Region," has provided different cases that show how local knowledge can serve as an effective tool in reducing risk from natural disasters. One of which reveals an effective result in combining local knowledge with scientific knowledge in the flood warning system in Dagupan City, Pangasinan, the Philippines. Dekens [44] conducted a literature review on local knowledge and provided a framework of local knowledge in disaster preparedness and its usage in disaster management. The study focalized on local knowledge composed of knowledge types (i.e., technical, ecological, historical, and cultural), practices (i.e., individual, household, and community levels, technical and nontechnical, and short-and long-term) and beliefs, values (i.e., respect, reciprocity, sharing, and humility), and worldviews. Inherent in the complexity of development problems and organizational environment are the large number and different interacting agents. These agents may have differences in terms of culture, belief, values, dogma, and worldview that can lead to multiplicity of knowledge. Recognizing this multiplicity of knowledge from various agents is highly important in the development process. According to Zirschky [47], various knowledge bases and perspectives are needed to be understood and integrated in knowledge management strategy. Moreover, Jones et al. [48] argued that multiple knowledge sources are crucial when it comes to informing policy.

3 DRRM and Social Networks

On November 8, 2013, Super Typhoon Haiyan (i.e., locally named Yolanda) wreaked unprecedented havoc in the Philippines. It claimed more than 6000 lives, displaced over four million Filipinos, and wiped out over 1.1 million homes, with the majority of damages to properties and loss of lives more pronounced in central Philippines [49]. It is considered one of the five strongest typhoons ever recorded and, undoubtedly, the most devastating weather-related disaster in the country with almost 40 billion pesos (i.e., nearly USD 1 billion) of total cost of damages in infrastructure and agriculture [49]. The complex nature of disaster risk reduction environment makes it important to understand the networks in which agents interact. The interconnectedness of actors within a network can influence knowledge sharing among them. Also, the frequency and degree of interaction may influence the strength of relationships among actors in a network. The study of networks is prominent in knowledge sharing research. Henderson [50] illustrated knowledge networks or communities of practice as the principal knowledge sharing approach of the United Nations Development Programme (UNDP). It was emphasized that these can be a strategic entry point in putting up KM initiatives within development agencies.

The networks, established in 1999, have become a tool for capacity-building, to connect country offices, and to promote South–South knowledge exchange. Also, the networks function as a knowledge sharing tool for best practices with discussion of its thematic areas available for internal and external UNDP staff globally. It has 20 knowledge networks, involving practice, sub-practice, knowledge, and cross-cutting types. These networks emphasized the connection rather than the collection of knowledge and considered as the entry point for UNDP's focus on KM. These networks also foster capacity development, especially the professional development as well as individual and organizational learning.

In studying networks and knowledge sharing, scholars usually employ social network analysis (SNA) as a theory, design, or methodology to illuminate behaviors and relationships of individuals in the organization. A social network is composed of a set of agents and their social relations. SNA gives emphasis on the patterns of interpersonal and social relationships among agents within the network. The agents can be individuals, groups, organizations, or institutions, and the analysis of ties or connections may vary on individual to organizational level. Several indicators are being used in the study of social networks. According to Chang and Hsieh [51], these include: (1) size which pertains to the number of nodes in the network, (2) density which is an indicator measuring the tightness among the actors in the network and defined as the ratio between the actual number of the linkages and the greatest number of linkages possible, and (3) centrality which is a measure of the linkage between actors and identifies the role of the actor in the network. Centrality looks into the degree, closeness, and betweenness of linkages. In a study of SNA on knowledge sharing of scientific groups, Lei and Xin [52] used self-administered questionnaires to collect information about the university members' individual behaviors and attitudes toward knowledge sharing as well as their interconnectedness. The indicators of measurement that were used included: (1) tie strength, (2) small group, (3) centrality, and (4) structural hole. Tie strength refers to the existing links between and among actors because of communication and interaction—it can be classified as strong (i.e., tight connections) or weak (i.e., loose connections). On the one hand, Lei and Xin [52] argued that strong ties can be important in knowledge sharing because it involves a high level of trust; however, one of its pitfalls is that knowledge can be repeated, and creation of new ones can be hampered. On the other hand, actors forming weak ties can have a wide range of knowledge that can be exchanged. This so-called strength of weak ties has been postulated by Granovetter [53] as cited in Prell [54]. Granovetter [53] aimed to know the nature and strength of a relationship between actors by measuring its tie strength in terms of frequency of contact (i.e., often, occasionally, or rarely). Specifically, the data was gathered among professionals in Massachusetts. The study tried to find out how these actors used their personal networks to gain information about new job opportunities. Results revealed that roughly 15% said they saw their contacts often, 55% said occasionally, and 29% said rarely. From these findings, weak ties are concluded to be a more important relation in finding information about new job opportunities than strong ties.

Centrality is one of the most popular concepts in SNA. It rests on the idea of who is at the core of social network structure. Furthermore, the analysis of centrality reveals that actors have different positions in the knowledge sharing network [52]. Centrality also unravels not just who is at the core, but also who are at the periphery of the network. This implies that central actors hold power to include those actors who are in the periphery to achieve more effective knowledge sharing. There are several measures of centrality (i.e., not just the number of in and out degree per actor or the degree centrality used by Marquez et al. [55]) that have been laid out by Prell [54]. These centrality measures include eigenvector centrality, betweenness centrality, closeness centrality, and beta-centrality. The eigenvector centrality is the sum of an actor's connection to other actors, weighted by degree centrality [54], while beta-centrality looks at the relationship between centrality and power (Bonacich [56] as cited by Prell [54]). According to Prell [54], eigenvector and beta-centrality take into account the degree centrality of other actors in the network, while degree, betweenness, and closeness centrality do not. However, according to a study by Chang and Hsieh [51], no matter what these centrality measures, those with higher centrality play a central role in the network, e.g., as knowledge contributors and holders of knowledge resources. SNA is a long-standing research interest, with a wide range of literature in social and behavioral sciences, among others. Over time, it became not just an analytical technique, but a paradigm which carries its own theoretical underpinnings, methodologies, and tools [54]. In conducting SNA, random sampling may not be suitable since SNA's unit of analysis is the relationship of actors in the network and not just merely the actors themselves.

4 Case Research from Marikina City

Marikina City in the Metro Manila area of the Philippines is one of the high-risk municipalities in Metro Manila when it comes to flooding, earthquake, and landslides. In 2009, Typhoon Ketsana (i.e., locally named Ondoy) ravaged 14 out of 16 barangays in Marikina City. The highest flood level, i.e., 22.6 m, in the city was recorded during the onslaught of this typhoon. With this, 80% of the residents were affected and 3059 families were evacuated. The Marikina City government estimated that some 55,926 houses were damaged, 1652 businesses were affected, and most of the government-owned equipment and vehicles as well as the city hall were damaged, leaving the city government operations momentarily paralyzed, based on data from the Marikina City DRRM authorities. Apparently, this shows that although Marikina City has institutionalized DRRM measures, its competence needs to be further improved so as to respond effectively to extreme weather-related hazards. The onslaught brought by Typhoon Ketsana served as tipping point to the Marikina City government to gear up more plans and actions toward resiliency. After the enactment of the Philippine Disaster Risk Reduction and Management Act in 2010, Marikina City established a DRRM Office, and the DRRM Council was reactivated in 2011. In 2013, the City developed its local DRRM plan (i.e., 2013–2019) and contingency plan detailing all the plans and actions needed to make Marikina City more resilient to disasters. It is now a continuing priority program of the City. Twenty-three members of the Marikina

City DRRM Council answered a standardized interview survey and participated in a series of in-depth interviews. The member cross-examinations included aspects of socio-demographic characteristics, views on effectiveness of various knowledge sharing channels, understanding of practice of adaptive knowledge sharing based on complexity concepts, and a component on SNA.

The interviews revealed that formal knowledge sharing channels are viewed as the most effective when sharing knowledge about DRRM within the Council due to the formal and bureaucratic nature of the organization. Meanwhile, all of the agents of the system scored "high" on adaptive knowledge sharing behavior—even though incentives and rewards are not remunerated. They regard knowledge sharing as integral to their work in DRRM. They also acknowledge the importance of knowledge of Marikina City residents. Likewise, they believe that various offices or sectors possess different knowledge that are valuable in DRRM. Lastly, the DRRM Council agents can freely share knowledge because the local government does not enforce restrictive rules governing informal interactions. Given that the DRRM landscape is complex and continuously changing, these sound practices allow the DRRM Council agents to learn and adapt to changes.

In the analysis of social networks, the metrics that were analyzed included network size and density, inclusiveness, tie strength, and centrality measures. Computer software called UCINET version 6.532 was used to analyze data on the said key indicators. The sociogram or visualization of the social networks on four thematic areas of DRRM was generated using Net Draw application in UCINET (Fig. 1). Key informant interviews were also conducted with selected participants to validate the results.

The DRRM Council's network was considered inclusive but not dense. If the network becomes too dense, intense knowledge sharing reduces individual variation and can lead to system collapse [57]. This means that if everyone is connected to one another, the same information can be repeated, and this dampens the capacity of the system to generate new knowledge, and thus, the system fails to innovate. The density of the networks also reflects its robustness or the vulnerability of the system to collapse [52]. If the central actors experienced problems or removed from the networks, the networks will be disconnected and may not withstand this fragmentation. The identified social networks generally have low frequency of interaction. These interactions happen mostly on monthly council meetings.

Moreover, the ratio of reciprocated ties is also low in the identified social networks. With these, the DRRM Council's agents have weak ties. Strong ties are important in knowledge sharing because this kind of tie involves a high level of trust; the same reason why this kind of tie is common with networks of close friends and family members. However, being in the network with strong ties, knowledge can be repeated, and creation of new ones may be hampered. In contrast to this, Granovetter [53] proposed the concept of strength of weak ties and argued that weakly tied agents tend to connect with other agents different from their background and will then have access to information separate from what they usually receive. By forming this kind of tie, agents can have a wider range of knowledge that can be exchanged, which

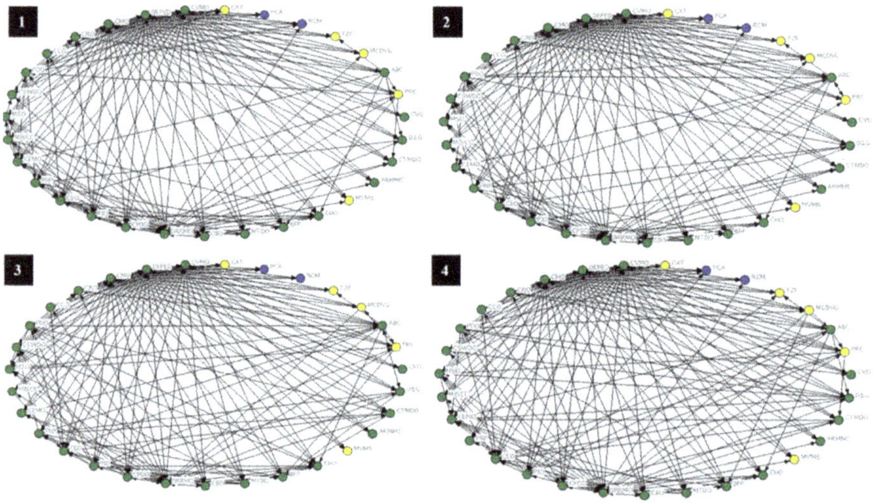

Fig. 1 Sociogram of the Marikina City DRRM Council social networks on (1) disaster prevention and hazard mitigation, (2) disaster preparedness, (3) disaster response, and (4) recovery and rehabilitation. The circles represent the agents (i.e., green = local government offices and national agencies, yellow = civil society organizations, and blue = private sector). The arrow heads pointing to agents mean that these agents were identified by other agent(s) as part of the social network. The sociograms depict strong centrality of local government offices in the network, but lacks dense connection with civil society groups and private sector

is more important in the knowledge construction process. By having weak ties, the agents in the DRRM Council can capitalize on this opportunity for the emergence of new knowledge about the four thematic areas in DRRM.

5 Conclusion: Key Features of an Adaptive Knowledge Sharing in a Complex DRRM System

The idea of adaptive knowledge sharing embedded in this chapter has a similarity to the concept of adaptive co-management. Armitage et al. [58] explained the principal features of adaptive co-management which include the following: (1) innovative institutional arrangements and incentives across socio-temporal scales and levels, (2) learning through complexity and change, (3) monitoring and assessment of interventions, (4) role of power, and (5) opportunities to link science with policy. Adaptive co-management is highly grounded in the paradigm of complexity that recognizes the uncertainty in systems, as well as to focus on adaptive capacity through learning. Moreover, Folke et al. [30] emphasized the importance of social networks in adaptive governance as it provides self-organizing systems that bridge local and scientific knowledge in supporting and understanding policies.

Based on the case research done with Marikina City DRRM Council, four key themes were developed in explaining adaptive knowledge sharing in complex DRRM system. First, flexibility is a key component in adaptive capacity. Flexibility can be fostered by tapping the potentials of various kinds of knowledge sharing channels, whether formal or informal, and using such depending on the context. Flexibility also becomes crucial when the internal and external contexts of the organization changes through time. Lastly, in social networks, flexibility is essential to allow knowledge flow and move across the network. Second, network density reflects robustness and resilience of the system. Less dense networks, as seen in the case of the Marikina City DRRM Council, enable members to assert their individual strengths that are necessary in balancing persistence and change. Third, skewed attention given to scientific knowledge undermines the importance of local knowledge in reducing uncertainties and in dealing with complexity. Civil society organizations and private sectors have unique knowledge that can contribute to DRRM initiatives. By engaging a broader base of external stakeholders in the knowledge sharing activities, multiple knowledge will be integrated, and this will enable the institution to reduce the uncertainties thereby increasing adaptive capacity of agents in DRRM Council. Moreover, the marginalized sector, youth, women, and religious groups, among others may have representatives in the Council to incorporate a more holistic and participatory DRRM. A re-examination of roles among the members of the Council may also be considered. For instance, the City Environment Office is the lead for recovery and rehabilitation program. However, being also the lead office in climate change adaptation in Marikina City, the City Environment Office can take a more active role in prevention and mitigation aspect of DRRM. Fourth, social networks are important in navigating DRRM organizations toward the edge of chaos. Local governments, being at the forefront of DRRM initiatives, are by nature formal organizations. Their structures, roles, and processes are guided by city ordinances and executive orders. However, with social networks, being a self-organized endeavor, DRRM organizations can counterbalance the traditional bureaucracies and operate in a more adaptive state. At length, local governments may be informed on how they should design knowledge sharing strategies that would effectively mitigate risk and save more lives. The culture of good governance plays a crucial part in valuing the importance of multi-stakeholder collaboration, local leadership, and community building. This chapter provides information to other local governments, particularly in the Philippines, which are vulnerable to disasters, and provides policy implications in developing collaboration mechanisms of agents in DRRM councils. This includes giving recommendation on who are the agents that need to be included in DRRM as well as streamlining an adaptive knowledge sharing framework on disaster risk reduction to strengthen communities as well as their resiliency in the complexity and uncertainty brought about by disasters.

References

1. Bellanet International Secretariat (2000) knowledge management for development organizations. Report of the proceedings of the knowledge management brighton workshop. University of Sussex, Brighton
2. World Bank (1999) World development report 1998/1999: knowledge for development. Oxford University Press, New York
3. Talisayon SD, Suministrado JP (2008) Community wealth rediscovered: knowledge for poverty alleviation. Center for Conscious Living Foundation, Manila
4. Van Den Hooff B, Ridder JA (2004) Knowledge sharing in context: the influence of organizational commitment, communication climate and CMC use on knowledge sharing. J Knowl Manag 8:117–130. https://doi.org/10.1108/13673270410567675
5. Christensen PH (2007) Knowledge sharing: Moving away from the obsession with best practices. J Knowl Manag 11:36–47. https://doi.org/10.1108/13673270710728222
6. Sherif K, Xing B (2006) Adaptive processes for knowledge creation in complex systems: the case of a global IT consulting firm. Inf Manag 43:530–540. https://doi.org/10.1016/j.im.2005.12.003
7. Metcalfe M (2005) Knowledge sharing, complex environments and small-worlds. Hum Syst Manag 24:185–195
8. Leach M, Scoones I, Stirling A (2010) Dynamic sustainabilities: technology, environment, and social justice. Earthscan, Bristol
9. Kauffman S (1993) The origins of order: self-organization and selection in evolution. Oxford University Press, New York
10. Levin S, Xepapadeas T, Crépin AS et al (2013) Social-ecological systems as complex adaptive systems: modeling and policy implications. Environ Dev Econ 18:111–132. https://doi.org/10.1017/S1355770X12000460
11. ADB (2004) Knowledge management in ADB. Asian development bank, Mandaluyong, The Philippines
12. World Bank (2007) Building knowledge economies. World Bank, Washington, DC
13. International Panel on Social Progress (2018) Rethinking society for the 21st century. Cambridge University Press, Cambridge
14. Park Y, Kim S (2006) Knowledge management system for fourth generation R&D: KNOWVATION. Technovation 26:595–602. https://doi.org/10.1016/j.technovation.2004.10.008
15. Ferguson J, Huysman M, Soekijad M (2010) Knowledge management in practice: pitfalls and potentials for development. World Dev 38:1797–1810. https://doi.org/10.1016/j.worlddev.2010.05.004
16. Cummings S, Regeer BJ, Ho WWS et al (2013) Proposing a fifth generation of knowledge management for development: investigating convergence between knowledge management for development and transdisciplinary research. Knowl Manag Dev J 9:10–36
17. Cummings J (2003) Knowledge sharing: a review of literature. OED working papers, world bank operations evaluation department, Washington, DC
18. Guzman MS (2007) Using ICTs for knowledge sharing and collaboration: an international experience based on Bellanet's work in the South. Knowl Manag Dev J 3:19–30
19. erguson J, Mchombu K, Cummings S (2008) Management of knowledge for development: meta-review and scoping study. IKM Emergent Res Programme, Bonn
20. Ramalingam B, Jones H, Reba T, Young J (2008) Exploring the science of complexity: ideas and implications for development and humanitarian efforts. Overseas Development Institute, London
21. Medd W (2004) Complexity science and the social world. In: Kempf-Leonard K (ed) Encyclopedia of social measurement. Elsevier, London, pp 399–403
22. Cilliers P (1998) Complexity and postmodernism: understanding complex systems. Routledge, London
23. Larsen-Freeman D (1997) Chaos/complexity science and second language acquisition. Appl Linguist 18:141–165. https://doi.org/10.1093/applin/18.2.141

24. Janus H, Paulo S (2013) Does aid stand on the age of chaos? The current curriculum. German Dev Inst, Bonn
25. McKelvey B, Andriani P (2010) Avoiding extreme risk before it occurs: a complexity science approach to incubation. Risk Manag 12:54–82. https://doi.org/10.1057/rm.2009.14
26. Scheffer MJ, Bascompte J, Brock WA et al (2009) Early-warning signals for critical transitions. Nature 461:53–59
27. Stirling A (2010) Keep it complex. Nature 468:1029
28. Sellnow TL, Seeger MW, Ulmer RR (2000) Chaos theory, informational needs, and natural disasters. Central states communication association, Detroit
29. Provitolo D, Dubos-Paillard E, Müller JP (2011) Emergent human behaviour during a disaster: thematic versus complex systems approaches. In: Proceedings of EPNACS 2011 within ECCS'11 emergent poperties in natural and artificial complex systems. European conference on complex systems, Vienna
30. Folke C, Hahn T, Olsson P, Norberg J (2005) Adaptive governance of social-ecological systems. Annu Rev Environ Resour 15:441–473. https://doi.org/10.1146/annurev.energy.30.050504.144511
31. Cutler RM (2002) Complexity science and knowledge-creation in international relations theory. Int Relations 1:1–7
32. Rivera-Vazquez JC, Ortiz-Fournier LV, Flores FR (2009) Overcoming cultural barriers for innovation and knowledge sharing. J Knowl Manag 13:257–270. https://doi.org/10.1108/13673270910988097
33. Pels J, Odhiambo F (2005) Design of and practical experiences with the Learn@WELL knowledge management module. Knowl Manag Dev J 1:4–18
34. Oye ND, Salleh M, Noorminshah (2011) Knowledge sharing in workplace: motivators and demotivators. Int J Manag Inf Technol 3:71–84. https://doi.org/10.5121/ijmit.2011.3406
35. Zin Aris AZ (2013) Requirement for knowledge sharing behavior: a review of empirical findings. J Asian Sci Res 3:517–526
36. Akram F, Bokhari DR (2007) The role of knowledge sharing on individual performance, considering the factor of motivation-the conceptual framework. Int J Multidiscip Sci Eng 11:22–42
37. Holste JS, Fields D (2010) Trust and tacit knowledge sharing and use. J Knowl Manag 14:128–140. https://doi.org/10.1108/13673271011015615
38. Rutten W, Blaas-Franken J, Martin H (2016) The impact of (low) trust on knowledge sharing. J Knowl Manag 20:199–214. https://doi.org/10.1108/JKM-10-2015-0391
39. Casimir G, Lee K, Loon M (2012) Knowledge sharing: influences of trust, commitment and cost. J Knowl Manag 16:740–753. https://doi.org/10.1108/13673271211262781
40. Al-Alawi AI, Al-Marzooqi NY, Mohammed YF (2007) Organizational culture and knowledge sharing: Critical success factors. J Knowl Manag 11:22–42. https://doi.org/10.1108/13673270710738898
41. Amayah AT (2013) Determinants of knowledge sharing in a public sector organization. J Knowl Manag 17:454–471. https://doi.org/10.1108/JKM-11-2012-0369
42. Ipe M (2003) Knowledge sharing in organizations: a conceptual framework. Hum Resour Dev Rev 2:337–359. https://doi.org/10.1177/1534484303257985
43. Bone C, Alessa L, Altaweel M et al (2011) Assessing the impacts of local knowledge and technology on climate change vulnerability in remote communities. Int J Environ Res Public Health 8:733–761. https://doi.org/10.3390/ijerph8030733
44. Dekens J (2007) Local knowledge for disaster preparedness: a literature review. International Centre for Integrated Mountain Development, Kathmandu
45. Beckford C, Barker D (2007) The role and value of local knowledge in Jamaican agriculture: adaptation and change in small-scale farming. Geogr J 173:118–128. https://doi.org/10.1111/j.1475-4959.2007.00238.x
46. UNDRR (2008) Indigenous knowledge for disaster risk reduction: good practices and lessons learned from experiences in the Asia-Pacific region. UNDRR, Bangkok

47. Zirschky P (2009) Knowledge management and multiple knowledges: a multi-case study within the development sector. KM Emergent Res Programme, Bonn
48. Jones N, Datta A, Jones H et al (2009) Knowledge, policy and power: six dimensions of the knowledge-development policy interface. Overseas Development Institute, London
49. NDRRMC (2016) NDRRMC Update: SiteRep No. 38 Effects of typhoon yolanda (Haiyan). In: Natl. Disaster Risk Reduction Management Council. http://ndrrmc.gov.ph/. Accessed 1 Feb 2021
50. Henderson K (2005) The knowledge sharing approach of the United Nations Development Programme. Knowl Manag Dev J 1:19–30
51. Chang CT, Hsieh PH (2012) The interaction patterns of knowledge sharing in design forums. PACIS 2012 Proc 171
52. Lei G, Xin G (2011) Social network analysis on knowledge sharing of scientific groups. J Syst Manag Sci 1:79–89
53. Granovetter MS (1973) The strength of weak ties. Am J Sociol 78:1360–1380
54. Prell C (2012) Social network analysis. Sage, London
55. Marques DV, Cardoso L, Zappalá S (2008) Knowledge sharing networks and performance. Comport Organ e Gestão 14:161–192
56. Bonacich P (1987) Power and centrality: a family of measures. Am J Sociol 92:1170–1182. https://doi.org/10.1086/228631
57. Webb C, Bodin Ö (2008) A network perspective on modularity and control of flow in robust systems. In: Norberg J, Cumming G (eds) Complexity theory for a sustainable future. Columbia Press, New York, pp 85–118
58. Armitage DR, Plummer R, Berkes F et al (2009) Adaptive co-management for social–ecological complexity. Front Ecol Environ 7:95–102. https://doi.org/10.1890/070089

Understanding the Disaster Risk of Human Settlements: Case Research

Ikrom Mustofa and Giuseppe T. Cirella

Abstract Risk from natural disasters is no longer defined as the natural occurrence. This chapter aims to understand the risk of such occurrences by analyzing certain socio-political situations in light of historical data via desk research and secondary analysis. The pressure and release model is adopted as a framework to distinguish the progression of vulnerability and safety. To analyze the data, a review of the literature (i.e., scientific journals, newspapers, and policy papers) is conducted. The analysis examined the historical conditions including physical, social, and political situations in two case research areas: Aceh and Haiti. Comparative research is applied to examine for factors in terms of similarities and differences that may have led to higher disaster risk. Moreover, solutions are presented in dealing with the multiple problems in terms of the progression of safety. The findings revealed that the case research areas suffered worsened impacts due to specific historical situations. Additionally, the progression of safety could be implemented as the solution to decrease the impact of disasters by human-related actions. Hence, the findings could have strong contributions for policy-makers to analyze the triggering factors of disasters as well as decrease the impact of disasters altogether.

Keywords PAR model · Disaster risk reduction · Progression of safety · Indonesia · Haiti

---

I. Mustofa (✉)
Piarea Environment and Technology, Bogor, Indonesia

Water System and Global Change Group, Wageningen University and Research, Wageningen, the Netherlands

G. T. Cirella
Faculty of Economics, University of Gdansk, Sopot, Poland
e-mail: gt.cirella@ug.edu.pl

G. T. Cirella (ed.), *Human Settlements*, Advances in 21st Century Human Settlements,
https://doi.org/10.1007/978-981-16-4031-5_3

1 Introduction

Natural disasters take place all around the world resulting in material loss and human casualties. The risk of natural disasters has been amplified by the impact of hazards, climate change, global change, population (i.e., moving closer to risk prone areas), and other social problems. Specifically, the trend of geological disasters (i.e., earthquakes, tsunamis, volcanic eruptions, and landslides) as well as hydrometeorological disasters (i.e., tropical cyclones such as typhoons and hurricanes, thunderstorms, hailstorms, tornados, blizzards, heavy snowfall, avalanches, coastal storm surges, floods including flash floods, drought, heatwaves, and cold spells) have been steadily on the increase in recent years [1–4]. For the most part, geological disasters are less numerous than hydrometeorological disasters; however, they are usually much more devastating. These extreme events have a high impact on the world. Identifying their frequency of occurrence must be mitigated as a high priority issue in all vulnerable countries. In certain developing countries, such as Indonesia and Haiti, such natural disasters have been followed by great devastation. It is essential to understand that risk from natural disasters is no longer defined as a natural occurrence [5, 6].

Different from hazards, disaster risk includes susceptibility and exposure among local people in the affected area. The negative effects of natural hazards can include the financial system through financial risk; such risk is higher in low-income countries than in high-income countries [7]. In this chapter, case research from two highly affected, low-income countries are examined. First, the 2004 Aceh tsunami in which a number of conditions lead to a high-risk situation, including: government instability, local conflict, lack of local institutions, and lack of preparedness. Second, the 2010 Haiti earthquake in which massive impacts were primarily affected due to government instability and absence of preparedness. The social production of vulnerability is also investigated by examining the degree of importance to understanding and addressing natural disasters as well as the hazards they pose. Thus, the aim will be to better understand the risk of such events—via the two cases—by way of socio-political circumstances in respect of historical precedents. The pressure and release (PAR) model is applied as a framework to distinguish the progression of vulnerability and progression of safety.

2 Conceptual Framework: PAR Model

As a causal type of disaster management tool, PAR model is used to suggest some underlying causes of disasters through the analysis of the nature of the hazard thus distinguishing a critical element in the research of the interaction of disaster stages. PAR is utilized by the development of two processes, which are the progression of vulnerability and progression of safety. The former gives insight into the development of vulnerability, while the latter provides how safety can be achieved [8]. PAR model is distinguished into several phases, which are root causes, dynamic pressure, and unsafe conditions. Specific root causes form vulnerability distinguished by economic, demographic, and political situations. These conditions, of course, impact the resources (i.e., allocation and distribution). They are a result of economic, social, and political structures, and also legal definitions and enforcement of the ideological order (i.e., right and gender). Root causes are also related to the function of the state, and with good governance, the law and administration. Root causes reveal the distribution of power in society [9, 10]. Dynamic pressure brings the activities and processes of root causes into unsafe conditions as the manifestation of general underlying economic, social, and political patterns. Dynamic pressure can form unsafe conditions that need to be considered with the different types of hazards—that are people-dependent. The processes of dynamic pressure operate to channel root cause into unsafe conditions to enable the analysis of how the forces play themselves out on the ground, in a strong spatial and temporal sense [11, 12]. Finally, unsafe conditions explain the vulnerability of local people that are expressed in time and space connected to a hazard. Moreover, they depend on the prosperity level among local people and how this level varies between government level (i.e., including households and individuals). It is essential to also consider the pattern of access to tangible resources (e.g., cash, shelter, food stocks, and agricultural equipment) and intangible things (e.g., support, knowledge and sources of assistance, morale, and ability to function in a crisis).

3 How Progression of Vulnerability Affects Losses: Two Cases

3.1 Aceh Tsunami

Thiscase research examines the effects of the 2004 tsunami in Aceh, Indonesia. On December 26, 2004, an earthquake extending over 1300 km and measuring between 9.1 and 9.3 on the Richter scale took place 30 km deep in the Indian Ocean. While this was one of the most massive measured earthquakes in human history, it led to the destruction of many structures, and it was, however, the following waves of the tsunami that did the most damage. The tsunami killed between 250,000 and 300,000 in countries bordering the Indian Ocean [13]. Aceh was hit the hardest of all, having between 130,000 and 150,000 killed and over half a million people displaced. The tsunami also resulted in around USD 4.5 billion in damages or losses [14]. The first of the three waves hits the east-side of Aceh just 10 min after the earthquake. It was, however, the second wave, that arrived 5 min later that was the most destructive. The tsunami wave varied in height between 10 and 30 m and reached several kilometers inland [15, 16]. Tsunamis are not a new phenomenon in Aceh. Over the centuries, several tsunamis have reached its shores. The difference in unpreparedness for a tsunami is visible between different local ethnicities. Several reports described the information of people picking up the fish left behind by the re-treating sea, unaware of the impending tsunami, while some indigenous peoples have been reported to flee in time and therefore have a much higher survivor rate [14, 17]. Unfortunately, the tsunami was not the only problem that Aceh had to face. Another problem the area has had to deal with is the conflict between the Free Aceh Movement (GAM) and the Indonesian state. Aceh as a region has seen various forms of conflict revolving around autonomy from the Dutch colonial government and later from the Indonesian state. This conflict, specifically in the last thirty years, has had a major impact on Aceh's vulnerability and was a significant reason the damage to the area, both physical and societal, was so extensive.

Figure 1 illustrates the PAR diagram for the Aceh tsunami. The assessment of vulnerability begins from the root causes which indicate the underlying problems. In this case, root causes utilize a typology similar to a triple bottom line of thinking (with a fourth bottom line acting as the political system) [18]. The bottom line systems are social, economic, political, and natural; these are the aspects that give rise to vulnerability and affect the distribution of resources among different parts and groups of people. The root causes form the basis for the next step in the model: dynamic pressure [19]. The dynamic pressure includes for instance a decreased access to resources, lack of local institutions, lack of community-based early warning system, and destruction of mangroves. The dynamic pressure then leads to several unsafe conditions which are described in the last phase of the PAR diagram. Such unsafe

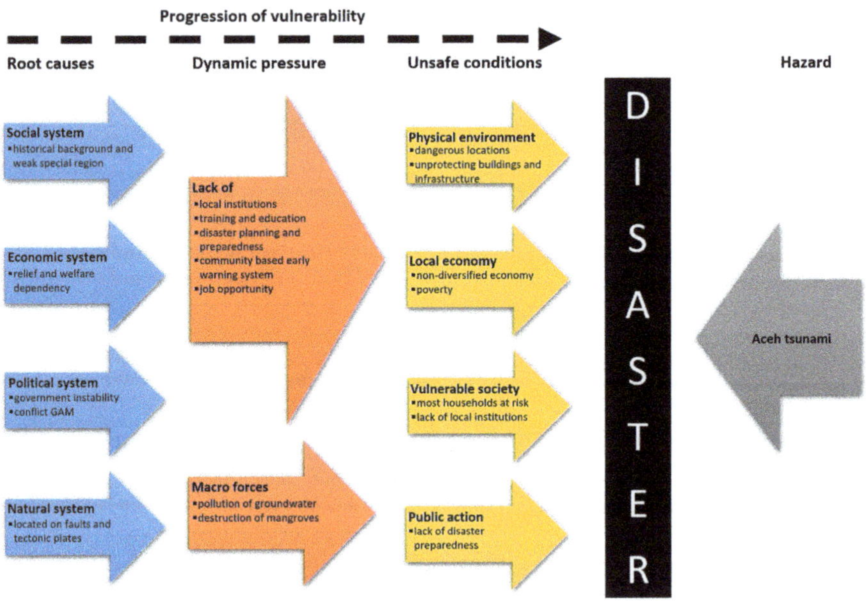

Fig. 1 PAR model progression of vulnerability for the Aceh tsunami

conditions could equate to weak and fragile local economic circumstances, problems in the physical environment, the creation of harmful livelihoods and issues in public actions.

3.1.1 Progression of Vulnerability: Aceh Tsunami

The progression of vulnerability for the Aceh tsunami incorporates six important factors: hydrology and ecology, spatial planning, physical infrastructure, social aspects, state and political issues, and management policy. First, in terms of hydrology and ecology, the coastal ecosystems and their importance regarding coastal protection are slowly gaining recognition in Aceh, but while some of these ecosystems are already supposed to be protected from human destruction, they are still under threat [20, 21]. Moreover, while mangroves were once very abundant, currently there are only a few patches of mangrove forest remaining. Most of them have been removed to make way for tambaks, i.e., fishponds or water basins built for shrimp cultivation. The diversity of the forests that remain is quite poor as well. Coastal reefs are under stress by pollution and sedimentation which leads to sometimes complete loss of these reefs [21, 22]. The same happens to seagrass beds that suffer under pollution, siltation, and mechanical damage (e.g., by trawling) [22].

Second, spatial planning was done through coastal zone management (CZM) and was carried out by the national government until 1999. Provincial and local government agencies had very little say about the management process. After 1999, the government became more decentralized, and the provincial and local government became active over their local resources. CZM, however, is still mostly managed by over 20 different agencies, of which only three are on a more local level [23]. Coordination between these agencies and between different programs is almost non-existent [23]. In Aceh, due to the conflict between the Indonesian government and GAM, some agencies had little to no control or barely functioning. These were, for instance, the National Land Agency, which regulates land ownership via the general courts [24]. Nonetheless, the call for local autonomy of Aceh has provided them with increasing authority of their resources [20].

Third, before the 2004 tsunami, the people of Aceh inhabited various buildings both engineered and non-engineered [25]. During the tsunami, almost all types of buildings were destroyed [25]. A significant number of non-engineered reinforced concrete structures were also structurally damaged, especially in first-floor columns. Previous research also explained that the damage done by the seismic disasters were not only to residential housings, but also government buildings which were multi-story reinforced concrete structures [25]. These buildings, unfortunately, were also built with poor seismic design. It was different with engineered reinforced real frames which appear to have had sufficient strength in facing the tsunami. Light timber frame buildings were extremely vulnerable to tsunami wave pressure. These building types were the most devastated infrastructure during the tsunami.

Fourth, the conflict between the Indonesian state and the GAM separatists has been going on for thirty years before the tsunami event [26]. The province of Aceh has known social conflict and unrest since the days of Dutch colonial rule; the most recent conflict could be seen as the contemporary version of these earlier struggles against the Java-based colonial government which continue against the independent Indonesian state. During the conflict, the separatist claims became not just territorial, but also religious, social, and economic [17]. About 90% of the 340,000 people living in Aceh are of ethnic Acehnese descent and have a cultural and historical heritage that they feel has been neglected and even disrespected by the Indonesian state. Thirty years of conflict left a profound impact on Acehnese society. Human rights abuses from both sides of the conflict and policies from the government such as martial law, the continued disrespect for Aceh's cultural, and religious customs. Also, the influx of Javanese via transmigration is seen by many Acehnese as an attempt to 'Indonesianize' Aceh had caused widespread poverty, distrust, and hatred toward the Indonesian government and sustained the fight for independence. It is argued that the conflict created a so-called 'conflict trap' where the violence, in turn, weakened security and the institutional capacities, reduced growth, lowered income, destroyed infrastructure, and redirected resources from development [26]—meaning that the conflict was detrimental to the resilience of communities. From this perspective, the reasons for the large number of victims lie not only in geographical closeness to the epicenter of the earthquake, but also in coastward population displacements, demanding access to land and resources, poverty, food insecurity, physical violence, and torture—as a result of the conflict [17].

Fifth, the United Nations Development Programme [27] explained that many areas which were damaged by the tsunami were also affected by conflict both vertical and horizontal. Originally, Aceh together with the Yogyakarta province were two regions among many areas in Indonesia which are so-called special regions of Aceh (i.e., Daerah Istimewa Aceh). Because of decentralization and also special autonomy, Aceh has been granted a larger part of the revenue from its natural resources, rather than being redistributed to other provinces in Indonesia. Before the tsunami, the coordinating agency at the provincial level was the intersectoral coordinating body (i.e., Satkorlak), which have full dependency on the provincial authority [28]. Under the provincial level, there are Satlak as the coordinating body of districts. Technically, Satkorlak and Satlak often create unclear circumstances by miscommunication and misconception in the implementation of mandates.

Sixth, disaster management is crucial—especially for a disaster-prone country such as Indonesia. Indonesia is vulnerable to natural hazards both geological and hydrometeorological [29]; unfortunately, Indonesia did not apply disaster risk reduction as the main priority to face various hazards before the 2004 tsunami event. There was no particular policy in place to prepare for all phases of disaster risk reduction, such as mitigation and prevention. This situation resulted in a high number of casualties, destroyed infrastructure, and emergencies when the tsunami hit in 2004.

3.2 *Haiti Earthquake*

On January 12, 2010, a 35 second earthquake struck Haiti measuring 7.0 on the Richter scale; it was one of the most devastating natural disasters of the last decade [30]. The epicenter of the earthquake was placed at 35 km from Port-au-Prince's metropolitan area and damaged the most populated areas of the country. In the following eight days, Haiti was struck by 50 aftershocks with a magnitude of an average of 4.0 further worsening the situation [31]. Haiti is situated in the island of Hispaniola between the Caribbean Sea and the North Atlantic Ocean with a total area of 27,750 km^2. It is comprised of one-third of the island, while the Dominican Republic covers the rest on it. The geographic location of Haiti also makes it prone to geological incidents as the Haitian part of the island of Hispaniola sits sandwiched between two fault lines known as strike-slip faults between the North American and Caribbean tectonic plates [32]. According to the International Federation of Red Cross and Red Crescent Societies [33], the disaster resulted in thousands of deaths and billions of USD worth of damage; at the same time, the country's rank lowered from 145 to 168th on United Nation's Human Development Index. The government capacity was also debilitated because many civil servants were killed, official buildings, main infrastructure, and energy sources were destroyed—inducing an overwhelming job loss [33].

Once called the Jewel of the Antilles, Haiti was the wealthiest colony in the world and provided approximately 50% of the national product of France by many natural exports. Meanwhile, France, refined and sold the exportation from Haiti to the rest of Europe throughout much of the 18th century [34]. One of the main reasons for this high productivity was the slave labor which was described as the most brutal in the Caribbean, by many documents of western slavery. The system of slavery had freedom earning possibility by the slaves which depended on their exceptional work, i.e., the reason why it worked well and had high productivity [34]. Subsequently, this led to a long slave struggle led by Jean-Jacques Dessalines, who brought Haiti's independence on January 1, 1804, resulting in the only successful slave revolution in the history [35]. This historic moment helped to build national pride and solidarity among social distinctions in Haiti, but despite this growing nationalism, Haiti's leaders and most of the population were set on different courses, resulting in the destruction of physical and human capital through revolutionary wars [36]. The revolution was seen as a dangerous precedent to other colonized states, which led to an international boycott of Haitian goods and commerce, pioneering the first blow to the Haitian economy [34]. The next blow to the economy was the 150 million Franc debt, including interest, to be paid to France for indemnities which took decades to payback [34]. A more recent overview of Haiti's history shows prevalent poverty levels that led it to be considered one of the poorest states in the Western Hemisphere [37]. This indicator reflects that half of the population is living in poverty, affecting their social conditions such as literacy, life expectancy, infant mortality, and child malnutrition—causing households to adopt several coping strategies [38]. The

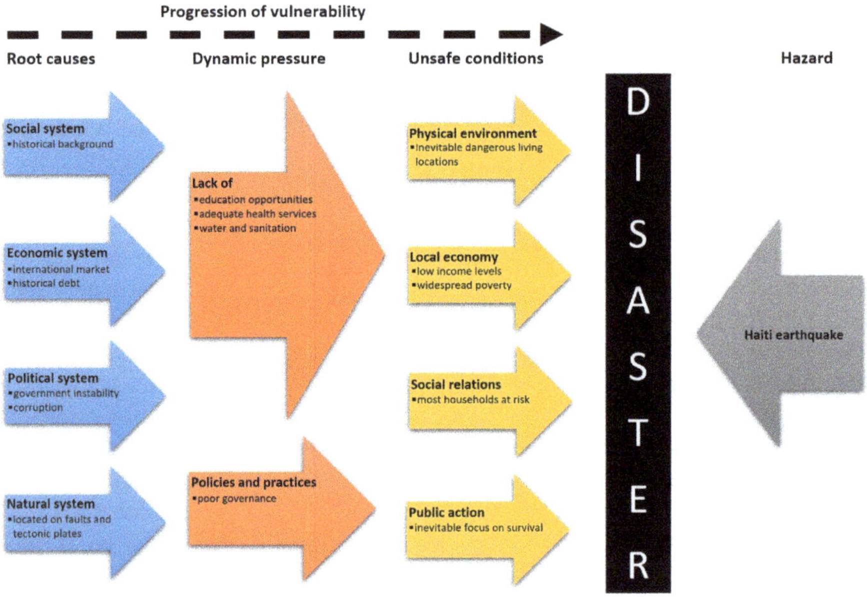

Fig. 2 PAR model progression of vulnerability for the Haiti earthquake

food restriction strategy affects hunger thus indirectly affecting productivity due to inadequate physical conditioning and development [38].

Throughout the occupation of Haiti by the USA (i.e., 1915–1934) up to the present day, the political system in Haiti has been known for its personal gain and, subsequently, high levels of corruption [36]. Before the earthquake, politics in Haiti was marked by uncertainty and a deep divide between the executive and opposition parties. When the earthquake occurred, i.e., in January, the country was in the middle of the presidential campaign with election being held in November, raising the question of whether these elections should be postponed [39]. In the face of these problems, the PAR model progression of vulnerability for the Haiti earthquake illustrates a valuable outlook into the state of the country when the earthquake hit (Fig. 2).

3.2.1 Progression of Vulnerability: Haiti Earthquake

The assessment of vulnerability begins from root causes, i.e., broken down into systems [18], including social, economic, political, and natural. It was clear that Haiti had straitened circumstances from almost all aspects, such as a severe international market, its location, i.e., on fault lines between two tectonic plates, political instability, and high levels of corruption. It caused failures, in terms of dynamic pressure, included lack of education, substandard water sanitation, and poor governance. In the last phase, PAR model illustrates additional cause and effect conditions that exacerbate the situations which are identified as unsafe. These unsafe conditions are problems with the physical environment, local economy, social relations, and public action. Suffering from these adverse conditions, while the earthquake came, it created a perfect storm which resulted in a devastating disaster. The progression of vulnerability for the Haiti earthquake includes four main factors: physical planning, engineering and construction (i.e., infrastructure), socioeconomics, and management and institution concerns.

First, in 2009, Haiti's urbanization rate was very high which resulted in a population density of more than 300 people per square kilometer, putting massive pressure on land and creating overcrowding. Inadequate income (i.e., with a gross domestic product of USD 2 per capita in 2009), lack of space, and unclear land tenure further consolidated low housing quality. Also, the construction sector in Haiti was under rapid and unregulated development before the earthquake. It, thus, can be established that neither the location of dwellings nor the various construction methods were considerate of multiple risks like earthquakes [40]. In this regard, it is complicated to conclude that Haiti's planning measures that make it apt for the prevention and mitigation stage were adequate or even tolerable.

Second, the aftermath of the devastating earthquake revealed that there was almost a complete absence of seismic detailing in construction. Seismic design was missing from the engineering curriculum—all together. It also appears there was a lack of building code and engineers who want to use proper seismic-friendly systems that provision earthquake design and safety, e.g., Beton Arme aux Etats Limites from France [41]. Despite the existence of laws on building permits and inspections, they were poorly regulated. The majority of the buildings were also constructed in non-ductile concrete, unreinforced masonry, and unconfined masonry which are all inadequate for earthquake-prone areas. Making matters worse is the utilization of poor quality materials, poor workmanship, poor maintenance, and use of corrosive materials all resulting in the bad performance in case of the occurrence of the earthquake [41]. As such, almost all the buildings in Haiti were mostly brittle and weak and not designed for earthquakes. Although many buildings were standing, nearly all did poorly [41]. All these steps show that the disaster risk reduction regarding mitigation was almost non-existent before the 2010 earthquake.

Third, the level of education especially for understanding such disasters in Haiti was insufficient. It was reflected by the unfamiliarity of residents of potential future hazards. Even engineers in Haiti were not educated on seismic designs [41]. In regards to education and awareness, environmental education was part of the curriculum and paves for inclusion to a disaster risk reduction plan; however, as of yet is not included. There are non-governmental organizations (NGOs) involved in bringing awareness of disaster risks and the disaster risk reduction to the public and schools in certain areas [42]. Though all these were promising activities, their significance is questionable considering the early stages the country was in before the earthquake hit which was not precisely of exceptional magnitude. As a result, only after the 2010, earthquake did Haiti start training seismologists and installing seismometers [43]. Hence, at that time, expertise and research were not a priority—though it could have greatly helped in prevention and mitigation. Moreover, Haiti's economy suffered as a result of the aftermath. The country was riddled with chronic poverty with 65% of its population under the international poverty line and ad hoc urbanization [44]. Modest disaster preparation (i.e., from predictive work and preventative action to overcoming possible catastrophic events) is just conducted for daily survival skills.

Fourth, it can be safe to assume that Haiti's sectoral laws contain provisions relevant to disaster risk management, but it cannot be concluded that these were properly or fairly implemented. Economic constraints have hindered the growth of institutional structures proposed under law especially at municipal and local levels [33, 42]. The National Risk and Disaster Management Plan and Emergency Response Plan which are Haiti's key national documents governing the disaster risk management system have only been in place since 2001. Though insignificant progress was made for the first few years after their implementation, Haiti had made progress in this regard since 2005 with development and installation of structures at all levels of government [42]. However, Haiti had no organized assessment team to assess various aspects of a disaster. It was also reflected in the lack of data on previous disasters which would have been beneficial in avoiding uncontrollable confusion once a disaster happened [45]. As a result, it explains the lack of a backup plan which was evident when various facilities were destroyed by the earthquake [46].

4 Problem Solution

From the analysis, a number of strategies as opportunities for change can be made. These opportunities have been conceptualized from the situation as it was and as it has been shaped post-disaster. There are several opportunities for change in which can be addressed to make present-day Aceh and Haiti more resilient toward future disasters. Two key solutions can be deduced: (1) a technical perspective and (2) a social valuation. To better explain these solutions and why we think they are essential, the use of the progression of safety, i.e., the release aspect from PAR model is applied. The release aspect can also be considered as recommendations in advancing safety and resilience (Table 1).

The occurrence of the Aceh tsunami and Haiti earthquake resolved problems contrary to the progression of vulnerability. It can be noted that both cases moved toward achieving safe conditions as the answer to vulnerable aftermath-based circumstances, e.g., improving the physical environment, local economy, and social relations, as well as public action. As a result, these improvements decrease dynamic pressure and encourage (and alleviate) the development of several sectors which may have been otherwise still managing older policies and practices. In the last phase, the progression of safety addresses with root causes such as the democratization of governance, development in the international community, and prevention of most critical locations from hazards. On the other hand, by using the PAR model for safety progression, it is clear that risks can be reduced by taking into account several actions.

After analyzing the two cases affected, there are certain similarities and differences between them. It is clear that both cases suffered greatly, and a number of conflicts

Table 1 Progression of safety from the case research

Progression of safety	Aceh tsunami	Haiti earthquake
Achieve safe conditions	Physical environment • resilient buildings and infrastructures • develop barriers Local economy and social relations • organized fishing • strengthen livelihoods • traditional knowledge to identify tsunami Public action • improve disaster preparedness through schools and increase local capacity	Physical environment • develop and use seismic building codes • change lands use planning Local economy and social relations • diversify income opportunities • strengthen livelihoods Public action • improve disaster preparedness through schools and increase local capacity
Reduce pressures	Development • training and education • build back better • adequate infrastructures and public institutions Policies and practices • develop disaster management policy for tsunami and other disasters	Development • training and education • adequate infrastructures and public institutions • ethical standard in public life Policies and practices • disaster management governmental structure • enhance framework for facilitation and regulation of international aid
Address root causes	Social and political • new organizations • increase the access of vulnerable groups to resources and structures • managing the conflict Economic • NGOs and donors to support livelihoods • market availability Natural • create buffer zone • avoid dangerous places	Social and political • democratization of governance • accesses of vulnerable groups to resources and structures Economic • develop voice in international community Natural • avoid most critical locations
Reduce hazard	Range of measures to reduce certain hazards	Range of measures to reduce certain hazards

were prevalent. Moreover, both Aceh and Haiti suffered low economic levels which also worsened by the impacts of the disasters. Collapsed and instability of governance were also deemed as a significant problem beforehand. International assistance and aid in terms of response were needed for the cases. In terms of differences, after the disasters, Aceh received more aid and international assistance than Haiti. In addition, political change in Aceh got much better, faster, and its national decrees to reduce the tensions and conflicts with the GAM successful. Haiti, on the other hand, continues to suffer more collapse.

It should be noted, however, the implementation of the framework in this chapter contains several limitations. The framework is unable to quantify the contribution of each factor affecting disaster (i.e., it has been noted that it is very difficult for policy-makers to decide on policy implementation without quantification scales). Some severe implications have been shown that impact of the uncertainties in knowledge concerning vulnerability can be explained via the underlying causes or pressures [47–50]. In practice, the lack of understanding and uncertainties impacted the decision-makers and also the policy as they suffered scarcity of resources [50–52]. Furthermore, the condition may address the pressures and unsafe conditions without the contribution from both social causes of vulnerability as well as the more distant root causes. Finally, it can be said that increase awareness among local people is unmistakably embedded—since the disasters. The historical problems have allowed for the development and societal acceptance of a framework that closely interrelates with community issues. Previously, before the disasters, local people were often not aware of their activities pre-, current, and post-disaster [49, 53]. By explaining the root causes of unsafe situations, community life has moved toward believing in a more systemized structure. The chapter has been an exercise in bettering human settlements and creating essential policy-based decisions for better possible measures.

5 Conclusion

The use of the PAR model framework revealed critical information regarding the historical components leading to disasters. It is clear that the case research areas suffered worsened impacts due to their specific historical situations. The 2004 tsunami was devastating for the province of Aceh. The high level of victims and damage done to society was not only due to the unique nature of the tsunami but the three decades of conflict between the GAM and the Indonesian state—significantly hampering Aceh's resilience. Its effects could be seen in both technical and societal aspects as physical infrastructure, livelihoods, and the local economy were all weakened. The 2010 earthquake in Haiti is known more for the catastrophic devastation it brought. This unmatched devastation can be accounted mostly to the nation's vulnerability. Haiti's vulnerability has been propagating for centuries now. It has its roots causes in international powers which Haiti could not compete with, and ultimately was rundown by poor governance which has characterized the country since its independence in the early nineteenth century. These factors resulted in the segregation of society which are distinguished by poverty, illiteracy, malnutrition, and so on—all contributing to the building up of each country's vulnerability. At length, the findings from this research could have strong contributions for policy-makers to analyze the triggering factors of disasters as well as to decrease the impact of the disaster itself by employing the progression of safety and recommendations after the disaster has occurred. It is clear that post-disaster situations often created the opportunity for change.

References

1. Tareghian R, Kashefipour SM (2007) Application of fuzzy systems and artificial neural networks for flood forecasting. J Appl Sci 7:3451–3459. https://doi.org/10.3923/jas.2007.3451.3459
2. Jayawardena AW (2015) Hydro-meteorological disasters: causes, effects and mitigation measures with special reference to early warning with data driven approaches of forecasting. In: Procedia IUTAM. Elsevier B.V., pp 3–12
3. Walch C (2018) Adaptive governance in the developing world: disaster risk reduction in the State of Odisha, India. Clim Dev 1–15. https://doi.org/10.1080/17565529.2018.1442794
4. UNDRR (2021) Building risk knowledge. In: United Nations official disaster risk reduction https://www.undrr.org/building-risk-knowledge. Accessed 2 Feb 2021
5. IPCC (2017) AR6 climate change 2021: the physical science basis—IPCC. Intergovernmental Panel on Climate Change, Geneva
6. Field CB, Barros V, Stocker TF et al (2012) Managing the risks of extreme events and disasters to advance climate change adaptation: special report of the intergovernmental panel on climate change
7. Chen X, Chang CP (2020) The shocks of natural hazards on financial systems. Nat Hazards 105:2327–2359. https://doi.org/10.1007/s11069-020-04402-0
8. Martin H, Lewis TM, Martin NM (2011) A critical review of haiti earthquake of 2010 : key development problems and focused solutions. In: Eighth annual conference on international institution for infrastructure, renewal and reconstruction. International Institute for Infrastructure, Renewal and Reconstruction, Kumamoto, pp 33–40
9. Bendito A, Barrios E (2016) Convergent agency: encouraging transdisciplinary approaches for effective climate change adaptation and disaster risk reduction. Int J Disaster Risk Sci 7:430–435. https://doi.org/10.1007/s13753-016-0102-9
10. Ndah AB, Odihi JO (2017) A systematic study of disaster risk in Brunei Darussalam and options for vulnerability-based disaster risk reduction. Int J Disaster Risk Sci 8:208–223. https://doi.org/10.1007/s13753-017-0125-x
11. Sh H (2018) Application of a PAR model for assessing vulnerability to drought hazard in Kondoa District. J Geogr Nat 8:1000232. https://doi.org/10.4172/2167-0587.1000232
12. Islam MS, Lim SH (2015) When "Nature" strikes: a sociology of climate change and disaster vulnerabilities in Asia. Nat Cult 10:57–80. https://doi.org/10.3167/nc.2015.100104
13. Cochard R, Ranamukhaarachchi SL, Shivakoti GP et al (2008) The 2004 tsunami in Aceh and Southern Thailand: a review on coastal ecosystems, wave hazards and vulnerability. Perspect Plant Ecol Evol Syst 10:3–40. https://doi.org/10.1016/j.ppees.2007.11.001
14. Wang F, Miyajima M, Li T et al (2012) Progress of geo-disaster mitigation technology in Asia. Springer, London
15. Paris R, Wassmer P, Sartohadi J et al (2009) Tsunamis as geomorphic crises: lessons from the December 26, 2004 tsunami in Lhok Nga, West Banda Aceh (Sumatra, Indonesia). Geomorphology 104:59–72. https://doi.org/10.1016/j.geomorph.2008.05.040
16. Borrero JC (2005) Field data and satellite imagery of tsunami effects in Banda Aceh. Science (80-) 308:1596. https://doi.org/10.1126/science.1110957
17. Gaillard JC, Clavé E, Vibert O et al (2008) Ethnic groups' response to the 26 December 2004 earthquake and tsunami in Aceh, Indonesia. Nat Hazards 47:17–38. https://doi.org/10.1007/s11069-007-9193-3

18. Elkington J (1998) Cannibals with forks: the triple bottom line of 21st century business. New Society Publishers, Stony Creek, CT
19. Awal MA (2015) Vulnerability to disaster: pressure and release model for climate change hazards in Bangladesh. Int J Environ Monit Prot 2:15–21
20. Peduzzi P, Allenbach K, Chatenoux B et al (2005) After the tsunami: rapid environmental assessment. United Nations Environment Programme, Nairobi
21. Laumonier Y, Uryu Y, Stüwe M et al (2010) Eco-floristic sectors and deforestation threats in Sumatra: identifying new conservation area network priorities for ecosystem-based land use planning. Biodivers Conserv 19:1153–1174. https://doi.org/10.1007/s10531-010-9784-2
22. Cochard R (2011) The 2004 tsunami in Aceh and Southern Thailand: coastal ecosystem services, damages and resilience. In: The Tsunami threat-research and technology. IntechOpen, London
23. Dahuri R (2006) Pre-and post-tsunami coastal planning and land-use policies and issues in Indonesia. In: Workshop on coastal area planning and management in Asian tsunami-affected countries. Food and Agriculture Organisation of the United Nations Regional Office for Asia and the Pacific, Bangkok, pp 111–129
24. Fitzpatrick D (2008) Managing conflict and sustaining recovery: land administration reform in tsunami-affected Aceh. Oxfam International, Oxford
25. Saatcioglu M, Ghobarah A, Nistor I (2005) Effects of the December 26, 2004 Sumatra earthquake and tsunami on physical infrastructure. ISET J Earthq Technol 42:79–94
26. Masyrafah H, Mckeon JMJA (2008) Post-tsunami aid effectiveness in Aceh. Brookings Global Economy and Development, Washington, DC
27. UNDP (2006) Local governance in tsunami recovery: lessons learned and emerging principles. United Nations Development Programme, Nairobi
28. Syam A, Dirhamsyah M, Priyono J (2010) Institutional and legislative system for disaster management in Aceh: past, present and the future. In: Fifth annual international workshop and expo on Sumatera Tsunami disaster and recovery. Tsunami and Disaster Mitigation Research Center, Banda Aceh, Indonesia, pp 80–83
29. Sari S (2009) Disaster risk reduction in Banda Aceh. Fourth annual international workshop and expo on Sumatra Tsunami disaster and recovery. Syiah Kuala University, Banda Aceh, Indonesia, pp 66–70
30. Meng L, Ampuero J, Sladen A, Rendon H (2012) High-resolution backprojection at regional distance: application to the Haiti M7.0 earthquake and comparisons with finite source studies. J Geophys Res 117:1–16. https://doi.org/10.1029/2011JB008702
31. Hou L, Shi P (2011) Haiti 2010 earthquake—how to explain such huge losses? Int J Disaster Risk Sci 2:25–33. https://doi.org/10.1007/s13753-011-0003-x
32. Hayes GP, Briggs RW, Sladen A et al (2010) Complex rupture during the 12 January 2010 Haiti earthquake. Nat Geosci 3:800–805. https://doi.org/10.1038/ngeo977
33. IFRC (2015) Haiti earthquake five-year progress report. International Federation of Red Cross and Red Crescent Societies, Geneva
34. Corbett B (2013) Why is Haiti so poor? People to people
35. Ghachem MW (2012) The old regime and the Haitian revolution. Cambridge University Press, Cambridge
36. Verner D, Egset W (2007) Social resilience and state fragility in Haiti: breaking the conflict-poverty trap. World Bank, Washington, DC
37. Risler EA, Kintzle S, Nackreud L (2015) Haiti and the earthquake: examining the experience of psychological stress and trauma. Res Soc Work Pract 25:251–256. https://doi.org/10.1177/1049731514530002
38. Échevin D (2011) Vulnerability and livelihoods before and after the Haiti earthquake. World Bank, Washington, DC
39. Herard D (2012) Disaster risk reduction and the action plan for national recovery and the development of Haiti. Florida International University, Miami
40. Bellerive JM (2010) Haiti earthquake PDNA: assessment of damage, losses, general and sectorial needs. UNEP, Port-au-Prince

41. Fierro E, Perry C (2010) Preliminary reconnaissance report: 12 January 2010 Haiti earthquake. Pacific Earthquake Engineering Research Center, UC Berkeley, Berkeley
42. Haiti Red Cross Society (2015) How law and regulation supports disaster risk reduction Haiti case—study report. International Federation of Red Cross and Red Crescent Societies, Geneva
43. Jones N (2010) Haiti to improve quake preparedness. Nature. https://doi.org/10.1038/news.2010.670
44. Commission E (2016) EU emergency support instrument. European Civil Protection and Humanitarian Aid Operations, Brussels
45. Cohen S (2011) Haiti earthquake January 2010: what actions and policies can the government of Haiti implement to improve emergency management response. San Jose State University, San Jose
46. Charles J (2008) Haiti's lax regulation yields unsafe buildings. McClatchy DC Bur
47. Glasson J, Therivel R, Chadwick A (2005) Introduction to environmental impact assessment, 3rd edn. Routledge, London
48. Government of Kenya (2013) Truth, justice and reconciliation report. Government Press, Nairobi, Kenya
49. Saputra A, Rahardianto T, Revindo MD et al (2017) Seismic vulnerability assessment of residential buildings using logistic regression and geographic information system (GIS) in Pleret Sub District (Yogyakarta, Indonesia). Geoenviron Disasters 4:11. https://doi.org/10.1186/s40677-017-0075-z
50. Yumarni T, Amaratunga D, Haigh R (2014) Assessing gender vulnerability within post-earthquake reconstruction: case study from Indonesia. Procedia Econ Financ 18:763–771. https://doi.org/10.1016/s2212-5671(14)01000-4
51. Koseki J, Yoshimine M, Hara T et al (2007) Damage survey report on May 27, 2006, mid java earthquake, Indonesia. Soils Found 47:973–989. https://doi.org/10.3208/sandf.47.973
52. Widyastuti H, Lumantarna E, Sofi M et al (2020) Earthquake vulnerability of port structures in Indonesia. In: IOP conference series: earth and environmental science. Institute of Physics Publishing, p 012001
53. Mardiatno D, Handayani T, Susanto D et al (2020) Earthquake vulnerability mapping in the at-risk Opak Fault, Sengon Village, Central Java, Indonesia. E3S Web Conf 200:01002. https://doi.org/10.1051/e3sconf/202020001002

Research and Development Within Public Transport Systems

Tran N. Anh, Ella Kozemko, and Giuseppe T. Cirella

Abstract In light of designing healthy, efficient, and safe human settlements, this chapter looks at the important aspect of developing a complete solution to public transport systems, i.e., to build them sustainably within the context of transport needs that are accessible, affordable, available, and acceptable. An examination of sustainable transport is reviewed with emphasis on environmental sustainability, stressing some of the current difficulties, i.e., congestion and its related costs via wasted time, impaired reliability, and exacerbated air pollution; environmental impacts, both at the global and local level; and health costs arising from air emissions and noise. Case research from Vietnam and Ukraine is investigated and elucidate public transportation trends as well as important needs facing these countries. Traffic congestion, environmental pollution, energy consumption, and the problem of people in traffic are major concerns that urban transport faces. To minimize the risks of vehicle emissions as well as to build green, clean, and sound settlements, important steps are needed in research and development within public transport systems. This chapter is a knowledge-based effort with recommendations to developing comprehensive public transport strategy and structure.

Keywords Transportation · Sustainability · Systematic review · Public survey · Vietnam · Ukraine

1 Introduction

Numerous studies launched by international entities and environmental researchers such as the International Council on Climate Change (IPPC) and the Stern Review on the Economics of Climate Change [1] have made clear notices that climate change is a global issue. In fact, atmospheric CO_2 concentration has increased dramatically from 280 million ppm prior to industrialization to 379 million ppm in 2005 and

T. N. Anh (✉) · E. Kozemko · G. T. Cirella
Faculty of Economics, University of Gdansk, Sopot, Poland

G. T. Cirella
e-mail: gt.cirella@ug.edu.pl

G. T. Cirella (ed.), *Human Settlements*, Advances in 21st Century Human Settlements,
https://doi.org/10.1007/978-981-16-4031-5_4

is expected to increase to 550 by 2030 [1, 2]. The IPPC [2, 3] has estimated that activities such as transportation play a huge role in environmental pollution since it predominately relies on fossil fuels such as oil and gas [4, 5]. One observation shows that in 2004, 23% of greenhouse gas (GHG) emissions were caused by transportation—paralleling a rise in CO_2 emissions. As such, the hazardous and fatal emissions from the transportation sector could seriously damage global health. The sector accounts for approximately 15% of overall GHG emissions and 23% of overall CO_2 emissions—the most prevalent of all GHGs. The global CO_2 emissions from the transportation sector grew by 45% from 1990 to 2007 and is expected to continue to grow by approximately 40% dating back from 2007 to 2030 [6] (Fig. 1). The extraordinary energy demand in the 2020s needs careful observation and prompt policy adjustments sector-wide.

Hrelja [7], Fiorio et al. [8], and Glazebrook [9] point out that policy-makers have developed public transportation, bicycle, and pedestrian facilities to their fullest capacity. Cultural actions have been done efficiently, and communities are motivated to use more active and environmental-friendly transport modes. Regulations toward limiting automobile use have been optimally deployed. Alternative cleaner burning fuels have been introduced, and vehicle manufacturers have employed top-notch technologies to produce first-class vehicles congruent with the greatest world standard emission protocols. Information technology has well-progressed, and the ease of online shopping has cut physical transportation activities—following the integration trends from developed and developing countries [10, 11]. Infrastructure construction is being promoted in which transport networks play an important role [12–14]. Although construction planning, construction investment in transport networks, and public passenger transport networks have been closely attended, many newly built streets in different countries lack capital, workloads are too large, and investment

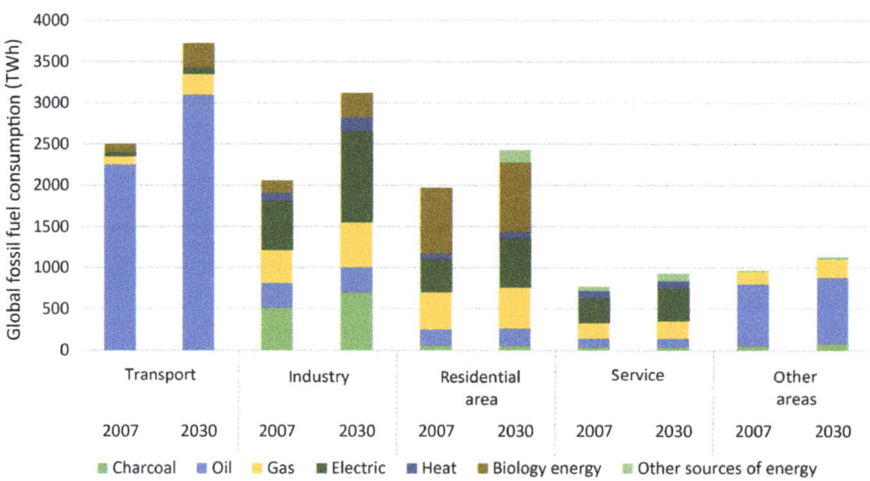

Fig. 1 World energy needs in terms of fuel usage per sector, adapted from the International Energy Agency [6]

in construction among countries is insufficient and inconsistent—hence, the effectiveness is limited. As a result, there are still many difficulties that do not meet the acknowledged requirements, including congestion and its related costs (i.e., wasted time, impaired reliability, and exacerbated air pollution); environmental impacts, both at the global level (e.g., GHG emissions) and the local level (i.e., noise and air pollution); and health costs arising from air and noise emissions. For these reasons, promoting the development of public transportation systems in both developed and developing countries will serve as an important tool to minimize human impacts on the environment and advocate sustainable development [15, 16].

In recent years, in every country, the problem of environmental pollution and traffic congestion is becoming a hot topic; one of the main causes are the number of vehicles which is constantly increasing and very high in comparison to the capacity of the transport infrastructure for motor vehicles (e.g., cars, motorbikes, etc.) [17, 18]. Although authorities have made great efforts to improve the traffic situation, traffic congestion in cities is very complicated. Moreover, there is an increasing risk of environmental pollute on, due to motor vehicle emissions at an alarming rate. In order to minimize urban congestion, governments need to come up with solutions to improve infrastructure, while encouraging people to use public transport to minimize the number of vehicles. Furthermore, instead of collecting a limited fee, a private vehicle could be charged when used to enter a large urban or city center. In order to reduce the number of cars, many priority bus routes with low prices should be introduced in a city that will facilitate and convince many people to give up their personal vehicles to go to the city via public transportation—saving time and money. This prioritization could include organizing priority bus routes for employees of an agency, company, or hospital, departing at convenient locations as well as running straight to work during peak hours [19].

Many major cities in the world such as Shanghai, Paris, Tokyo, and Bangkok have succeeded in developing viable public transport systems [20, 21]. By mobilizing, encouraging public officials to go to work by bus to significantly reduce the number of private cars, they will reduce traffic congestion and create safety for people while commuting in and out of an urban center [19]. Likewise, the density of emissions and waste in the air will be reduced to ensure the health of the people as well as to make the environment green, clean, and authentically pleasing [17]. Experience in some developed countries indicates that, for the cities to develop sustainably, the optimal solution chosen is to develop a public transport network with large transport capacity to prevent congestion—ensuring necessary areas for traffic, apply modern and civilized forms of transport minimizes urban pollution [21]. This approach will have a positive impact in contributing to the improvement of the technical infrastructure system for the whole of society as well as promote trade and transportation of the people in the city, create conditions for minimizing environmental pollution, and traffic congestion during peak hours—all hallmarks of an effective public transport arrangement [11, 21]. This chapter will investigate the important factor in the make-up of human settlements by examining research and development within public transport systems within the scope of sustainable development.

2 Sustainable Transport: Background

A review of the relevant literature covering the issue of public transport and its impact on sustainable development was conducted via case research. Sustainable development is a concept that is currently under very high focus worldwide. The development of humanity cannot only focus on economic development but must also respect the indispensable needs of social and environmental impacts [22–24]. Sustainable development is development that can meet current needs without compromising the ability to meet the needs of future generations [25, 26]. For a simpler understanding, Fig. 2 illustrates a complete breakdown of the term. In other words, sustainable development must ensure economic efficiency, social welfare, and environmental protection and preservation. Sustainable transport is used to describe vehicles and transport planning systems that are relevant to the concerns of sustainable development. The concept of a sustainable transport system proposed by the Council of Transport of European Union countries outlines three main factors: (1) access and develop the basic needs of individuals, businesses, and society in a safe and appropriate manner for human health and ecosystems; (2) affordable, fair, efficient operation, diverse transportation options, and support for economic development while creating a balance in local development; and (3) limit emissions and emissions within the planet's ability to absorb them, use renewable resources equal to or below the generation, and use non-renewable resources equal or lower development of alternative renewable energy, while minimizing land use and noise. According to Sindakis et al. [27], public transport could become the backbone of new integrated knowledge services that are provided on a cooperative basis and framework. Since

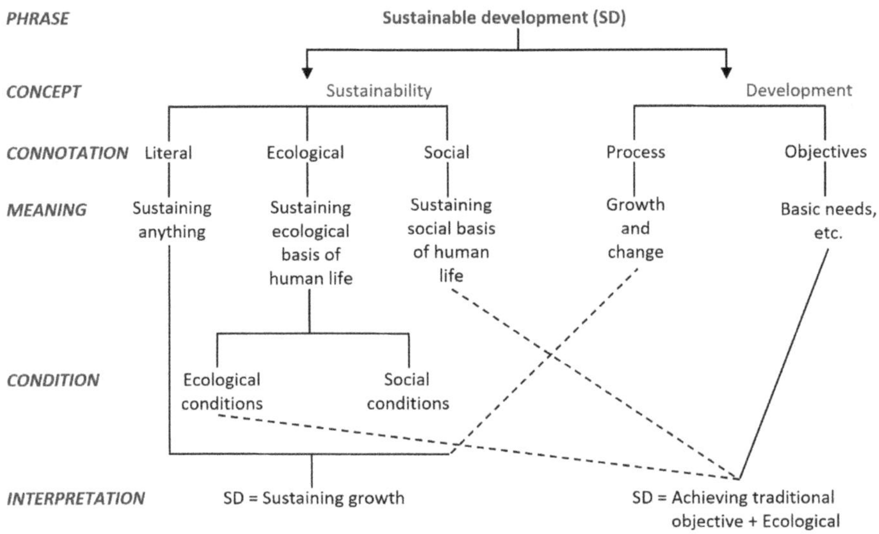

Fig. 2 Definitional breakdown of sustainable development, adapted from Lélé [26] and Elkington [24]

the world's population has increased significantly, there is greater need for mobility in all aspects of social life—especially in transport [28, 29]; hence, the development of public transport services is an important element of social quality [30]. Sustainability of transport, health, and environmental quality as well as economic conditions can be enhanced by transfers from private to public transport [31]; however, these ideas will only occur when a mobility-oriented public transport system exists. Gershon [32] stated that public transport adds to the society a host of additional benefits from an employee perspective that provides employees with the means to work and reduce their reliance on unemployment benefits, to workers who are more likely to go to work if they have easy and affordable commutable transport modes. As a result of such change, the regional economy will save car parking space and expenses related to increased environmental vehicle emissions [27].

Although there have been many changes in technology and infrastructure, the public transportation sector still reveals many shortcomings in terms of business culture. The public transportation industry needs to attract more customers, especially those who use private transport, as well as policies and knowledge to ensure customer movement is smooth [33]. In large and small cities, activities such as labor, investment, transportation, education, and finance are held and relative to whether people have access to them. In order to travel to these areas, we need to provide an efficient and modern transport system inclusive to minimizing travel time between different points, reducing traffic congestion, accidents, and environmental pollution. Maintaining an efficient public transport system is not easy, but it is possible if properly thought out. This requires the authorities in the field of land reclamation and planning to use available funding to upgrade infrastructure, invest more in public transport (i.e., transportation without or minimizing the burning of fossil fuels), and ease transfers between them to meet mobility demands. In this chapter, high quality service for public transport is discussed alongside how to operate it, in so that, access is more effective and efficient then private transportation, i.e., via fast accessibility, comfort, safety, and high reliable and mobility. Mbara and Pisa [33] state the important relationships in transport as traffic and traffic safety, public transport, and environmental contamination and smog. Through a number of specific studies on sustainable urban development, Pojani and Stead [11] have identified nine factors that promote sustainable urban transport in developed cities, noting: road infrastructure, rail-based public transport, road-based public transport, support for non-motorized modes, technological solutions, awareness-raising campaigns, pricing mechanisms, vehicle access restrictions, and control of land uses as vital to the research and development of the topic.

3 Systematic Review of Public Transportation Systems

To assess the current situation, strengths, and weaknesses of public transport systems, this chapter builds upon a number of theoretical and practical nodes by proposing solutions to develop system trends, improve the quality of road networks to meet travel needs, formulate cleaner urban areas and facilities, and modernize sustainably. The scope identifies public transport systems and the impacts of it via a sustainable development, i.e., triple bottom line [34], paradigm and examines case research from selected Asian and European countries. Articles covering issues related to the type of public transport, its activities, impact on the environment, and society were reviewed and put into typologies to best categorize the state of the art. Information from the collected data sources, i.e., scientific databases, classified the research as follows: (1) article type, (2) definition of sustainability, (3) key findings in terms public transport impact, and (4) vehicle mode and characteristics.

First, the article type was classified by its nature, i.e., its conceptual versus practical experience. Conceptual articles are those that are based on deductive reasoning without factual research. In contrast, the articles considered experience-based contained reports on gathered knowledge from various surveys and research methods. The review coded the approach of the articles as: interpretive (i.e., meaning how sustainability could be interpreted in the context of public transportation), descriptive (i.e., meaning how sustainability is interpreted in the context of public transportation), and prescriptive (i.e., meaning how sustainability should be integrated into projects from a moral or logical point of view). Second, the definition of sustainability reviewed and expressed what sustainability is in various publications, especially in accordance with the concept of the triple bottom line [23, 24, 34]. The three-pillar approach of environment, social, and economic stances was adopted. Third, key findings on the impact of public transport on sustainable development was based on the intersection between public transportation and sustainable development. In this regard, a summary of all the relevant publications and research papers was used to draw conclusions and information related to public transport in the context of contemporary sustainable development theory. Fourth, vehicle mode and their characteristics (i.e., design and manufacturing) summarize the types of transport, studied and reviewed, including similarities and differences between vehicle types in different areas and contexts—noting sustainability can be somewhat varied for this category [35]. The modes of the public transportation systems included bus, rail, metro, airplane, and car sharing, while other relevant factors were the policies, objectives, infrastructure, and convenience of the public transport system (Table 1). The research aspect of the chapter highlights the transport recommendations policy-makers can utilize in advancing safety and resilience in the field. As such, systemic review methods were utilized to identify and analytically investigate 57 articles and reports.

An exploratory, deterministic effort to study public transportation services was looked at in terms of the impact it has on society as well as on sustainable development. The research identified which services are trusted by the public—excluding the following: non-long-distance vehicles, walking, bicycles, scooters, and electric scooters. Five public modes of transport were considered: bus, rail, metro, car sharing, and airplane (Fig. 3). Additional literature explored sustainability and its relationship within the transport sector. From this standpoint, the satisfaction as well as the utility of each type of public transport mode was investigated to provide incentives and change services in-line with sustainable development principles.

Table 1 Modes and reviewed literature of public transportation systems

Category	Description	No	References
Bus	A bus is a large wheeled vehicle, powered by a motor and built to carry several people at the same time. Buses typically run shorter distances than other types of passenger cars and bus routes often connect between urban locations	8	Soh et al. [36], Lyu et al. [37], Ayuningtyas et al. [38], Nguyen and Kajita [39], Boren [40], Dyr et al. [41], Diab and El-Geneidy [42], Rohani et al. [43]
Rail	A train is a type of public transport, including a train locomotive and reconnected carriages. Trains run on railroads and are used to transport people, equipment or other supplies. Train is a central element of rail transport—a popular and relatively inexpensive form of transport	7	Sharma and Newman [44], Tan et al. [45], Kasraian et al. [46], Polom et al. [47], Fremdling [48], Smith [49], van de Velde [50]
Metro	Metro is a vast system of transporting passengers in an urban area, often running on rails. These routes can be underground, or suspended by bridges. Unlike street trains (i.e., tramway), subway can reach high speed because there is a separate path, not shared roads with other means of transportation	5	Derrible [51], Chang and Murakami [52], Pojani and Stead [11], Lyu et al. [53], Stepanik et al. [54]
Car sharing	Car sharing is a car rental service by minute or hour. Each member after registration will be granted a use card. Via a service application, users rent a car and then simply book their location and time to use it	10	Hui et al. [55], Munzel et al. [56], Turon et al. [57], Laurino and Grimaldi [58], Shaheen and Cohen [59, 60], Machado et al. [61], Couzineau-Zegwaard and Meier [62], Morency et al. [63], Turoń et al. [57]
Airplane	Civil aircraft is the main means of transport for civil aviation transportation. This means of transport can travel very long distances—quickly, requires an airport, and utilizes a lot more fuel than other transport modes	9	Sarkar [64], Chen et al. [65], Rajiani and Kot [66], Alaeerad and Khoshnood [67], Lutte and Bartle [68], Elofsson et al. [69], Gillen [70], Thomas [71], Karsner [72]
Other	Some issues related to environment, investment policy on transport infrastructure	18	IPCC [4, 73], OECD [74], Stone and Mees [75], International Energy Agency [6], Hrelja [7], Fiorio et al. [8], Glazebrook [9], Hernandez et al. [76], Jia et al. [12], Bickel and Friedrich [77], Müller et al. [14], EEA [17], Kelly and Fussell [18], Suzuki et al. [19], World Bank [20], Kujala et al. [21], Zhou et al. [78]

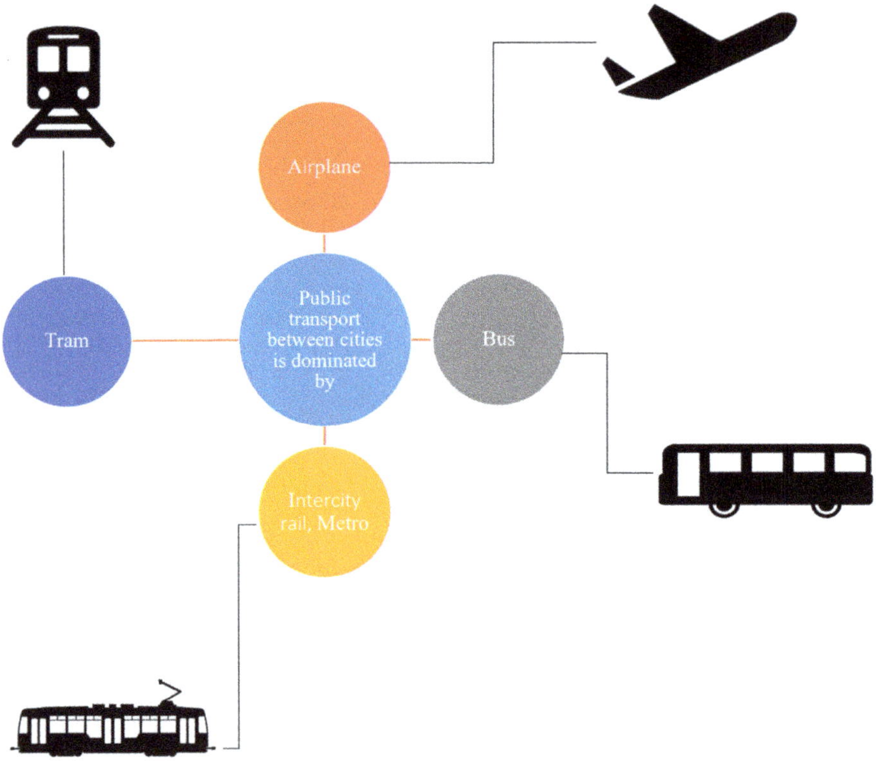

Fig. 3 Modes of public transportation

3.1 Public Survey

Public opinion surveys are a very important method of gathering information needed to supplement and amend problems. In this review, a public survey on transportation satisfaction was administered to examine individual viewpoints toward the utility of public transportation systems in regards to sustainable development whether more professional public transportation systems are needed or desired. The survey was conducted online, using Google Forms, in two countries: Vietnam and Ukraine. The conducted survey was spread over a number of social networking sites (e.g., Facebook, Instagram, and Twitter) from October 2019 to November 2019. The amount of valid responses with IP addresses from those countries number 2980, i.e., 1590 from Vietnam and 1390 from Ukraine. Survey data was analyzed with Microsoft Excel 2019 and StatPlus version 6. In combination with the literature review, the public survey allowed for the collection, integration, and merging of data for analysis.

3.2 Behavior Patterns Applied to Different Types of Public Transport to the Environment and Society

In reference to the study by Saif et al. [79], a sufficient level of positive impact of public transportation, i.e., affecting all aspects of social life, includes health, employment rates, social exclusion, traffic mobility, stability,economy, spatial, and temporal efficiency. Utilizing these indicators, Table 2 develops some strong categorical impacts, with 38 referenced works, that look at the socio-environmental viability of public transportation.

Utilizing the public survey on transportation satisfaction, a series of behavioral patterns were formulated. They show that nearly 63% of the respondents thought that it was necessary to restrict the circulation of cars and motorcycles in order to advance the use of public transportation (i.e., buses, subways, trams, etc.) as well as reduce smog, traffic congestion, reduce the gap between rich and poor, and make it easier for people to meet their daily transportation needs. Moreover, more than 80% suggested adjusting school hours and working hours to reduce traffic congestion and nearly 70% supported the collection of automobile fees in the central area—85.5% supported the application of information technology in traffic management and operational tasks (e.g., automatic fee collection and fining). Among the surveyed respondents, they all recorded they regularly use public transport in various forms. It was found that most people were satisfied with using public transport, and the average travel time was from 30 to 60 min per day.

The survey results indicated that passengers were also supportive of using public transport as a form of daily transportation instead of private motorization. As such, public transportation, e.g., buses, that can accommodate 50 people—equates to 50 personal vehicles being taken off the road during peak hours—thus reducing traffic congestion and emissions to the environment. The surveyed countries, i.e., Vietnam and Ukraine, showed a willingness and desire from commuters to use public transportation—if it was readily available. It can be inferred and is reasonable to suggest; these findings can be reciprocated to other nearby countries that operate comparable public transportation systems in a similar manner.

4 Conclusion

Global energy demand has increased significantly in these recent decades. Between 1973 and 2007, demand for energy global primary intake has doubled [6]. Fossil fuels remain the main source of energy for the world. As such, the demand for renewable energy will only increase. Road traffic using approximately 70% of the energy is used in the system. Inside of that, private passenger road transport occupies 50% of that energy consumed [6]. It has been established from the case research in this chapter that if traffic continues to become more difficult it will not be sustainable. Public transport is an important urban infrastructure and an essential need of human

Table 2 Public transportation—environment and societal impacts

Category	Impact of public transportation	No	References
Health	NO_x is the common name for oxide nitrogen (i.e., including: NO, NO_2, N_2O, N_2O_3, N_2O_4, and N_2O_5). Accordingly, the use of public transport will minimize NO_X emissions to the environment as well as black carbon substances, which directly harm the human respiratory system. In addition, studies have shown that people who are obese have the opportunity to lose weight by daily physical activities such as walking to the bus station or subway station	4	Basagaña et al. [80], MacDonald et al. [81], Glazener and Khreis [82], Leirião et al. [83]
Employment rate	Public transport, with reduced travel time and costs, can move commuters to different locations (i.e., including work), thereby increasing personal income and reducing unemployment rates as well as augment business activity and local economics	8	Gannon and Liu [84], Christidis et al. [85], Lavee [86], Jędrzejczak-Gas and Wyrwa [87], Tyndall [88], (Bastiaanssen et al. [89], Talbot et al. [90], Phillips [91]
Social exclusion	In rapidly growing cities, deprived populations usually have to settle in outer suburbs where rent is more affordable. As this ever-growing part of the population relies solely on walking and public transport for its mobility, public transport can contribute both to social inclusion and economic development by providing access to jobs. A key challenge for public transport is to effectively play the role of connecting poor neighborhoods at the fringes of cities with areas where jobs opportunities exist	6	Fransen et al. [92], Stępniak et al. [93], El-Geneidy et al. [94], Kamruzzaman et al. [95], Boisjoly and El-Geneidy [96, 97]

(continued)

Table 2 (continued)

Category	Impact of public transportation	No	References
Traffic mobility	A reasonable public transport network that meets the mobility needs of the people is a public transportation system that is accessible to the people in the shortest distance and saves time. Mobility, in terms of public transport, is measured by actual movement or supply of transit form of number of trips taken or total kilometers traveled	9	Handy [98], Bok and Kwon [15, 99], (Setton et al. [100], Yigitcanlar et al. [101], Cedar [102], Bertolini [103], Martinez and Viegas [104], de Almeida Correia and Menendez [105]
Stability	Presently, the biggest challenge for society is environmental pollution—in which transportation emissions are one of the main factors causing pollution. If personal vehicle wages can be reduced and instead stimulating people to use public transport increased, the sustainability of transportation and thus improved economic, health, and environment of people can be achieved	7	Elias and Shiftan [31], Kujala et al. [21], (Muñoz and Cohen [106], Cohen and Muñoz [107], Ammann [108], Weiland [109], Hangelbroek [110]
Economy, spatial, and temporal efficiency	In terms of spatial efficiency, the reduction of a large number of private vehicles will contribute to expanding the land use area and limiting traffic congestion in urban areas. Public transport is an arterial system of a city, closely related to urban economic growth. Public transport is the basis to ensure that urban economic activities can proceed normally, as a necessary condition of urban economic growth. At the same time, the scale, speed, and level of development of public transport have a certain relationship with the scale, speed, and level of development of the urban economy	4	Dadashpoor and Rostami [111], Mononen et al. [112], Solomon et al. [113], Vihervaara et al. [114]

settlements [15, 115–117]. In recent years, public transport has been focused on developing, but has not been able to meet actual needs [14, 77]. Traffic congestion, environmental pollution, energy consumption, and the problem of people in traffic are major problems that urban transport is facing [118]. To minimize the risks of vehicle emissions as well as to building green, clean, and pleasing looking human settlements, a reasonable public transportation system, with the support of the community, is needed. The system should ensure the right of transportation for all people and, at the same time, develop the local economy. Developing transport infrastructure on the basis of preserving green space and landscapes associated with regulations on urban expansion, as well as, enhancing traffic safety and creating a convenient transportation system (i.e., highly appreciated by its users) will add and ensure a quality living environment [11, 119].

Recommendations for public transport systems need to ensure the following basic principles. First, meet the travel needs of people quickly and conveniently as well as ensure traffic safety—hence, sustainable transport must first and foremost ensure it is functional. A sustainable transportation system is a transportation system that satisfies the right for all to travel including the elderly, children, and people with disabilities [10, 120]. Second, build a modern and synchronous transportation system to meet current and future traffic needs. The transport system in particular and the infrastructure system in general must be one step ahead to create a premise for socio-economic development [121]. Third, combine the construction of transport infrastructure with environmental protection, on the basis of rational use of economic resources. This will include the development of associated environmental protection as a basis of sustainable transport. Traffic for sustainable development needs to be closely linked with environmental protection [14]. Fourth, control the growth of cars by formulating a transport development strategy and structure [75].

From the perspective of various studies and reports on public transportation, which can be seen as one of the modern pillars of globalization, the proper use of public transportation not only helps reduce traffic congestion, but also helps develop the environment, social concerns, and economic issues—society-wide [24]. Being able to propagate people's awareness to the benefits of using public transport will reduce private transport vehicles. This not only requires the responsibility of the unit that manages and operates the public transport route, but also requires the participation of the information and communication systems at all levels where the systems are in place. The practical significance of this chapter, in retrospect of human settlements, depicts an important picture of developing a complete solution to public transport systems, i.e., to build it sustainably within the context of transport needs, via accessibility, affordability, availability, and acceptability.

References

1. Stern N (2006) Stern review on the economics of climate change. Government of the United Kingdom, London
2. Santalla E, Córdoba V, Blanco G (2013) Greenhouse gas emissions from the waste sector in Argentina in business-as-usual and mitigation scenarios. J Air Waste Manage Assoc 63:909–917. https://doi.org/10.1080/10962247.2013.800167
3. IPCC (2014) Climate change 2014: synthesis report. Contribution of working groups I, II and III to the fifth assessment report of the intergovernmental panel on climate change. Intergovernmental Panel on Climate Change, Geneva
4. IPCC (2017) AR6 climate change 2021: the physical science basis—IPCC. Intergovernmental Panel on Climate Change, Geneva
5. IPCC (2008) AR5 synthesis report: climate change 2014. https://www.ipcc.ch/report/ar5/syr/. Accessed 28 Jan 2021
6. International Energy Agency (2009) World energy outlook 2009. OECD, Geneva
7. Hrelja R (2015) Integrating transport and land-use planning? How steering cultures in local authorities affect implementation of integrated public transport and land-use planning. Transp Res Part A Policy Pract 74:1–13. https://doi.org/10.1016/j.tra.2015.01.003
8. Fiorio CV, Florio M, Perucca G (2013) User satisfaction and the organization of local public transport: evidence from European cities. Transp Policy 29:209–218. https://doi.org/10.1016/j.tranpol.2013.06.004
9. Glazebrook G (2009) Taking the con out of convenience: the true cost of transport modes in Sydney. Urban Policy Res 27:5–24. https://doi.org/10.1080/08111140802369820
10. Cirella GT, Bąk M, Kozlak A et al (2019) Transport innovations for elderly people. Res Transp Bus Manag 30:100381. https://doi.org/10.1016/j.rtbm.2019.100381
11. Pojani D, Stead D (2015) Sustainable urban transport in the developing world: beyond megacities. Sustainability 7:7784–7805. https://doi.org/10.3390/su7067784
12. Jia GL, Ma RG, Hu ZH (2019) Review of urban transportation network design problems based on citespace. Math Probl Eng 2019:1–22. https://doi.org/10.1155/2019/5735702
13. Bickel P, Friedrich R, Droste-Franke B et al (2005) ExternE: externalities of energy, methodology 2005 update. Office for Official Publications of the European Communities, Luxembourg
14. Müller DB, Liu G, Løvik AN et al (2013) Carbon emissions of infrastructure development. Environ Sci Technol 47:11739–11746. https://doi.org/10.1021/es402618m
15. Bok J, Kwon Y (2016) Comparable measures of accessibility to public transport using the general transit feed specification. Sustainability 8:224. https://doi.org/10.3390/su8030224
16. Bureau B, Glachant M (2011) Distributional effects of public transport policies in the Paris Region. Transp Policy 18:745–754. https://doi.org/10.1016/j.tranpol.2011.01.010
17. EEA (2018) Air quality in Europe—2018 report. European Environment Agency, Luxembourg
18. Kelly FJ, Fussell JC (2015) Air pollution and public health: emerging hazards and improved understanding of risk. Environ Geochem Health 37:631–649. https://doi.org/10.1007/s10653-015-9720-1
19. Suzuki H, Cervero R, Iuchi K (2013) Transforming cities with transit: transit and land-use integration for sustainable urban development. World Bank, Washington, DC
20. World Bank (2015) East Asia's changing urban landscape: measuring a decade of spatial growth. World Bank, Washington, DC
21. Kujala R, Weckstrom C, Darst RK et al (2018) Data descriptor: a collection of public transport network data sets for 25 cities. Sci Data 5:1–14. https://doi.org/10.1038/sdata.2018.89
22. IUCN (2019) Nature-based solutions. Int. Union Conserv. Nat. https://www.iucn.org/commissions/commission-ecosystem-management/our-work/nature-based-solutions. Accessed 30 Oct 2019
23. Elkington J (2018) 25 years ago i coined the phrase "triple bottom line." here's why it's time to rethink it. Harv Bus Rev

24. Elkington J (1997) Cannibals with forks: the triple bottom line of 21st century business. Capstone, Oxford
25. UN (1987) Our common future: report of the world commission on environment and development. General Assembly Resolution 42/187, United Nations
26. Lélé SM (1991) Sustainable development: a critical review. World Dev 19:607–621. https://doi.org/10.1016/0305-750X(91)90197-P
27. Sindakis S, Depeige A, Anoyrkati E (2015) Customer-centered knowledge management: challenges and implications for knowledge-based innovation in the public transport sector. J Knowl Manag 19:559–578. https://doi.org/10.1108/JKM-02-2015-0046
28. Buhaug H, Urdal H (2013) An urbanization bomb? Population growth and social disorder in cities. Glob Environ Chang 23:1–10. https://doi.org/10.1016/j.gloenvcha.2012.10.016
29. Cepeda EP, Galilea P, Raveau S (2018) How much do we value improvements on the accessibility to public transport for people with reduced mobility or disability? Res Transp Econ 69:445–452. https://doi.org/10.1016/J.RETREC.2018.08.009
30. Yatskiv I, Budilovich E, Gromule V (2017) Accessibility to Riga public transport services for transit passengers. Procedia Eng 187:82–88. https://doi.org/10.1016/j.proeng.2017.04.353
31. Elias W, Shiftan Y (2012) The influence of individual's risk perception and attitudes on travel behavior. Transp Res Part A Policy Pract 46:1241–1251. https://doi.org/10.1016/j.tra.2012.05.013
32. Gershon RRM (2005) Public transportation: advantages and challenges. J Urban Heal 82:7–9. https://doi.org/10.1093/jurban/jti003
33. Mbara T, Pisa N (2019) An analysis of impediments to deliver sustainable transport in cities of developing countries: the case of Harare, Zimbabwe. WIT Trans Built Environ 182:241–252. https://doi.org/10.2495/UT180231
34. Elkington J (1994) Towards the sustainable corporation: win-win-win business strategies for sustainable development. Calif Manage Rev 36:90–100. https://doi.org/10.2307/41165746
35. Van Marrewijk M (2003) Concepts and definitions of csr and corporate sustainability: between agency and communion. Journal of business ethics. Springer, Netherlands, pp 95–105
36. Soh KL, Wong WP, Le CC, Hiew YH (2014) Improving traffic infrastructure in a developing country: an investigation into the usage of public bus transport in Malaysia. Ind Eng Manag Syst 13:172–184. https://doi.org/10.7232/iems.2014.13.2.172
37. Lyu J, Xiang J, Zhao J, Ming (2019) Design of bus station based on sustainable development and humanization design principles-take two design plans for example. MATEC Web Conf 278:05002. https://doi.org/10.1051/matecconf/201927805002
38. Ayuningtyas KNS, Frazila RB, Wibowo SS, Farda M (2019) Network analysis of intercity bus terminal and inner-city toll road development—the case of Bandung City. Int J Sustain Transp Technol 2:8–18. https://doi.org/10.31427/ijstt.2019.2.1.2
39. Nguyen DT, Kajita Y (2018) Traffic congestion and impact on the environment in Vietnam: development of public transport system—experience from actual operation of bus in Hanoi. J Civ Environ Eng 08:317. https://doi.org/10.4172/2165-784x.1000317
40. Borén S (2020) Electric buses' sustainability effects, noise, energy use, and costs. Int J Sustain Transp 14:956–971. https://doi.org/10.1080/15568318.2019.1666324
41. Dyr T, Misiurski P, Ziółkowska K (2019) Costs and benefits of using buses fuelled by natural gas in public transport. J Clean Prod 225:1134–1146. https://doi.org/10.1016/j.jclepro.2019.03.317
42. Diab EI, El-Geneidy AM (2015) The farside story: measuring the benefits of bus stop location on transit performance. Transp Res Rec J Transp Res Board 2538:1–10. https://doi.org/10.3141/2538-01
43. Rohani MM, Wijeyesekera DC, Karim ATA (2013) Bus operation, quality service and the role of bus provider and driver. Procedia Eng 53:167–178. https://doi.org/10.1016/j.proeng.2013.02.022
44. Sharma R, Newman P (2017) Urban rail and sustainable development key lessons from Hong Kong, New York, London and India for emerging cities. Transp Res Procedia 26:92–105. https://doi.org/10.1016/j.trpro.2017.07.011

45. Tan P, Ma J, Zhou J, Fang Y (2016) Sustainability development strategy of China's high speed rail. J Zhejiang Univ Sci A 17:923–932. https://doi.org/10.1631/jzus.A1600747
46. Kasraian D, Maat K, van Wee B (2016) Development of rail infrastructure and its impact on urbanization in the Randstad, the Netherlands. J Transp Land Use 9:151–170. https://doi.org/10.5198/jtlu.2015.665
47. Połom M, Tarkowski M, Puzdrakiewicz K (2018) Urban transformation in the context of rail transport development: the case of a newly built railway line in Gdańsk (Poland). J Adv Transp 2018:1–15. https://doi.org/10.1155/2018/1218041
48. Fremdling R (1977) Railroads and German economic growth: a leading sector analysis with a comparison to the United States and Great Britain. J Econ Hist 37:583–604. https://doi.org/10.1017/S0022050700095395
49. Smith RA (2003) The Japanese Shinkansen. J Transp Hist 24:222–237. https://doi.org/10.7227/TJTH.24.2.6
50. van de Velde DM (2013) Learning from the Japanese railways: experience in the Netherlands. Policy Soc 32:143–161. https://doi.org/10.1016/j.polsoc.2013.05.003
51. Derrible S (2012) Network centrality of metro systems. PLoS One 7:e40575. https://doi.org/10.1371/journal.pone.0040575
52. Chang Z, Murakami J (2019) Transferring land-use rights with transportation infrastructure extensions: evidence on spatiotemporal price formation in Shanghai. J Transp Land Use 12:1–19. https://doi.org/10.5198/jtlu.2019.1357
53. Lyu G, Bertolini L, Pfeffer K (2016) Developing a TOD typology for Beijing metro station areas. J Transp Geogr 55:40–50. https://doi.org/10.1016/j.jtrangeo.2016.07.002
54. Stepanik M, Vebr L, Delgado JPM (2016) Metro as urban public transport and sustainable urban mobility in the Europe and Brazil. J Eng Archit 4. https://doi.org/10.15640/jea.v4n2a9
55. Hui Y, Wang Y, Sun Q, Tang L (2019) The impact of car-sharing on the willingness to postpone a car purchase: a case study in Hangzhou, China. J Adv Transp 2019:1–11. https://doi.org/10.1155/2019/9348496
56. Münzel K, Boon W, Frenken K et al (2020) Explaining carsharing supply across Western European cities. Int J Sustain Transp 14:243–254. https://doi.org/10.1080/15568318.2018.1542756
57. Turoń K, Kubik A, Łazarz Bogusław B et al (2018) Car-sharing in the context of car operation. IOP Conf Ser Mater Sci Eng 421:032027. https://doi.org/10.1088/1757-899X/421/3/032027
58. Laurino A, Grimaldi R (2012) The Italian way to carsharing. TeMA J L Use Mobil Environ 5:77–90. https://doi.org/10.6092/1970-9870/942
59. Shaheen SA, Cohen AP (2007) Growth in worldwide carsharing. Transp Res Rec J Transp Res Board 1992:81–89. https://doi.org/10.3141/1992-10
60. Shaheen SA, Cohen AP (2012) Carsharing and personal vehicle services: worldwide market developments and emerging trends. Int J Sustain Transp 7:5–34. https://doi.org/10.1080/15568318.2012.660103
61. Machado C, de Salles HN, Berssaneti F, Quintanilha J (2018) An overview of shared mobility. Sustainability 10:4342. https://doi.org/10.3390/su10124342
62. Couzineau-Zegwaard E, Meier O (2018) Carsharing and innovation through the prism of actor network theory. J Innov Econ 27:37–67
63. Morency C, Habib KMN, Grasset V, Islam MT (2012) Understanding members' carsharing (activity) persistency by using econometric model. J Adv Transp 46:26–38. https://doi.org/10.1002/atr.142
64. Sarkar AN (2012) Evolving green aviation transport system: a hoilistic approah to sustainable green market development. Am J Clim Chang 1:164–180. https://doi.org/10.4236/ajcc.2012.13014
65. Chen Y, Yu J, Li L et al (2018) An empirical study of the impact of the air transportation industry energy conservation and emission reduction projects on the local economy in China. Int J Environ Res Public Health 15:812. https://doi.org/10.3390/ijerph15040812
66. Rajiani I, Kot S (2018) The prospective consumers of the Indonesian green aviation initiative for sustainable development in air transportation. Sustainability 10:1772. https://doi.org/10.3390/su10061772

67. Alaeerad E, Khoshnood M (2016) Airport development impact on the sustainable development of the national tourism industry. Turkish Online J Des Art Commun 661–669. https://doi.org/10.7456/1060JSE/034
68. Lutte RK, Bartle JR (2017) Sustainability in the air: the modernization of international air navigation. Public Work Manag Policy 22:322–334. https://doi.org/10.1177/1087724X16679845
69. Elofsson A, Smedby N, Larsson J, Nässén J (2018) Local governance of greenhouse gas emissions from air travel. J Environ Policy Plan 20:578–594. https://doi.org/10.1080/1523908X.2018.1473152
70. Gillen D (2009) Airline business models and networks: regulation, competition and evolution in aviation markets. Rev Netw Econ 5. https://doi.org/10.2202/1446-9022.1103
71. Thomas C (2016) Future development of the air transport sector—implications for sustainable development. QScience Proc 2016:16. https://doi.org/10.5339/qproc.2016.qulss.16
72. Karsner D (1997) Aviation and airports: the impact on the economic and geographic structure of American Cities, 1940s–1980s. J Urban Hist 23:406–436. https://doi.org/10.1177/009614429702300402
73. IPCC (2007) A report of working group I of the Intergovernmental Panel on Climate Change—summary for policymakers. IPCC. http://www.ipcc.ch/pdf/assessment-report/ar4/wg1/ar4-wg1-spm.pdf
74. OECD (2017) Gaps and governance standards of public infrastructure in Chile. OECD, Paris
75. Stone J, Mees P (2010) Planning public transport networks in the post-petroleum era. Aust Plan 47:263–271. https://doi.org/10.1080/07293682.2010.526550
76. Hernandez S, Monzon A, de Oña R (2016) Urban transport interchanges: a methodology for evaluating perceived quality. Transp Res Part A Policy Pract 84:31–43. https://doi.org/10.1016/j.tra.2015.08.008
77. Bickel P, Friedrich R (2001) Environmental external costs of transport. Springer, Berlin
78. Zhou C, Li S, Wang S (2018) Examining the impacts of urban form on air pollution in developing countries: a case study of China's megacities. Int J Environ Res Public Health 15:1565. https://doi.org/10.3390/ijerph15081565
79. Saif MA, Zefreh MM, Torok A (2019) Public transport accessibility: a literature review. Period Polytech Transp Eng 47:36–43. https://doi.org/10.3311/PPtr.12072
80. Basagaña X, Triguero-Mas M, Agis D et al (2018) Effect of public transport strikes on air pollution levels in Barcelona (Spain). Sci Total Environ 610–611:1076–1082. https://doi.org/10.1016/j.scitotenv.2017.07.263
81. MacDonald JM, Stokes RJ, Cohen DA et al (2010) The effect of light rail transit on body mass index and physical activity. Am J Prev Med 39:105–112. https://doi.org/10.1016/j.amepre.2010.03.016
82. Glazener A, Khreis H (2019) Transforming our cities: best practices towards clean air and active transportation. Curr Environ Heal Reports 6:22–37
83. Leirião LFL, Debone D, Pauliquevis T et al (2020) Environmental and public health effects of vehicle emissions in a large metropolis: case study of a truck driver strike in Sao Paulo, Brazil. Atmos Pollut Res 11:24–31. https://doi.org/10.1016/j.apr.2020.02.020
84. Gannon C, Liu Z (1997) Poverty and transport. World Bank, Washington, DC
85. Christidis P, Navajas E, Brons M et al (2014) Future employment in transport: analysis of labour supply and demand. European Commission, Joint Research Centre, Seville
86. Lavee D (2019) The impact of investment in transport infrastructure on employment: the case of Israel. Giordano Ed Eur Transp 74:1–21
87. Jędrzejczak-Gas J, Wyrwa J (2020) Determinants of job satisfaction in a transport company: a Polish case study. Equilibrium 15:565–593. https://doi.org/10.24136/eq.2020.025
88. Tyndall J (2017) Waiting for the R train: public transportation and employment. Urban Stud 54:520–537. https://doi.org/10.1177/0042098015594079
89. Bastiaanssen J, Johnson D, Lucas K (2020) Does transport help people to gain employment? A systematic review and meta-analysis of the empirical evidence. Transp Rev 40:607–628. https://doi.org/10.1080/01441647.2020.1747569

90. Talbot R, Rackliff L, Nicolle C et al (2016) Journey to work: exploring difficulties, solutions, and the impact of aging. Int J Sustain Transp 10:541–551. https://doi.org/10.1080/15568318.2015.1014529
91. Phillips DC (2014) Getting to work: experimental evidence on job search and transportation costs. Labour Econ 29:72–82. https://doi.org/10.1016/j.labeco.2014.07.005
92. Fransen K, Neutens T, Farber S et al (2015) Identifying public transport gaps using time-dependent accessibility levels. J Transp Geogr 48:176–187. https://doi.org/10.1016/j.jtrangeo.2015.09.008
93. Stępniak M, Pritchard JP, Geurs KT, Goliszek S (2019) The impact of temporal resolution on public transport accessibility measurement: review and case study in Poland. J Transp Geogr 75:8–24. https://doi.org/10.1016/j.jtrangeo.2019.01.007
94. El-Geneidy A, Levinson D, Diab E et al (2016) The cost of equity: Assessing transit accessibility and social disparity using total travel cost. Transp Res Part A Policy Pract 91:302–316. https://doi.org/10.1016/j.tra.2016.07.003
95. Kamruzzaman M, Yigitcanlar T, Yang J, Mohamed MA (2016) Measures of transport-related social exclusion: a critical review of the literature. Sustainability 8:1–30
96. Boisjoly G, El-Geneidy AM (2017) The insider: a planners' perspective on accessibility. J Transp Geogr 64:33–43. https://doi.org/10.1016/j.jtrangeo.2017.08.006
97. Boisjoly G, El-Geneidy A (2016) Daily fluctuations in transit and job availability: a comparative assessment of time-sensitive accessibility measures. J Transp Geogr 52:73–81. https://doi.org/10.1016/j.jtrangeo.2016.03.004
98. Handy S (2005) Smart growth and the transportation-land use connection: what does the research tell us? Int Reg Sci Rev 28:146–167. https://doi.org/10.1177/0160017604273626
99. Satariano WA, Guralnik JM, Jackson RJ et al (2012) Mobility and aging: new directions for public health action. Am J Public Health 102:1508–1515. https://doi.org/10.2105/AJPH.2011.300631
100. Setton E, Marshall JD, Brauer M et al (2011) The impact of daily mobility on exposure to traffic-related air pollution and health effect estimates. J Expo Sci Environ Epidemiol 21:42–48. https://doi.org/10.1038/jes.2010.14
101. Yigitcanlar T, Foth M, Kamruzzaman M (2019) Towards post-anthropocentric cities: reconceptualizing smart cities to evade urban ecocide. J Urban Technol 26:147–152. https://doi.org/10.1080/10630732.2018.1524249
102. Ceder A (Avi) (2020) Urban mobility and public transport: future perspectives and review. Int J Urban Sci 1–25. https://doi.org/10.1080/12265934.2020.1799846
103. Bertolini L (2020) From "streets for traffic" to "streets for people": can street experiments transform urban mobility? Transp Rev 40:734–753. https://doi.org/10.1080/01441647.2020.1761907
104. Martinez LM, Viegas JM (2017) Assessing the impacts of deploying a shared self-driving urban mobility system: an agent-based model applied to the city of Lisbon, Portugal. Int J Transp Sci Technol 6:13–27. https://doi.org/10.1016/j.ijtst.2017.05.005
105. de Almeida Correia GH, Menendez M (2017) Automated and connected vehicles: effects on traffic, mobility and urban design. Int J Transp Sci Technol 6:iii–iv
106. Muñoz P, Cohen B (2017) Mapping out the sharing economy: a configurational approach to sharing business modeling. Technol Forecast Soc Change 125:21–37. https://doi.org/10.1016/j.techfore.2017.03.035
107. Cohen B, Muñoz P (2016) Sharing cities and sustainable consumption and production: towards an integrated framework. J Clean Prod 134:87–97. https://doi.org/10.1016/j.jclepro.2015.07.133
108. Ammann C (2011) Stability studies needed to define the handling and transport conditions of sensitive pharmaceutical or biotechnological products. AAPS PharmSciTech 12:1264–1275
109. Weiland J (2012) Stability and transport in magnetic confinement systems. Springer, New York
110. Hangelbroek RJ (1985) On the stability of the transport equation. Integr Equations Oper Theory 8:1–12. https://doi.org/10.1007/BF01199979

111. Dadashpoor H, Rostami F (2017) Measuring spatial proportionality between service availability, accessibility and mobility: empirical evidence using spatial equity approach in Iran. J Transp Geogr 65:44–55. https://doi.org/10.1016/j.jtrangeo.2017.10.002
112. Mononen L, Auvinen AP, Ahokumpu AL et al (2016) National ecosystem service indicators: measures of social-ecological sustainability. Ecol Indic 61:27–37. https://doi.org/10.1016/j.ecolind.2015.03.041
113. Solomon S, Qin D, Manning M et al (2007) IPCC, 2007: Climate change 2007: the physical science basis. Contribution of working group I to the fourth assessment report of the intergovernmental panel on climate change. Cambridge University Press
114. Vihervaara P, Viinikka A, Brander L et al (2019) Methodological interlinkages for mapping ecosystem services—from data to analysis and decision-support. One Ecosyst 4:1–17
115. Guihaire V, Hao JK (2008) Transit network design and scheduling: a global review. Transp Res Part A Policy Pract 42:1251–1273. https://doi.org/10.1016/j.tra.2008.03.011
116. Hensher DA (2007) Sustainable public transport systems: Moving towards a value for money and network-based approach and away from blind commitment. Transp Policy 14:98–102. https://doi.org/10.1016/j.tranpol.2006.10.004
117. Mees P (2014) TOD and multi-modal public transport. Plan Pract Res 29:461–470. https://doi.org/10.1080/02697459.2014.977633
118. Ackerman F (2009) Stern advice for Copenhagen. Nat Clim Chang 1:62–63. https://doi.org/10.1038/climate.2009.34
119. Aarhaug J, Elvebakk B (2015) The impact of Universally accessible public transport—a before and after study. Transp Policy 44:143–150. https://doi.org/10.1016/J.TRANPOL.2015.08.003
120. Mackett RL, Thoreau R (2015) Transport, social exclusion and health. J Transp Heal 2:610–617. https://doi.org/10.1016/j.jth.2015.07.006
121. Jiang X, Zhang L, Xiong C, Wang R (2016) Transportation and regional economic development: analysis of spatial spillovers in china provincial regions. Netw Spat Econ 16:769–790. https://doi.org/10.1007/s11067-015-9298-2

Perspectives in Energy Efficiency and Agriculture Security

Energy Transition in Maritime Transport: Solutions and Costs

Ernest Czermański and Giuseppe T. Cirella

Abstract Modern human settlements rely on shipping and maritime-related services to sustain key provisions of modern life. Energy transition in maritime transport is an important factor in developing sound and healthy delivery of such services. Exploratory research as well as costs are examined in the implementation of hybrid and electric propulsion engines in ships as well as their use of on-shore power. The use of renewable energy sources is considered, and the electrification of sea ports noted as a key piece to the energy transition puzzle. The chapter utilizes a number of example technologies and illustrates a schematic of a typical pure electric ship propulsion system. The capital costs function of hybrid propulsion is illustrated in relation to the change of CO_2 emission reduction costs—indicating its unsuitability for long haul shipping. The methods described rear toward reducing the carbon footprint of the entire transshipment service and, thus, developing a cleaner and more energy efficient maritime transport sector.

Keywords Shipping · Renewable energy · Hybrid propulsion · Electric propulsion · On-shore power · Electrification of ports

1 Introduction

Human settlements comprise of a number of fundamental infrastructural necessities for their success. In modern times, the path toward sustainable development has led these necessities toward a number of structural changes consistent with energy origin and achieving a 100% share of renewable sources (e.g., solar, wind, water, and geothermal). On land, this is possible by building appropriate model-based structures that generate energy as well as the distribution network. These systems require storage and return-oriented capability during periods of temporary energy reduction.

E. Czermański (✉) · G. T. Cirella
Faculty of Economics, University of Gdansk, Sopot, Poland
e-mail: ernest.czermanski@ug.edu.pl

G. T. Cirella
e-mail: gt.cirella@ug.edu.pl

G. T. Cirella (ed.), *Human Settlements*, Advances in 21st Century Human Settlements,
https://doi.org/10.1007/978-981-16-4031-5_5

The most far-reaching concept is that of smart grids which distribute and store electricity in an international system, balance production, and transmit consumption on an ongoing basis [1–5]. Electricity storage is based on different kinds of batteries, capacitors, and supercapacitors; however, they have limited capacity and lifespan in terms of energy demand [6–9]. Energy consumption itself is highly variable in time and space and does not coincide with the seasonality of energy production from renewable sources. As such, energy storage systems should cover the difference [10–13]. The currently known technologies are based on lithium and cobalt, and according to the state of the art, the world's resources of these elements, i.e., in the form of minerals, are too little to cover the potential demand for battery production for all energy storage the world over [6, 14, 15]. Renewable energy, on the other hand, is certainly not in short supply. Solar energy alone can provide 3.1×10^{17} kWh annually, with the total worldwide annual energy demand measured at 1.6×10^{14} kWh [16–18]. This fact alone entertains further analysis of the possibility of transforming energy sources needed for the proper functioning and development of modern life. One basic requirement of a successful human settlement is the transport of cargo, i.e., its movement, in and out of cities. An important mode, that has lived the ages, is maritime transport. The faculty to ship cargo in a sustainable manner and, moreover, in accordance with the far-reaching goal of the International Maritime Organization (IMO) calls for the significant reduction of shipping emissions—i.e., until it reaches a full zero-emission status. This chapter will explore energy transition in shipping and elucidate possible solutions and costs of its implementation.

2 Hybrid and Electric Ship Propulsion

Among the many types of alternative fuels and alternative propulsion systems for ships, hybrid propulsion and increasingly pure electric propulsion are frequently mentioned in the literature [19–23]. Both of these types of propulsion systems are inhomogeneous in every respect; while they have in common the fact that they are technological innovations, in ship propulsion and that they are at an early stage of development, they are represented very narrowly in shipping practice. All existing cases of using pure electric propulsion can be met in linear shipping where a very short distance is the subject of the service, i.e., an estimated for 20–40 Nm [19, 21, 24]. Hybrid propulsion is the combination of two or more alternative sources of energy used for both the propulsion of the ship and the operation of onboard equipment. Hybrid in this respect includes: (1) a combination of two or more energy sources put together to form a propulsion system, where they may operate alternately or together and (2) a combination of two systems in which one produces energy that is transferred to the other, which in turn converts it into mechanical energy for ship propulsion. A classic example is a diesel-electric propulsion system based on a diesel engine and electric motors together with a set of batteries as energy storage onboard. In such a system, the internal combustion engine (i.e., diesel) operates at a constant load with minimal fuel consumption and exhaust emissions, i.e., generating electricity transferred to the electric motors, which in turn, the appropriate transmission systems drive the propeller shaft. The onboard battery system allows for quick and flexible response with the need to change, i.e., increase, the shaft speed and to store excess energy in moments of overproduction. This gives the ship sufficient redundancy for emergency events at sea much faster than traditionally oil-based engines. Such a system is predestined for shipping based on a high proportion of low-speed port operations (i.e., entry, exit, and berthing), hence suitable for short-distance transport operated by ferries or RoPax vessels. Very often, this type of shipping is currently based on powering the ship in port from the shore or, alternatively, generating its own electricity while stationary from low-emission diesel generators. At the same time, this is an area of great potential for further innovation based on renewable energy by replacing these generators with photovoltaic installations or wind turbines [25–29]. An example of this technology, introduced in 2020, is the small feeder-class container ship Yara Birkeland from Norway, where one internal combustion engine drives two electric motors that power two azimuthal propulsion units. According to the IMO, Yara Birkeland, registered as No. 9865049, offers 3120 deadweight tonnage, 3000 gross tonnage, 120 twenty-foot equivalent unit (TEU) cargo capacity, 13.0 kn maximum speed, and 6.0 kn operational speed. It is an example of a fully autonomous ship.

The second type of electric propulsion are the so-called pure electric ships, which are completely powered by electricity, which is not generated onboard but is stored within the hull in the form of battery cells. They are charged while in port and provide energy to the propulsion systems and onboard installations during the voyage (Fig. 1). The most suitable type of shipping which can adopt such solutions is very short sea distances with simultaneous proportionally long stopovers in port. There are already examples of such lines equipped with all-electric ships, mainly in Scandinavia (e.g., the Larvik-Oppedal line), and increasingly, there are already diverse companies building larger vessels, e.g., China. At present, the shipbuilding industry has seen strong development of electric ships for China's transportation system with the introduction of general cargo vessels starting at a length and width of 70.5 m and 13.9 m, 3.3 m draft, and 2000 deadweight tonnage. Energy supply is provided by a set of supercapacitors and batteries with a total capacity of 2.4 MWh. In general, the efficiency of such a propulsion system for deep sea shipping is estimated to be very low, mainly due to the relatively low battery capacity compared to ship demand. In this regard, shipowners are still waiting on new technologies that will allow the carrying capacity and ship capacity the allocated space and tonnage for the battery system for suitable ship propulsion.

Fully electric ships are only a marginal part of maritime navigation with specific navigational characteristics. There are currently a small fleet of 89 such ships with an average of 536 deadweight tonnage and length of 70.7 m [31]. Nonetheless, zero-emission ships insofar as they are powered by renewable sources already exist. In the case of hybrid propulsion, the energy is obtained from the combustion of fossil fuel, which is a source of emissions. As a result, for the purposes of further consideration, it should be assumed that unless renewable energy sources replace combustion engines, or at least reduce them to some extent, low-emission or, even more so, zero-emission

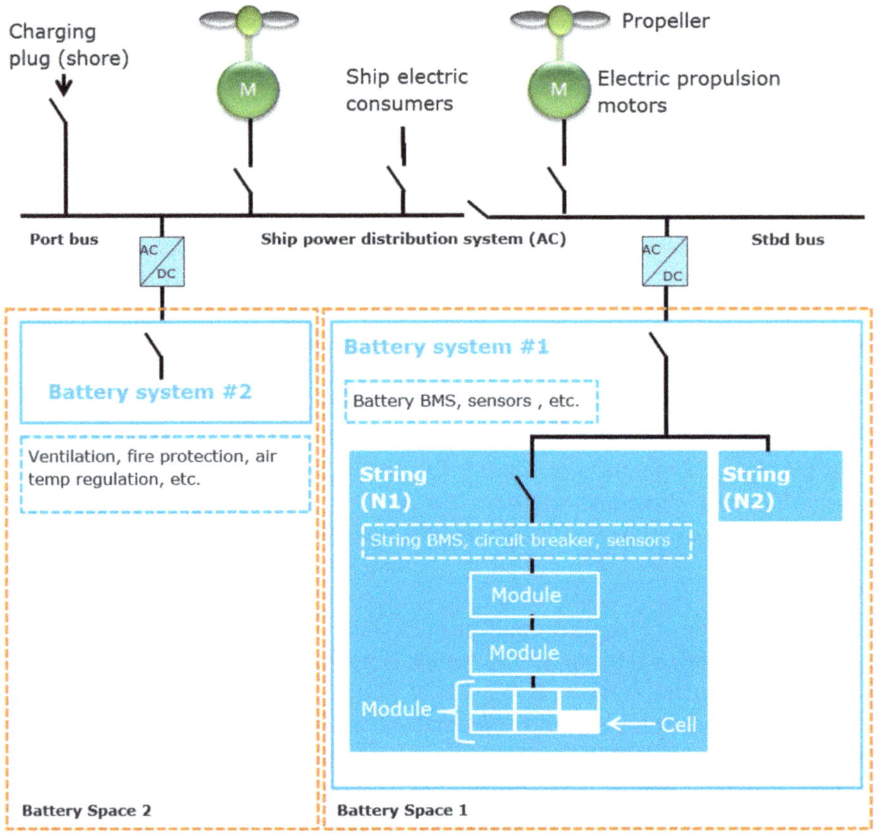

Fig. 1 Schematic of a typical pure electric ship propulsion system, adopted from DNV GL [30]

hybrid ships should not be considered. As such, the technologies available on the market do not yet make it possible to develop a system that could fully replace internal combustion, with the exception of fuel cells using liquid hydrogen to produce onboard electricity [32, 33]. This solution, however, is still in its infancy and conceptualization stage.

3 Costs of Implementing Hybrid Propulsion

Hybrid propulsion, i.e., technology combining electric propulsion motors with onboard power supply from diesel generators as well as a battery system providing power reserve in case of the need for rapid acceleration is generally considered to be a solution which pays off only in the case of new ships and not for the modernization of existing ones [34–36]. Valuing the components of a complete hybrid propulsion system is very complicated because it has to cover practically the entire propulsion system, from the main engine room to the generators as well as to the propulsion shaft itself. For the purpose of the unified analysis system adopted in this chapter, the average values of the costs found in various case studies [19, 22, 23, 34–36], mainly as finished products offered by suppliers of these propulsion systems or even ship repair yards specializing in this type of ship conversion, were adopted. The potential CO_2 emissions reduction of the hybrid propulsion system has been determined to be between 2 and 45%, with a median of 6% [37]. To achieve these levels of reduction, 10, 20, and 40% of the value of the new vessel must be invested. These are, therefore, already significant costs. For a typical USD 60 million ferry for full hybrid propulsion, it would equate to an additional USD 24 million for a propulsion conversion. However, as Fig. 2 illustrates, there is very little investment in terms of fuel saving versus cost to the shipowner. As a result, the cost of reducing 1 ton of CO_2 for the lowest environmental performance threshold is as high as USD 451.00/t-CO_2 for an investment of USD 5–10 million. For a reduction level of 6%, this cost drops to USD 256.00/t-CO_2. Further reductions up to the maximum efficiency limit of 45% would lead to a very low level of profit of USD 31.00/t-CO_2, which marks the end of the shipowner's financial cost zone. The maximum reduction of CO_2 emissions is, however, financially viable despite the very high outlays necessary for its implementation—ranging from USD 24 to 40 million [37].

According to Stefan Krüger, from the Institute of Ship Constructing and Shipping Safety of the Technische Universität Hamburg, hybrid propulsion systems are not suitable for ships operating over long distances, both in irregular and liner shipping, i.e., ocean container shipping. Energy loss is mainly in the form of heat, of up to 10%, which is the sum of the losses generated at the stage of, firstly, combustion of marine gasoil in generators, then during conversion into electrical energy and, thirdly, during conversion of electrical energy into mechanical energy transferred to the propulsion shaft. Hybrid propulsion systems, on the other hand, are suitable for short sea shipping and ferry services where the vessel often accelerates and decelerates and then stands for a long time in port for loading. This gives time to recharge the batteries and rebuild the power potential. In this context, it seems necessary to use another technology in

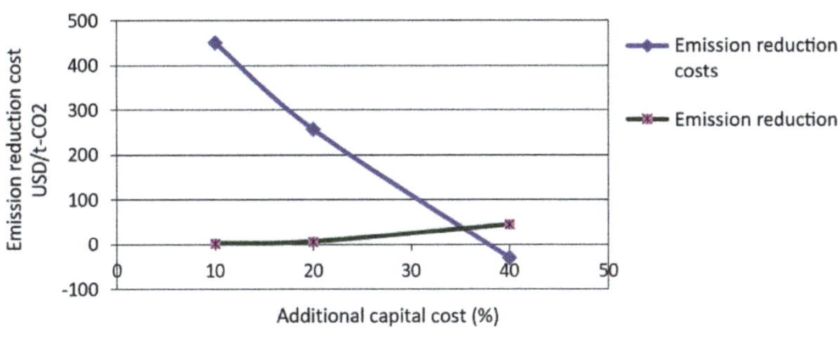

Fig. 2 Additional capital costs function of hybrid propulsion in regards to the change of CO_2 emission reduction costs [37]

close combination—on-shore power for ships in sea ports. Both of these solutions can have a positive effect on each other and, through synergy, lead to a minimum of fuel consumption by ship generators.

4 Cost of Implementing On-Shore Power

The potential for reducing CO_2 emissions by switching ships in port to on-shore power is estimated to be between 3 and 10% with a median of 5.3% [38]. Compared to other methods, it is one of the least efficient solutions; nonetheless, it is attracting a lot of interest and is the subject of numerous studies and analyses. The profitability of investment in on-shore connection (i.e., shore-to-ship power or cold-ironing) must be considered in two ways—for the port (i.e., which also acts for the port city) and for the shipowner (i.e., since the two groups of cold-ironing users have different considerations). Market observations suggest that the main interest for this solution comes from the hinterland, from decision-makers at both international and national level, as well as locally [39–41]. Factors motivating port authorities to invest in port infrastructure to enable shore-side ship power include: clean air in the port and port city, ability to produce electricity in-house (i.e., wind, solar, wave, or tidal power) using equipment installed within the administrative area of the port, integration of offshore wind energy into the grid via an interconnector located in the port area and connection of the harbor grid to the transmission grid from the offshore wind farm, favorable structure of origin of electricity in the local distribution network coming from on-shore renewable sources (i.e., energy of rivers, wind, sun), and a port development strategy toward a green port concept implementation [42, 43].

It should be stressed that the total cost of creating an on-shore power supply system in a port per berth, together with the entire service system, represents a very high cost for the port's infrastructure. It would be calculated at several of USD millions per unit. This cost, however, decreases rapidly as the total number of berths in a single system increases and, independent of shipowners, does not take into account further calculations of the effectiveness of shore-side power in reducing CO_2 emissions. In contrast, the factors motivating shipowners to invest in ship readiness for connectedness to the on-shore grid include: strategic sustainability objectives adopted by the shipowner in its operations (i.e., sometimes falling within the scope of the shipowner's corporate social responsibility), lower costs of purchasing electricity in port compared to the cost of generating it on land, and restrictions by some sea ports to connect to the on-shore power supply network, while the shipowner is willing to handle cargo generated (i.e., particularly in the case in liner shipping where there are economies of scale where a single ship's call does not give rise to major considerations or demands on the part of the port authority to adapt the ship to shore-side power). A search among known cases of ship investments in shore power installation, the average cost of such an installation corresponds to about 0.25% of the value of building a new ship and does not change with increasing ship size. This is a constant correlation. In value terms, it can vary most often between USD 100,000 and

300,000 [44]. Various calculations show that for small container ships, cold-ironing is uneconomic in any fuel price range. The size limit is 1000 TEU per unit. However, for vessels above 1000 TEU, cascading profitability can be determined annually, as such: 3000 TEU (up to USD 175,000), 4250 TEU (USD 220,000–250,000), 5500 TEU (USD 350,000–420,000), 8160 TEU (USD 550,000–560,000), and 9075 TEU (USD 1.1 million) [45]. Evidently, the latter thresholds cause significant increments in the profitability of cold-ironing and are not profitable.

5 Conclusion: Electrification of Ports

The implementation of hybrid and electric ship propulsion as well as on-shore power supplies is expensive and not necessarily cost effective or up to par with current energy expectations. As such, the electrification of port equipment, broadly defined as the process of modernizing existing loading and unloading equipment used within port handling terminals, can be another alternative in maritime transport energy saving. The most vulnerable parts of this type of process are mobile transshipment equipment, such as rubber tire gantry cranes, straddle carriers, and reach stackers or lifting vehicles. The conversion of a drive to electric is limited to the replacement of the motor and power transmission gears. This can be carried out on site, without having to move the equipment to the manufacturer of the new technology, and therefore minimizes the period of time that such equipment is out of production. The capacities that are needed for them are much lower than for ship power plants, and therefore, the investment costs are relatively lower. In addition to the cost of the technology, the electrification of port equipment has numerous advantages, including: (1) increased reliability of electrical equipment, (2) lower energy costs, (3) reduction of risk associated with the use of liquid fuels thus increasing traffic safety in the port, (4) large potential for developing own energy sources from renewable technologies (e.g., photovoltaics, wind energy, and wave energy), (5) greater independence of the equipment's operation from the amount of energy consumed and elimination of fuel level control and refueling (and thus longer equipment operation or disposal time), (6) greater independence from weather conditions, and (7) greater susceptibility of electric port equipment to increase efficiency via automation. As result, this alternative can develop a greener-based port in which economic advantages, in tangent with environmental benefits, aids in rectifying fundamental infrastructural necessities for modern maritime transport.

The primarily elimination of local exhaust emissions that occur with internal combustion drives (i.e., mainly diesel engines) is important to developing human settlements that are environmentally friendly and mindful human health and habitation. The use of electric-only drives powered entirely from renewable sources has not been fully implemented and no such examples exists to date. However, there are many countries aiming to change their electricity origin structure toward renewable sources. The possibility of storing electricity in ports as large economic facilities where not only the production of cargo handling services, but also, increasingly, development and implementation work takes place is promising. The green port policy which is being put into practice in many European ports (i.e., notably Bremen [46]) indicates possible energy transition solutions in maritime transport thinking and, to a larger degree, human settlements and how human beings organize port-to-port transport. Moreover, methods and systems described in this chapter should be further explored to also reduce the electrical energy consumption of port equipment through specific technological systems (e.g., Alasali et al. [47]). Utilizing such methods, it is possible to reduce the carbon footprint of the entire transshipment service and, thus, complete the door-to-door delivery service in a much cleaner and sounder manner.

References

1. Aikhuele DO, Ighravwe DE, Akinyele D (2019) Evaluation of renewable energy technology based on reliability attributes using hybrid fuzzy dynamic decision-making model. Technol Econ Smart Grids Sustain Energy 4. https://doi.org/10.1007/s40866-019-0072-2
2. Hattam L, Greetham DV (2017) Green neighbourhoods in low voltage networks: measuring impact of electric vehicles and photovoltaics on load profiles. J Mod Power Syst Clean Energy 5:105–116. https://doi.org/10.1007/s40565-016-0253-0
3. El-Hawary ME (2014) The smart grid—state-of-the-art and future trends. Electr Power Components Syst 42:239–250. https://doi.org/10.1080/15325008.2013.868558
4. Pahwa A, Venkata SS (2011) Preparing the workforce for smart distribution systems. In: 2011 IEEE PES innovative smart grid technologies, ISGT Asia 2011 conference: smarter grid for sustainable and affordable energy future
5. Majeed Butt O, Zulqarnain M, Majeed Butt T (2020) Recent advancement in smart grid technology: future prospects in the electrical power network. Ain Shams Eng J (In Press). https://doi.org/10.1016/j.asej.2020.05.004
6. Trahey L, Brushett FR, Balsara NP et al (2020) Energy storage emerging: a perspective from the joint center for energy storage research. Proc Natl Acad Sci USA 117:12550–12557. https://doi.org/10.1073/pnas.1821672117
7. Zerrahn A, Schill WP, Kemfert C (2018) On the economics of electrical storage for variable renewable energy sources. Eur Econ Rev 108:259–279. https://doi.org/10.1016/j.euroecorev.2018.07.004
8. Staffell I, Rustomji M (2016) Maximising the value of electricity storage. J Energy Storage 8:212–225. https://doi.org/10.1016/j.est.2016.08.010
9. Genoese M, Genoese F, Möst D, Fichtner W (2010) Price spreads in electricity markets: what are fundamental drivers? In: 2010 7th international conference on the european energy market, EEM 2010
10. Azad AK, Khan MMK, Ahasan T, Ahmed SF (2014) Energy scenario: production, consumption and prospect of renewable energy in Australia. J Power Energy Eng 02:19–25. https://doi.org/10.4236/jpee.2014.24004
11. Saidur R, Rahim NA, Islam MR, Solangi KH (2011) Environmental impact of wind energy. Renew Sustain Energy Rev 15:2423–2430. https://doi.org/10.1016/j.rser.2011.02.024
12. Czermański E, Pawłowska B, Oniszczuk-Jastrząbek A, Cirella GT (2020) Decarbonization of maritime transport: analysis of external costs. Front Energy Res 8:28. https://doi.org/10.3389/fenrg.2020.00028
13. Czechowski PO, Dabrowiecki P, Oniszczuk-Jastrzabek A et al (2020) A preliminary attempt at the identification and financial estimation of the negative health effects of urban and industrial air pollution based on the agglomeration of Gdańsk. Sustainability 12:1–28. https://doi.org/10.3390/SU12010042

14. Saeedmanesh A, Mac Kinnon MA, Brouwer J (2018) Hydrogen is essential for sustainability. Curr Opin Electrochem 12:166–181. https://doi.org/10.1016/j.coelec.2018.11.009
15. Vymazal J, Kröpfelová L, Švehla J et al (2009) Trace elements in Phragmites australis growing in constructed wetlands for treatment of municipal wastewater. Ecol Eng 35:303–309
16. Hayat MB, Ali D, Monyake KC et al (2019) Solar energy—a look into power generation, challenges, and a solar-powered future. Int J Energy Res 43:1049–1067. https://doi.org/10.1002/er.4252
17. Delucchi MA, Jacobson MZ (2011) Providing all global energy with wind, water, and solar power, part II: reliability, system and transmission costs, and policies. Energy Policy 39:1170–1190. https://doi.org/10.1016/j.enpol.2010.11.045
18. Deng YY, Haigh M, Pouwels W et al (2015) Quantifying a realistic, worldwide wind and solar electricity supply. Glob Environ Chang 31:239–252. https://doi.org/10.1016/j.gloenvcha.2015.01.005
19. Geertsma RD, Negenborn RR, Visser K, Hopman JJ (2017) Design and control of hybrid power and propulsion systems for smart ships: a review of developments. Appl Energy 194:30–54. https://doi.org/10.1016/j.apenergy.2017.02.060
20. Kim S, Jeon H, Kim J (2020) Trend analysis of domestic and international regulations for electric propulsion system. J Int Marit Saf Environ Aff Shipp 4:113–121. https://doi.org/10.1080/25725084.2020.1809949
21. Jeon H, Kim J, Yoon K (2019) Large-scale electric propulsion systems in ships using an active front-end rectifier. J Mar Sci Eng 7:168. https://doi.org/10.3390/jmse7060168
22. Geertsma RD, Negenborn RR, Visser K, Hopman JJ (2017) Parallel control for hybrid propulsion of multifunction ships. IFAC-PapersOnLine 50:2296–2303. https://doi.org/10.1016/j.ifacol.2017.08.229
23. Capasso C, Notti E, Veneri O (2019) Design of a hybrid propulsion architecture for midsize boats. Energy Proc 158:2954–2959. https://doi.org/10.1016/j.egypro.2019.01.958
24. Solem S, Fagerholt K, Erikstad SO, Patricksson Ø (2015) Optimization of diesel electric machinery system configuration in conceptual ship design. J Mar Sci Technol 20:406–416. https://doi.org/10.1007/s00773-015-0307-4
25. Gul M, Kotak Y, Muneer T (2016) Review on recent trend of solar photovoltaic technology. Energy Explor Exploit 34:485–526. https://doi.org/10.1177/0144598716650552
26. Kobougias I, Tatakis E, Prousalidis J (2013) PV systems installed in marine vessels: technologies and specifications. Adv Power Electron. https://doi.org/10.1155/2013/831560
27. Kjaer SB, Pedersen JK, Blaabjerg F (2005) A review of single-phase grid-connected inverters for photovoltaic modules. IEEE Trans Ind Appl 41:1292–1306. https://doi.org/10.1109/TIA.2005.853371
28. Spagnolo GS, Papalilo D, Martocchia A (2011) Eco friendly electric propulsion boat. In: 2011 10th international conference on environment and electrical engineering, EEEIC.EU 2011—conference proceedings
29. Oğuz E, Şentürk AE (2019) Selection of the most sustainable renewable energy system for Bozcaada Island: wind versus photovoltaic. Sustainability 11:4098. https://doi.org/10.3390/su11154098
30. DNV GL (2016) DNV GL handbook for maritime and offshore battery systems, report no. 2016–1056, Document No.: 15DJV2L-2. DNV GL, Olso
31. IHS (2021) Online database. In: IHS Marit. Portal. https://maritime.ihs.com/EntitlementPortal/Home/Index. Accessed 14 Feb 2021
32. Hansson J, Månsson S, Brynolf S, Grahn M (2019) Alternative marine fuels: prospects based on multi-criteria decision analysis involving Swedish stakeholders. Biomass Bioenerg 126:159–173. https://doi.org/10.1016/J.BIOMBIOE.2019.05.008
33. Blanco H, Nijs W, Ruf J, Faaij A (2018) Potential for hydrogen and power-to-liquid in a low-carbon EU energy system using cost optimization. Appl Energy 232:617–639. https://doi.org/10.1016/j.apenergy.2018.09.216
34. Völker T (2013) Hybrid propulsion concepts on ships. Sci J Gdynia Marit Univ 3:66–76

35. Bennabi N, Charpentier JF, Menana H et al (2016) Hybrid propulsion systems for small ships: Context and challenges. In: Proceedings—2016 22nd international conference on electrical machines, ICEM 2016. Institute of Electrical and Electronics Engineers Inc., pp 2948–2954
36. Xiao N, Xu X, Chen B (2020) Research on simulation and experiment of ship complex diesel-electric hybrid propulsion system. J Sh Res 64:171–184. https://doi.org/10.5957/jsr.2020.64.2.171
37. Czermański E (2019) Morska żegluga kontenerowa a zrównoważony rozwój transportu. Institute of Maritime Transport and Seaborne Trade Press, Gdansk
38. Bouman EA, Lindstad E, Rialland AI, Strømman AH (2017) State-of-the-art technologies, measures, and potential for reducing GHG emissions from shipping—a review. Transp Res Part D Transp Environ 52:408–421. https://doi.org/10.1016/J.TRD.2017.03.022
39. Sisson M, Mcbride K (2011) The economics of cold ironing. AECOM, Los Angeles
40. Theodorus PG (2012) A cold ironing study on modern ports, implementation and benefits thriving for worldwide ports. School of Naval Architecture and Marine Engineering, National Technical University of Athens, Athens
41. Klopott M, Maśnicki R (2010) Cold-ironing jako innowacja w transporcie morskim—aspekty techniczne i ekonomiczne—Zeszyty Naukowe Uniwersytetu Szczecińskiego. Problemy Transportu i Logistyki—Numer nr 10 Innowacje w transporcie technologie i procesy (2010)—Biblioteka Nauki—Yadda. Zesz Nauk Uniw Szczecińskiego Probl Transp i Logistyk 10:99–109
42. Bergqvist R, Monios M (2019) Green ports: inland and seaside sustainable transportation strategies. Elsevier, London
43. Gonzalez-Aregall M, Bergqvist R (2019) Green port initiatives for a more sustainable port-city interaction: the case study of Barcelona. Maritime transport and regional sustainability. Elsevier, Amsterdam, pp 109–132
44. Weidenholm C (2009) Cold ironings cuts costs. Swedish Club Lett 3:6–7
45. Weiner PH (2012) Cold-ironing as a cost-effective tool for improving sustainability in the shipping industry. Nicholas School of the Environment, Durham
46. Krämer I, Czermański E (2020) Onshore power one option to reduce air emissions in ports. Sustain Manag Forum 28:13–20. https://doi.org/10.1007/s00550-020-00497-y
47. Alasali F, Haben S, Holderbaum W (2019) Energy management systems for a network of electrified cranes with energy storage. Int J Electr Power Energy Syst 106:210–222. https://doi.org/10.1016/j.ijepes.2018.10.001

Sustainability and Renewable Energy Education: Children of the Next Generation

Hen Friman, Yafa Sitbon, Ifaa Banner, Yulia Einav, and Giuseppe T. Cirella

Abstract Creating a clean energy future is a worldwide goal. At present, world energy consumption relies heavily on fossil fuels, which are non-renewable and eventually finite, such as coal, oil, and natural gas. These resources have already become too expensive and too environmentally damaging. The importance of renewable energy and energy efficient technologies is constitutively rising. How should children be taught about renewable energy? What is air pollution and how does harming the natural habitat affect us? Students from the Holon Institute of Technology participate in the course "Green ambassadors" which looks at these questions by integrating practical work. As part of the course requirements, students are asked to conduct lessons within the topic of preserving the environment to fifth and sixth grade pupils in the Arabic elementary school "El Omariya" situated in the city of Ramla. During meetings held within the school, the students taught the pupils, via games and activities, what renewable energy means, how to turn waste into a resource, what energy conversion and renewable energy mean, etc. In order to illustrate the topics studied by the pupils, the students used a moveable laboratory containing demonstrations, experiments and creative activities. Environmental education, with an emphasis on renewables, is the focus of this chapter.

H. Friman (✉) · Y. Einav
Faculty of Engineering, Holon Institute of Technology, Holon, Israel
e-mail: henf@hit.ac.il

Y. Einav
e-mail: yulia_e@hit.ac.il

Y. Sitbon · I. Banner
Holon Institute of Technology, Holon, Israel
e-mail: yafasit@hit.ac.il

I. Banner
e-mail: ifaab@hit.ac.il

G. T. Cirella
Faculty of Economics, University of Gdansk, Sopot, Poland
e-mail: gt.cirella@ug.edu.pl

G. T. Cirella (ed.), *Human Settlements*, Advances in 21st Century Human Settlements,
https://doi.org/10.1007/978-981-16-4031-5_6

Keywords Environmental education · Clean energy · Elementary school · Green ambassador · Israel

1 Introduction

The development of renewable energy sources is spreading rapidly. Today, as a result of the energy crisis and growing concern over climate change, national security, and the health effects of poor air quality, renewable energy is gaining the attention of the general public. Fortunately, renewable energy technologies have also been vastly improved in the past two decades [1–4]. Scientific advances and global market growth have transformed solar, wind, and bioenergy from backyard novelties into practical systems capable of fueling vehicles and heating as well as lighting homes, schools, and businesses. Costs have also fallen as the technology has improved. The cost of generating electricity from wind and solar power has decreased by 90% over the past 30 years [5, 6]. Renewables are cleaner and safer than non-renewables (e.g., coal, oil, or nuclear power), and their use helps improve public health and energy security, as well as reduce the emissions of CO_2. Education is an important first step in making this transition. Renewable energy technologies are ready to be implemented, but increased public confidence, regulatory reforms, and a system of economic incentives for development of these resources are needed to make large-scale use a reality. An understanding of renewable energy will be a crucial part of scientific literacy for the future. When current middle and high school students reach adulthood, many of them will be commuting from solar energy-powered homes in biomass-fueled cars. Much of the electricity they use at home and in the workplace will come from solar, wind, biomass, and geothermal power.

How should children be taught about renewable energy? Last year students from the Holon Institute of Technology (HIT) participate in the course called "Green ambassadors" which combines various practical exercises into the curriculum. This course integrates work run by the Dean of the Students, Social Involvement Unit, and "Israeli Hope" in the academia activated at HIT. Among the goals of the course is to engage students with ecological and environmental issues, enlighten them about the importance of the quality of the environment, and educate (i.e., the next generation) about basic energy efficiency. As part of the course requirements, students are asked to conduct lessons within the topic of preserving the environment to fifth and sixth grade pupils in the Arabic elementary school "El Omariya" situated in the city of Ramla. During meetings held within the school, the students taught the pupils, via games and activities, what renewable energy means, how to turn waste into a resource, what energy conversion and renewable energy mean, etc. In order to illustrate the topics studied by the pupils, the students used a moveable laboratory containing demonstrations, experiments, and creative activities. In this chapter, descriptive research pieces together the content of the course and how knowledge obtained was passed on to young pupils using a creative, hands on teaching approach which is more effective than traditional teaching.

2 Environmental Education: Children of the Next Generation

Educative arguments for change are needed. At present, young people's interest in choosing a scientific career is declining, scientific ignorance in the general population is extensive, economic importance of scientific knowledge is inclining, and last, but not least, student opinion that science in school is boring and not relevant for them is prevalent. In the past decade, there has been mounting evidence that this problem has become more acute. Studies by Lyons [7] and Sjøberg and Schreiner [8], have indicated that most youth expressed positive attitudes on the importance of scientific and technological issues to society but also that new strategies for increasing young peoples' interest and knowledge in science and their ability to use science outside school are needed. Learning in out-of-school environments is common worldwide. Students get to visit science, natural history, and art museums. There is much evidence in the research literature that out-of-school learning has many positive impacts on learning outcomes of various sorts [9, 10]. Students learn scientific content, develop positive attitudes toward science, interact with each other while being engaged in learning meaningful things, and gain opportunities to use all senses to experience phenomena in real-contexts [9].

In Israel, the Arab public-school system, legally obliged to provide a level of education equal to that offered to Jewish citizens, is in fact inferior, on average, to the Jewish public-school system due, in part, to the unequal budgets and resources allocated by Israel's government. This often creates gaps in knowledge in a range of subjects [11–13]. Arab students from Israel also find it more difficult than their Jewish peers to meet the demands of an academic system that requires critical discourse, as Arab schools tend to allow less room for expressing opinions and encourage more passive instruction [14]. Therefore, many Arab students are at a disadvantage especially relative to their Jewish counterparts. Many of them lack both the cultural capital [15] and the type of academic skills required for coping with Western-influenced Israeli culture, making it difficult for them to match the achievements of their Jewish peers. Henry Giroux asserts that "while the hidden curriculum cannot be entirely eliminated, its structural properties can be identified and modified" thus enabling the teacher to develop new pedagogical methods [16]. These methods, which Giroux and others identify as critical pedagogy, may partially and temporarily transform the power relations in class and may develop critical and political consciousness among participants, students, and teachers. However, critical pedagogy, claims Norman Denzin, "requires citizens and citizen-scholars committed to taking risks, persons willing to act in situations where the outcome cannot be predicted in advance" [15]. Bell Hooks [17] has accurately acknowledged the reluctance of many teachers "to see the classroom change, to allow for shifts in relations between students" and their teacher. Hooks and Denzin use methods of critical pedagogy that can transform the educational power structure and aspire the classroom towards persons jointly working together to develop new lines of action, new stories, new narratives in a collaborative effort. To create these changes teachers should overcome their fears, work harder, be adventurous, imaginative, and spontaneous [18–20].

3 Academy—Holon Institute of Technology

HIT was established in 1969 and became an independent public academic institution of higher education in 1999, certified by the Council of Higher Education of Israel. HIT focuses on the teaching of sciences, engineering, computer science and technology, and management of technology and design. It also emphasizes multidisciplinary theoretical and practical research of innovative technologies from a professional scientific, economic, and cultural perspective. HIT trains highly qualified students in the realms of science, engineering, management and design, and plays an important role in their integration upon graduation into key positions within the industry. The Institute aspires to quality and excellence in teaching and innovative research, and strives to introduce novel and unique cutting-edge teaching and research technologies. It also prides itself on its advanced academic achievements, application of innovative techniques, and interdisciplinary professionalism that lead to creative teaching and new technologies. HIT aims to utilize the intellectual and professional potential of each and every student, so that they can fully integrate into the fast-paced technological world of today. Providing superior technological and scientific education enables HIT graduates to enter key leadership positions in both the private and public sectors.

3.1 Faculty of Engineering

The aims and goals of the Faculty of Engineering are to keep the study programs updated to meet the ever-changing requirements for engineers of the future, enrich student theoretical knowledge as well as teach practical and design skills. The Faculty objectives are to provide the students with a rich and comprehensive study program, adapt teaching methodologies and techniques—focusing on understanding as a goal, enable students to achieve skills such as self-learning and acquire expertise via practice, constantly update the teaching methods and the study programs, maintain relationships with the various relevant industry sectors, introduce the students to state of the art equipment and facilities, and promote research in the various fields and explore cooperation with other institutes in Israel and abroad.

3.2 Social Involvement Unit at the Dean of Students Office

One of the many goals of the Social Involvement Unit is to promote social involvement of students and staff in the community. It also promotes weak applicants and students at the Institute by offering mentoring, tutoring, emotional support guidance to learning, and adjustments in school. Over the years, the Unit has worked in many education and welfare arenas to promote immigrants, youth, and more. The Social

Involvement Unit serves as a professional center to encourage and promote the social impact of students and staff and to leverage knowledge, expertise, and human capital for the benefit of the community through social involvement projects and course actions involving meaningful activities.

4 Action Learning Course

The action learning course is an academic course which involves academic learning and social activities. The courses deal with processes and social challenges, reveal different ideologies, and develop critical thinking and pragmatic ideas. Students receive course credits and a grade for being part of such courses. Participating students enroll in courses that involve action and activities to engage in the experiential learning process, thereby creating a dialogue and cross-fertilization between being taught in the classroom and experiencing the reality in the real world [19]. The learning experience includes meeting with social organizations, institutions, and state authorities and carrying out practical work with diverse populations. Through experience, students strengthen their academic skills, formulate ethical attitudes toward reality, develop professional and civilian perspectives, and realize how they can influence their surrounding in the present and the hereafter.

Under the guidance and supervision of Hen Friman, HIT has set up an innovative course which puts together learning of students in the classroom and social doing-and-passing on of material learnt to elementary pupils in an enjoyable way to boost awareness and accessibility to knowledge and information regarding environmentalism. The ultimate goal has been for young pupils to become green-oriented ambassadors—i.e., children with environmental awareness. In the frontal part of the course, students received an introduction to Arabic society and acquired knowledge on environmental issues, such as ecology, electricity production, air pollution, renewable energy, water sector, waste, and recycling. The students learnt how an effective and enjoyable method of teaching is based on the knowledge acquired and were divided into working groups. In the practical part of the course (i.e., each group plans, purchases, and builds sample as well as forms a teaching model exemplifying one scientific doctrine studied in the course). The students presented their workshops to fifth and sixth graders from "El Omariya" School in Ramla.

4.1 Solar Energy

Solar energy is clean energy. It produces no hazardous solid, liquid, or gas waste [21, 22]. It does not create water or air pollution. Direct production of electricity using sunlight is accomplished using photovoltaic cells, also called solar cells. They have no moving parts and are "clean" energy [22, 23]. A major limitation is cost, which greatly exceeds the cost of producing electricity using fossil fuels or nuclear power.

Fig. 1 (left) Solar car race lesson, (right) the pupils are participating in starting them via a moveable light source. Photographs were taken by H. Friman on 21 November 2017

The best solar cells are only 20% efficient and only provide 50 W of electricity per square meter of cell size [24, 25]. The group which engaged in solar energy set itself a goal to present the pupils the possibility to produce energy in a way which does not pollute the environment. In order to incorporate all the pupils and make the activity competitive, the pupils were required to divide into groups, each having to start a solar vehicle by a moveable light source (Fig. 1); thus, facilitating a solar car race for the lesson.

4.2 Wind Energy

Wind energy has been utilized for thousands of years. The wind is free, commonly available, and can provide clean, pollution-free energy. Today's wind-turbines are very high tech. In most places, the cost of commercial wind power on a large scale is not economically competitive with conventionally generated electricity [26, 27]. One important factor is that with a doubling of wind speed, power output increases by a factor of eight. The United States remains the world leader in wind energy, but Europe has embarked on a very ambitious wind-power development program. It is predicted that by 2030, wind energy will supply at least twice the electricity it does now [28]. Comparative studies of wind energy have shown it to be a viable energy source that can reduce environmental pollution and water consumption. Negative backlash, however, includes noise pollution, visual interference, and potential impacts to wildlife [26, 29].

4.3 Waste and Recycling

Another problem in Israel is solid waste. The Ministry of Environmental Protection is continually working to reduce the amount of waste accumulating at landfills. One the best approaches to reduce waste is a recycling process, in which appropriate materials are being removed from the waste stream and used as raw materials for creating new materials [30–32]. Most recycling in Israel is carried out in sorting and separation plants [33, 34]. A waste and recycling group of students and pupils had the goal to raise awareness about the poor situation in the country on waste issues with emphasis on the recycling process as the best solution (i.e., equivalent to the best solution to date). During the course of the activity, the pupils prepared a page of newspaper scraps and turned an old tire into an armchair.

4.4 Soil Contamination

The group responsible for this topic explained to the pupils the layered structure of the soil. The pupils built a sample of soil profiles and discussed the overall impact of soil pollution, agriculture, ecology, and so on. The pupils were challenged to try to sprout beans on fruitful ground versus contaminated soil (Fig. 2). The results were conclusive; they were presented to the pupils' parents and siblings.

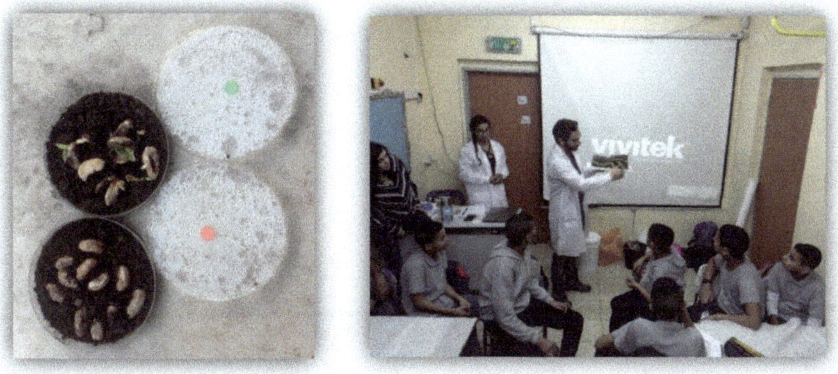

Fig. 2 (left) An experiment from the soil pollution group carried out home sprouting of beans on fruitful ground as opposed to contaminated soil. (right) Pupils are taught about the layered structure of the soil profile to help elucidate with the exercise. Photographs were taken by H. Friman on 21 November 2017

5 "Green Ambassadors" Methodology

The course methodology included mutual fertilization (Fig. 3); the pupils acquired knowledge while the students got familiar with Arab society in an informal and unprejudiced manner (Fig. 4). All the groups conducted theoretical knowledge lessons accompanied by a presentation including pictures and videos illustrating the material with a significant part being comprised of performing scientific experiments in class so as the pupils would internalize it. Throughout the activity, the students helped raise the motivation among the pupils. Even with the quietest pupils, they were encouraged to ask questions and take an active part in the lesson. By working in small

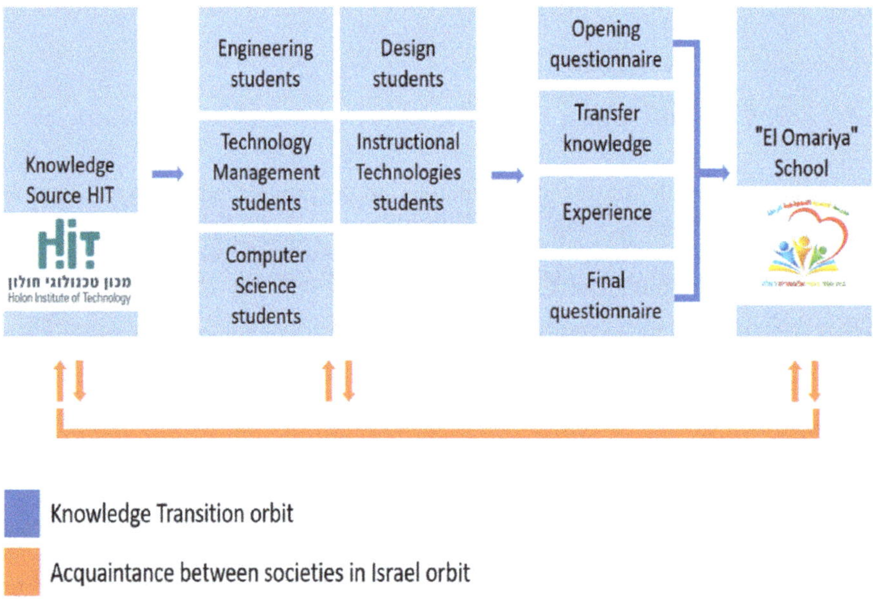

Fig. 3 Mutual fertilization framework of the course

Fig. 4 **a** HIT students visited "El Omariya" School in Ramla and **b** learn from the school's staff about Arabic society. Photographs were taken by H. Friman on 7 November 2017

groups, pupils learnt teamwork, how to give and receive criticism, and how to plan, monitor, and evaluate their individual and joint activities with others. It appears that modern workplaces increasingly require such partial delegation of authority, group management, and cooperative skills. The pupils understood the purpose of the course and the importance of the issues studied. Based on the responses from pupils and teachers at the end of the activity, the students learnt that the pupils are interested in preserving the environment and making a better future for everyone—agreeing that they would become ambassadors of preserving the environment even by doing small things, which in turn is meaningful (e.g., not leaving garbage after a picnic and turning the lights off when leaving a room). As a result, pupils become aware and confident of their ability to make a significant and positive change as green-friendly ambassadors.

6 Conclusion

In the "Green ambassador" community-based course, i.e., action learning, participating students engaged in an experiential learning process that included creating a dialogue and cross-fertilization between what was being studied in the classroom and reality. Through the experience, students strengthened their academic skills and values, formulated ethical attitudes toward reality, developed a professional and civilian approach, and understood how they could affect and influence their surroundings. Thanks to all these, the pupils became more aware and learnt an important lesson about the ways to preserve the environment.

Acknowledgements We would like to thank the Higher Education Council for budgeting and supporting this course. To the Social Involvement Unit, thank you for the support and for allowing such a course to take place at HIT, as well as helping us contribute to society and future generations. Last, but not least, we want to thank "El Omariya" School for the opportunity to take part in the next generation of education for a better and cleaner environment.

References

1. Owusu PA, Asumadu-Sarkodie S (2016) A review of renewable energy sources, sustainability issues and climate change mitigation. Cogent Eng 3:1167990. https://doi.org/10.1080/23311916.2016.1167990
2. Buonocore JJ, Hughes EJ, Michanowicz DR et al (2019) Climate and health benefits of increasing renewable energy deployment in the United States. Environ Res Lett 14:114010. https://doi.org/10.1088/1748-9326/ab49bc
3. Gielen D, Boshell F, Saygin D et al (2019) The role of renewable energy in the global energy transformation. Energy Strateg Rev 24:38–50. https://doi.org/10.1016/j.esr.2019.01.006
4. Abdelwahab H, Moussaid L, Moutaouakkil F, Medromi H (2018) Energy efficiency: improving the renewable energy penetration in a smart and green community. Proc Comput Sci 134:352–357. https://doi.org/10.1016/j.procs.2018.07.199

5. Bilgen S, Kaygusuz K, Sari A (2004) Renewable energy for a clean and sustainable future. Energy Sources 26:1119–1129. https://doi.org/10.1080/00908310490441421
6. Kaltschmitt M, Streicher W, Wiese A (2007) Renewable energy: technology, and environment economics. Springer, Berlin Heidelberg, Berlin
7. Lyons T (2006) Different countries, same science classes: students' experiences of school science in their own words. Int J Sci Educ 28:591–613. https://doi.org/10.1080/09500690500339621
8. Sjøberg S, Schreiner C (2005) How do learners in different cultures related to science and technology? Results and perspectives from the project rose. Asia-Pacific Forum Sci Learn Teach 6:1–17
9. Falk JH, Dierking LD (2000) Learning from museums: visitor experiences and the making of meaning. AltaMira Press, Walnut Creek, CA
10. Bamberger Y, Tal T (2008) Multiple outcomes of class visits to natural history museums: the students' view. J Sci Educ Technol 17:274–284. https://doi.org/10.1007/s10956-008-9097-3
11. Al-Haj M (1995) Education among the Arabs in Israel: control and social change. Magnes Press, Jerusalem
12. Al-Haj M (2003) Higher education among the Arabs in Israel: formal policy between empowerment and control. High Educ Policy 16:351–368. https://doi.org/10.1057/palgrave.hep.8300025
13. Al-Haj M (1988) Arab graduates in Israel: main characteristics, potential of graduates, and employment situation (Hebrew)
14. Barak M, Peleg R, Avrahami M (2000) Scientific technological education as a focus of the renewal of the Arab-Druze school in Israel. Iyunim B'khinukh 4:51–76
15. Denzin N (2007) The politics and ethics of performance pedagogy: toward a pedagogy of hope. The SAGE handbook of performance studies. SAGE Publications, Thousand Oaks, CA, pp 325–338
16. Golan-Agnon D (2005) Inequality in education. In: Champagne D, Abu-Saad I (eds) Indigenous and minority education: International perspectives on empowerment. Negev Center for Regional Development, Beer-Sheva, pp 201–214
17. Hooks B (1994) Teaching to transgress: education as the practice of freedom. Routledge
18. Olneck M (2000) Can multicultural education change what counts as cultural capital? Am Educ Res J 37:317–348. https://doi.org/10.3102/00028312037002317
19. Jabareen Y, Agbaria A (2011) Israeli Government policy and civil society initiatives to improve Arab Education in Israel education on hold. DIRASAT The Arab Center for Law and Policy, and The Arab Minority Rights Clinic, Faculty of Law, University of Haifa, Haifa
20. Arar K (2012) Israeli education policy since 1948 and the state of Arab education in Israel. Ital J Sociol Educ 1:113–145
21. Lewis NS (2007) Toward cost-effective solar energy use. Science (80-)315:798–801
22. Kannan N, Vakeesan D (2016) Solar energy for future world: a review. Renew Sustain Energy Rev 62:1092–1105
23. Lu Y, Khan ZA, Alvarez-Alvarado MS et al (2020) A critical review of sustainable energy policies for the promotion of renewable energy sources. Sustainability 12:5078. https://doi.org/10.3390/su12125078
24. Messenger RA, Abtahi A (2017) Photovoltaic Systems Engineering, 4th edn. CRC Press, Boca Raton, FL
25. Wenham SR, Green MA, Watt ME, Corkish R (2007) Applied photovoltaics, 3rd edn. Earthscan, London
26. Baerwald MR, Feyrer F, May B (2008) Distribution of genetically differentiated splittail populations during the nonspawning season. Trans Am Fish Soc 137:1335–1345. https://doi.org/10.1577/t07-097.1
27. Zahedi A (2014) Sustainable power supply using solar energy and wind power combined with energy storage. Energy Proc 52:642–650. https://doi.org/10.1016/j.egypro.2014.07.120
28. Manwell JF, McGowan JG, Rogeres AL (2010) Wind energy explained: theory, design and application, 2nd edn. Wiley, West Sussex

29. Knopper LD, Ollson CA (2011) Health effects and wind turbines: a review of the literature. Environ Health A Glob Access Sci Source 10:78
30. Sharp V, Giorgi S, Wilson DC (2010) Methods to monitor and evaluate household waste prevention. Waste Manag Res 28:269–280. https://doi.org/10.1177/0734242X10361508
31. Korica P, Cirman A, Žgajnar Gotvajn A (2016) Decomposition analysis of the waste generation and management in 30 European countries. Waste Manag Res 34:1109–1116. https://doi.org/10.1177/0734242X16670488
32. Bartl A (2014) Moving from recycling to waste prevention: a review of barriers and enables. Waste Manag Res 32:3–18
33. Nissim I, Shohat T, Inbar Y (2005) From dumping to sanitary landfills—solid waste management in Israel. Waste Manag 25:323–327. https://doi.org/10.1016/j.wasman.2004.06.004
34. Daskal S, Ayalon O, Shechter M (2018) The state of municipal solid waste management in Israel. Waste Manag Res 36:527–534. https://doi.org/10.1177/0734242X18770248

Fostering Sustainable Development: Green Energy Policy in the European Union and the United States

Chelsea R. Spring and Giuseppe T. Cirella

Abstract Climate and environmental pollution have a long-term effect on world economics. Evidence has shown that climatic events have had drastic impacts since pre-industrial times. Industrialization has played a key role in pollution-based emissions. Most industrialized countries come from the developed world. Mass industrial and economic development has burdened less developed countries by exposing them to harmful methods of emission development. This chapter examines the need for developed countries to reverse harmful environmental emissions by creating green energy-based policies that, in effect, reduce greenhouse gases and environmental pollution. A focalized breakdown of the European Union and the United States, i.e., two parties that produce substantial amounts of polluting emissions, is assessed by looking at effective green energy policy to foster stronger sustainable development and ecologically friendly human settlements for the timeframe 2005–2011. This historical-environmental perspective is valuable to understand where current green deal policy originates from and recognize the political and economic elements policy-makers should consider for future policy development.

Keywords Green taxes · Energy socioeconomics · Environmental pollution · Policy development · Developed countries

1 Introduction

Environmental taxes, commonly known as green taxes, are levies on activities that cause environmental pollution. These taxes are created to place financial burdens on industries that produce pollutants while accelerating the advancement and implantation of cost-effective smart methods of production and consumption. In an ideal world, the implementation of green taxes should reduce environmental pollution in industries and foster the rationale for sustainable development [1–6]. Globally,

C. R. Spring (✉) · G. T. Cirella
Faculty of Economics, University of Gdansk, Sopot, Poland

G. T. Cirella
e-mail: gt.cirella@ug.edu.pl

G. T. Cirella (ed.), *Human Settlements*, Advances in 21st Century Human Settlements,
https://doi.org/10.1007/978-981-16-4031-5_7

numerous governments and regulatory agencies within various industries have vigorously created and implemented a variety of taxes, duties, and licensure fees that fall under the category of green regulations and policies [1, 2]. A major aim in the creation of green-oriented protocol is to reduce pollution emitting agents, e.g., particulate matter and CO_2 emissions, which are produced when industries use fossil fuels that are synthetically produced or natural. A huge debate within the global scientific community is climate concerns—with long-term influence on world economics. Within the science community, a consensus on keeping risks at acceptable levels, i.e., to best deal with climate change, involves a comprehensive understanding of both energy usage and its sourcing [7–9].

Regional effects of climate change will differ; as such, the global economic burden of green energy-based policies will require an optimal transitional knowhow as well as a compliant and well-informed citizenry. As a result, the benefits (i.e., cost) of emissions in each country are impartial of whether it is currently contributing to greenhouse gases (GHGs) [10]. For instance, CO_2 emissions, a type of GHGs, are being distributed on an unequal bias, i.e., as the US and other developed countries act as key polluters, while less developed countries inadvertently are sharing the burden [10, 11]. This, in essence, affects how countries can contribute, i.e., in terms of costs, since underdeveloped countries are not capable of making a dramatic contribution. In other words, the US and other developed countries—principle to the source—should lead by example and reform their regulations and policies to aim for a reduction in GHG emission levels in conjunction with environmental pollution. This chapter is broken into three segments that introduce the subject matter by reviewing the contextual causes of climate, green policies currently set in place within the European Union and the US, and analytical findings for the timeframe 2005–2011. Further comparative research look at how these parties can re-evaluate their laws and guidelines to work toward sustainable development.

2 Climatological Rationale

Climate shifts and naturally occurring events pose a universal risk on society. According to the Organization for Economic Cooperation and Development (OECD), such an event "it threatens the basic elements of life for all people: access to water, food production, health, use of land, and physical and natural capital" [12]. If actions are not taken to slow or reverse the emission of polluting gases, negative consequences will affect human well-being, hinder economic growth, and intensify the risk of unanticipated and large-scale changes to the planet's ecological systems [12, 13]. In this context, a disparity exists between developed and less developed countries—namely that the former generates a significantly higher level. Hence, the responsibility of developed countries to contribute more to preventing polluting emissions comes down to the fact that these developed countries play a bigger role GHG production and can afford the cost of making economical changes toward climate-resilient growth paths [10, 12]. As such, one of the key concerns of the climate issue is how to effectively regulate and implement policy-based decisions without collapsing a country's energy needs or its economy.

Throughout history, earth's climate has always changed, e.g., with several cycles of glacier advances and retreats over the millennia. According to the National Aeronautics and Space Administration (NASA) [14], the average surface temperature of the planet has increased 1.18 °C since the late-nineteenth century. Linked to this change, natural disasters are common weather-related events that are carefully tracked and monitored. The increase in frequency and magnitude of extreme weather events, such as superstorms, can have very devastating effects [15–17]. Examples of superstorms include Hurricane Katrina (i.e., a category 5 hurricane) which hit the Gulf of Mexico in August 2005—impacting over 250,000 km^2 of the US (i.e., mainly Louisiana, Alabama, and Mississippi). Hurricane Katrina's devastation resulted in 80% of New Orleans, Louisiana, being under water, approximately 2000 people were killed, 600,000 households were displaced, and aside from the human costs, nearly USD 150 billion of damage was done [18, 19]. That same season, three other category 5 hurricanes, Emily, Rita, and Wilma, affected the same region. Three years later, another superstorm, Hurricane Ike, hit the Caribbean and Gulf region [19]. This hurricane, classified as a category 4, caused damage across the Caribbean and the southern US (i.e., Texas, Arkansas, and Mississippi). Numerous areas were flooded and landslides occurred in many affected regions causing an overall death toll of 128, with approximately 34 people missing. The estimated damage from Hurricane Ike was about USD 20.3 billion [18, 19]. Climatologically speaking, the reasons for these extensive and frequent weather events in the subtropical Pacific are due to the increasing temperature of the ocean's surface layer. This layer is closely coupled with atmospheric temperature. The increase in temperature is caused by changes in concentration of the cooler La Niña weather patterns offset by alternating irregular El Niño ones in recent years [15, 20–26]. These cooler weather patterns cause trade winds and cooler currents that transport warmer surface water deep into the ocean. This phenomenon is believed to be the reason for the superstorms that are frequently occurring throughout this part of the world. Other significant weather events with a devastating impact on the environment were the recording breaking heatwaves throughout the US within the past decade including, prolonged droughts in California, the heatwave and drought in Russia during 2010, the heatwave in India in 2015 which killed approximately 2500 people and Typhoon Haiyan in 2013 [15, 22]. According to NASA [27], CO_2 emissions are currently 420 ppm (a record high) and up from 300 ppm in the 1950s [10]. Since the 1990s, CO_2 emissions from fossil fuels increased from 20.9 gt in 1990 to 28.8 gt in 2007 [28]. It is projected that by 2030, fossil fuel consumption will increase by another 80%. To this end, the world transport sector, for instance, is 95% dependent on oil—contributing up to 60% of all oil consumption. Road transportation, i.e., mainly automobiles, is the main cause for these emissions compared to air and rail. Road vehicles account for three-fourths of energy in transport—concluding the direr need for countries to rethink, reduce, and pollute less.

2.1 Regulations and Policies

There are various types of regulations and policies that can be implemented to reduce emissions in industries. Implementation can be geared toward energy consumption or emission reduction from industrial processes. Direct emission from the industry in Annex I countries accounted for 2108 Mt CO_2 (i.e., 15.4%) of the total Annex I emissions from fossil fuels [29]. Similarly, there is an indirect relationship between emission release from industry and electricity usage. Emissions that were produced by the production process (i.e., production emissions) account for an additional 5% of total emissions from 23 countries [29, 30]. To combat emissions within industry, varying approaches such as taxes and trading policies can be implemented domestically and internationally. The most common form of policy instruments among Annex I countries is the creation of policies that address issues related to energy consumption and its sourcing. The classification of "voluntary non-binding agreements on reporting emissions and progress to self-defined targets, [i.e.,] to negotiated agreements that are legally binding, have benchmarking and performance assessment and contain sanctions in the case of non-compliance" [30]. For varying approaches to be successful, policies need to be economically efficient and feasible to implement realistic outcomes, minimize impacts on competitiveness, and invest in climate-resilient planning [29, 30]. Other approaches include soft measures of success which create awareness of a specific climate change concerns within industry or with alleviation-based actions—and can measure success by assessing areas for future improvement or progress. In general, this can include a production line change or improvement to create fewer emissions and pollution. Additionally, partnerships or agreements within industry can be executed to measure success, e.g., between "government and industry, industry and non-governmental organizations, within groupings of industry, or individual declarations by particular companies" [30].

The US and many countries within the European Union have implemented a variety of environmental taxes that have targeted CO_2 emissions in hope that industries would switch to low-carbon or renewable energy resources. Taxes are the most common instrument used in the European Union's market-based approach [29]. Most countries within the European Union target emissions by implementing a form of consumption tax that targets these emitting energy sources [31]. Several countries have applied supplemental taxes that indirectly effect emissions in a positive way as well as taxes whose purposes are to create financial revenue for funding and researching for certain renewable energy sources [29]. These various consumption taxes indirectly create additional taxes on the substances that cause emissions from non-energy sources such as a motor vehicle tax on car usage [31]. Countries including Denmark, Sweden, Finland, and the Netherlands have implemented taxes on specific GHGs such as CO_2 and SO_4 [29, 31]. These taxes target transportation and mineral oils. More controversial than the abovementioned specific taxes is the carbon tax, i.e., an explicit tax for which the rate is linked to the "carbon content of the fuel, irrespective of whether the resulting carbon price is uniform across fuels and uses." At present, the carbon tax is not commonly used in greening policies [32]. It is being debated among economies globally and currently only 20 countries worldwide have implemented it.

The current environmental taxing concept is based on British economist Arthur Pigou's concept that environment and social concepts are not included in the goods or services which they generate, therefore, obliging the government to use its authority to establish costs values by implementing taxes [33–35]. Carbon taxes can be implemented in various ways, such as "imposing levies on the production, distribution, and use of fossil fuels" [36]. For instance, by implementing a carbon tax, the goal would be to decrease the overall demand for energy by increasing the cost of energy; this in turn, would create an increasing cost to high-carbon or carbon-intensive fuels, e.g., coal and oil—making low-carbon fuels, e.g., gas and renewables, the main fuel source. As a result, this would cause renewables to gain a proportional competitive advantage over non-renewables [37]. The concept of the carbon tax seems modeled for global economies; however, since only a small portion have moved forward with it, it is worth discussing some of the reservations and opponents' claims [33].

One of the main reasons for reluctance to implement a carbon tax is the cost or the adverse impact it may have on overall economies including gross domestic product (GDP) and social welfare [33, 37]. Factors that would affect whether the carbon tax would have a negative or positive impact on an economy, include: (1) how the economy recycles its revenues, (2) if an economy is high-carbon or low-carbon driven within industries, and (3) competitiveness and emission leakage [33]. How an economy recycles its revenues from the carbon tax will determine if GDP and social welfare programs are positively or negatively affected. Ways revenues can be recycled into the economy are by lumpsum transfer to households using revenue to cut existing taxes, that being: (1) corporate income tax, taxes on goods and services, and capital gains tax; (2) subsiding technology that focuses on greening; and (3) "increasing public expenditure on infrastructure or welfare programs" [33]. Countries that focus on high-carbon fuels would be expected to have a negative impact on GDP and social welfare programs than countries who focus on low-carbon fuels and renewables. This would be because high-carbon countries are more prone to competitiveness than countries who have low-carbon fuels since implementing a carbon tax could cause enterprises to move production to other countries where there is a smaller carbon tax (or none at all) [34]. This competitiveness can be countered, in part, by using a varying approach system within the economy to allow tax breaks within certain industries or by using a sort of emissions trading system (ETS) in conjunction with the carbon tax [37]. The negative impacts to taxing sources of emissions seem effective in reducing global climate change; however, many exemptions exist, for example, numerous subsectors of industries where CO_2 emission and energy taxes are implemented, certain enterprises are exempt from paying the tax or pay an extremely reduced tax due to competitiveness concerns, especially when the enterprise is subject to internationalization [30]. As a result, these exemptions could have a negative impact on greening policies and prevent targeted emissions and pollutants levels from being reached.

2.2 Emissions Trading

The scheme of ETS is to create an artificial market that cap the amount of emissions of a certain substance that can be released into the atmosphere within a set time period within an industry [37–39]. Emissions trading can be split into two categories: cap-and-trade or baseline and credit. Cap-and-trade schemes are "market-based approaches to controlling pollutions within industries by providing economic incentives for reaching reductions in emissions of pollutions" [38]. There are two forms of emissions trading in the cap-and-trade scheme: rate-based and market-based. Baseline and credit schemes put a cap on emissions by implementing a trading mechanism. The cap-and-trade system is the predominate type of emissions scheme—categorized as statutory (i.e., mandatory) or non-statutory (i.e., voluntarily). Statutory emissions schemes are government initiated, whereas voluntarily schemes are at the discretion of the enterprises in the industry to decide if they want to participate. There are no government mandates in voluntarily trading schemes [38, 40–42]. Regardless of the legal status of the cap-and-trade system, the two systems are similar. Within a statutory scheme, a government passes a law that establishes emissions trading with a goal to limit the amount of emissions from a specific type of GHG within a certain territory. This law allows for a transfer of ability from the emitting source to the government otherwise known as the allocation of credits [38, 39]. After authorization of the law or voluntary agreement, enterprises must apply for permits to emit and carry out activities within the restriction of the scheme; during this time, enterprises within the industry will be allocated credits. In order to ensure the scheme is being properly executed, and adhered to, a regulatory body monitors and imposes negative actions, as needed, per regulatory laws of the scheme. For example, at the end of the trading period, firms must turn-in the amount of emissions they have allocated. The units are canceled from their account, and their allocations are counted [38, 40, 41]. Enterprises who successfully reduced emissions more than the allocated amount receive a surplus of emissions for the next round, while enterprises who emitted more than allowed must pay a certain amount for their overage [38]. It should be noted that a "permit does not act as a mechanism to control the cap of emissions, it only creates a population that can emit; therefore, an incentive, such as allowances, must be implemented" [38]. Stipends must offset participant emissions and be limited by the cap's overall emission goal in order to be successful at reducing emissions. Typically, enterprises will sell their surplus allowances after the turn-in period is over, and just before the start of a new trading period, as this is when most regulatory agencies allow for trading to be done [38].

Baseline and credit schemes are similar to cap-and-trade as both have a law established on emission caps within a certain jurisdiction, that is to say (i.e., the law allows the ability to freely transfer emissions from emitting sources to the government in statutory systems) [38]. Once a law is enacted, scheme participants apply for a permit to emit in order to carry out regulating activities. The difference between baseline and credit schemes and cap-and-trade schemes is that they implement a different trading mechanism. "Unlike cap-and-trade systems that create transferable allowances up to the overall cap and allocate allowances to eligible participants, baseline and credit schemes assign a baseline of emissions to regulated sources of emissions" [38]. The baselines are linked to a certain type of emissions source, meaning it is harder to freely trade baselines among enterprises. Baselines are like allowances in a cap-and-trade scheme since a baseline establishes the amount of allowable emissions which a member may emit without being penalized [38]. The trading scheme works by assessing an enterprise's emission output within a set period and establishes the trading mechanism by issuing credits to sources whose emissions remained below their associated baselines within the set period [38]. A source that has emitted below the baseline will receive the difference in credits. Credits are transferable and may be sold during the trading period or banked for use in future compliance periods. However, if a source exceeds the baseline of emissions allowed, they must use their credits to pay back the regulatory body for exceeding their allowable emissions [37–40, 42, 43]. Overall, within the cap-and-trade system, sources can trade their allowances among each other within a set period, while in the baseline and credit system, credits are only given after the set period has ended and can only then be banked or sold. The International Emissions Trading Association, a non-profit business organization created in 1999, established a functional international framework for trading in GHG emission reductions and includes membership from leading international companies from across the carbon trading cycle [44].

2.3 OECD Climate Focus

The OECD, an international economic organization consisting of 36-member countries, works to stimulate economic growth and world trade. Founded in 1961, it is a forum of countries whose goals are to stay committed to democracy and the market economy in order to be a platform that "compares policy experience, seek answers to common problems, identify good practices, and coordinate domestic and international policies of its members" [45]. A majority of OECD members come from high income economies and have extremely high Human Development Index scores—they are internationally recognized as developed countries. Its member countries consist of Australia, Austria, Belgium, Canada, Chile, Czech Republic, Denmark, Estonia, Finland, France, Germany, Greece, Hungary, Iceland, Ireland, Israel, Italy, Japan, Korea, Luxembourg, Mexico, the Netherlands, New Zealand, Norway, Poland, Portugal, Slovak Republic, Slovenia, Spain, Sweden, Switzerland, Turkey, the United Kingdom, and the US. The platform the OECD stood behind to promote its goal of combating climate change was the Kyoto Protocol signed on December 11, 1997. Numerous parties of the OECD—including all of the European Union and the US—were signatory parties with the Clinton Administration's Vice President Al Gore as a main participant in putting the Protocol together. In the case of the US, President Bill Clinton signed the agreement in November 1998, but the US Senate refused to ratify it, citing potential damage to the American economy. The Kyoto Protocol applied to the six GHGs: CO_2, CH_4, N_2O, HFCs, PFCs, and SF_6. For the purpose of this chapter, the focus GHG is CO_2. The Kyoto Protocol extended the 1992 United Nations Framework Convention on Climate Change (UNFCCC) that committed its parties to reduce GHG emissions to 1990 levels [46]. Based on scientific evidence from the Intergovernmental Panel on Climate Change that believed industrialization lead to a 30% CO_2 increase from 1750 to 1992 [43, 47, 48], the UNFCCC created the Protocol on the principle that different countries have different responsibilities in combating climate change—in terms of their respective development and economic success. Hence, it stresses the responsibility of developed countries to adapt their policies to be on a path of low-carbon sustainability, as stated in Article 2 of the convention. Specifically, stabilizing the concentration of GHGs in the atmosphere to a level would prevent dangerous anthropogenic interference with the climate system [46]. The convention did not take into consideration underlying national differences in GHG emissions, wealth, and capacity would affect how parties would adapt their policies to achieve this goal. The treaty which went into effect on February 16, 2005, positioned first round commitments by Annex I countries, i.e., inclusive of OECD countries, from 2006 to December 31, 2012. Utilizing this timeframe, this chapter will explore the European Union and United States' energy policies to see which party was able to decrease their CO2 emissions effectively and if they were able to sustain their energy requirements and emissions levels based on the 2018 data.

3 European Regulations and Policies Related to Decreasing Emissions

Green energy policies are methods and tools implemented by governments or governing bodies with the intent to reduce CO_2 emissions. Reducing emissions must be a broad-based approach to encourage behavioral adjustments throughout the entire economy in replacement of fossil-based energy [29]. The overall goal of policy-makers is to create environmentally friendly policies with minimal economic cost. The European Union adopted the 20-20-20 formula, which sets climate and energy targets. The goals of this plan were to slash GHGs 20% by 2020 compared to 1990 levels, achieve 20% of primary energy from renewable energy, and improve energy efficiency by 2020. Within the Union, strategies to accomplish these goals were taxes and subsides, regulations, renewable feed-in tariffs, and ETS policies. Table 1 expresses the main points of the 20-20-20 formula in terms of sustainable development.

The most widely applied market-based instruments European Union governments have implemented are environmental taxes and charges, which are implemented in the sectors of air pollution, emissions, transport, energy, and mineral oil [29]. To enhance energy efficiency in the areas of buildings, transport, and households, many European Union governments have granted subsidies or tax credits. Another instrument that has been implemented in the energy sector are renewable feed-in tariffs. Table 2 pieces together key green energy and climate focused policy action each European member

Table 1 European Union 20-20-20 formula for greening policy goals by 2020, adapted from ILO [29]

Type of growth	Target	Incentive
Smart growth	Compared to 1990 levels, GHG emissions will be decreased 20% by 2020	Europe will have resource efficiency
Sustainable growth	Sharing renewable energy in the final energy consumption will increase 20% by 2020	There will be an industrialized policy in effect for the globalization era
Inclusive growth	Energy efficiency will increase 20% by 2020	

state has implemented as of 2011—broken down via tax; duty, fee, and charge; and subsidy.

The European Union emissions trading system (EU ETS) is currently the largest cap-and-trade program in the world [39]. With its launch in 2005, it allocated tradeable emissions permits to over 14,000 power stations and industrial plants within 31 countries—accounting for 40% of the European Union's total GHG emissions. The goal of the EU ETS was to reduce carbon emissions in a cost-effective manner and aid in the creation of low-carbon technologies [29, 39]. The system allows participants to buy and sell allowances within the trading period decided by the ETS. Holders of the allowance have the right to emit one ton of CO_2 or equivalent amount of another GHG. The EU ETS "covers the CO_2 emissions in the power sector, including all fossil fuel generators over 20 MW, iron and steel manufacturing, oil refining, cement, glass, ceramics, and paper and pulp production" [29]. After 2005, members could opt-in to smaller systems within sectors. In the second phase of this system, 2008–2012, new sectors that did not include CO_2 GHG emissions were added. Between 2005–2007, members of the European Union could opt-out of specified sectors, but by 2008, all eligible installations were required to be covered by ETS.

Within the first two trading periods of the scheme, parties had to create National Allocation Plans (NAPs); these determined the total quantity GHG emission allowances companies in member states would be granted. Countries had to take the total number of allowances for the specific trading periods and allocate the allowance among ETS installations. However, the third trading period that took place in 2013 was set by the European Union regulatory body instead of the individual nations, which suspended the use of individual allocation plans [29]. The first trading period (2005–2007) was marked by March 31, 2004, and May 1, 2004, as acceptable dates NAPs were to be conferred, they included: Austria, Belgium, Cyprus, Denmark, Estonia, Finland, France, Hungary, Germany, Ireland, Latvia, Lithuania, Luxembourg, Malta, the Netherlands, Portugal, Spain, the Slovak Republic, Slovenia, Sweden, and the UK. The NAPs had to comply with the Kyoto Protocol. NAPs for the Czech Republic, Greece, Italy, and Poland were not accepted until 2005. The NAPS that were accepted in this trading period created the core infrastructure of the carbon market. The second trading period (2008–2012) compared the NAPs of

Table 2 European Union member states green energy and climate policy, 2011

Country	Tax	Duty, fee, and charge	Subsidy[a]
Austria	– Energy tax for the consumption of natural gas, coal, and electricity (1996) – Mineral oil tax for the consumption of petrol, diesel, and heavy and light fuel oils – Motor vehicle tax (2007) – Tax on motor vehicle insurance for – Vehicles less than 3.5 tons (2000) – Vehicle registration tax (1992, 2008) – Air transportation tax includes all routes, no matter distance (2011) – Renewable energy and energy efficiency partnership (2002)	– Road transportation duty for lorries and trailers (1981) – Charge only in Vienna for parking cars in "limited parking zones" – Toll for alpine roads usage – Vignette for highway usage by motorcycles (1997)	– Combined heat and power subsidy program to obligate grid companies to purchase electricity from those plants, provided they served public districts (2002) – New housing subsidization scheme to reduce GHG in the building sector (2009) – Federal environmental fund (2001)

(continued)

Table 2 (continued)

Country	Tax	Duty, fee, and charge	Subsidy[a]
Belgium			– Subsidies for renewable energy investment—Wallonia (2005) – Subsidies for passive house construction which abylids to low-energy renovations to meet current policies (2007) – Financial support for demonstration projects that would allow subsidies to be provided for demonstration projects in the sectors of rational use of energy and renewables (1992) – Tax deduction for enterprises who invested in energy efficiency and renewable energy (1992) – Walloon invested in car sharing incentive program (1998) – Tax deductions for homeowners who renovate their homes to be energy efficient and use renewable energy (2003) – Free public train service for civil servants (2004) – Awarded grants to installations of micro-cogeneration systems and high-efficiency wood-burning furnaces and heating boilers to promote small-scale heat generations (2005) – Brussels—personnel have the option of giving up their old vehicles, and in exchange, the regional government would provide subsidies in the form of a one– or two-year subscription to a car share programming (2006)

(continued)

Table 2 (continued)

Country	Tax	Duty, fee, and charge	Subsidy[a]
Bulgaria	– Fuel excise tax (1991, 2006) – Motor vehicle import tax (2003) – Passenger car excise tax and motor vehicle tax (engine power) (1994, 2006) – Road tax based on weight of vehicle (2002) – Vehicle, vessel, and aircraft tax and tax on agricultural, construction, and passenger cars (1998, 2006)	– Air pollution non-compliance fees (1993) – Fuel road charge on petrol, diesel, liquified oil, and gas (2000, 2002) – Liquid fuel product charge (2000)	– Feed-in tariffs for hydro, wind power, and biomass (2007)
Croatia	None		
Cyprus	– Fuel excise tax on consumption of natural gas, coal, and electricity	– Excise duty on motor vehicle usage	– Energy savings and promotion of alternative energy sources, grant scheme (2003) -allowances for environmentally friendly vehicles (2000)

(continued)

Table 2 (continued)

Country	Tax	Duty, fee, and charge	Subsidy[a]
Czech Republic	– Air pollution fee of major stationary sources (1967, 2008) and air pollution fee of small stationary sources (1991, 2008) – Electricity tax (2008) – Natural gas tax (2008) – Road tax for the use of motor vehicles and passenger cars (1993, 2008) – Solid fuels tax (2008)	– Fuel excise duty on motor petrol, LPG, heavy fuel oils, and other mineral oils (1993, 2007) – Electronic road toll fee– fee for vehicles above the allowed 3.5t (2007, 2010) – Highway fee for vehicles greater than 3.5t	– International financial cooperation subsidy – Share programs, energy performance contracting, and funds of small projects – State environmental fund of the Czech Republic, various subsidy programs, such as, preparation of regional strategies (2005) – Promotion of environmental education and enlightenment subsidy (2005) – Promotion of renewable energy sources subsidy (2005) – Protection of air subsidy (2005); protection of water subsidy 2005 – Green investment scheme, to provide homeowners and apartment owners grants to make their household energy efficient (2009) – Building retrofit subsidies, panel program that provides incentives and subsidies as support for the repair, reconstruction, and modernization of apartment building (2004, 2009) – Act No. 180/2005 promotion of the use of renewable energy sources through incentives and subsidies (2005, 2010)

(continued)

Table 2 (continued)

Country	Tax	Duty, fee, and charge	Subsidy[a]
Denmark	– 2009 reform to cutting taxes on labor, and raising taxes on energy, climate, and transportation by EUR 1.1 billion – CO_2 tax on fuels (1982) – Mineral oil production, consumption, and sales tax (1977) – Petrol tax (1950) – Waste disposal and battery tax (2005) – Tax and fee reductions for fuel efficient or electric cars – Motor vehicle weight tax (2007) – Passenger car consumption fuel tax of passenger vehicles that use petrol (1997) – Sustainable transport tax (2009) to reduce CO_2 emissions from road transport, through initiatives on green car taxes, investment in public transport, intelligent traffic systems, and new roads	– Energy consumption duty (1980s) – Duties on coal, electricity, and natural gas (1995) – passenger car petrol duty (1977) – duty on natural gas (1995) – Annual fee on CO_2 emissions and smog from vehicles – Heavy good vehicle fees (late 1990s) – Fines for personnel who lack vehicle insurance – Road usage charge – First time registration for motor vehicles (2007) – Environmental duty for passenger vans and cars up to 9 persons max (2000) – Duty on CO_2 emissions from petrol, gasoil, diesel oil, fuel oil, and electricity (1998, 2010) – Duty on coal (1982) – Duty on electricity consumption for heating and insurances on pleasure boats – Duty on motor vehicle compulsory insurance of private cars, motorcycles, and mopeds (2002) – Duty on the use of natural gas for motor vehicles fuel and town gas usage (1995, 2010) – Duty on leaded and unleaded petrol (1950, 2008)	– Wind energy feed-in tariffs – Energy development and demonstration program (1976) – Subsidized energy and electricity efficiency related research and development (1999) – Climate change strategy (2003) – Emissions trading in the electric sector (2001) – CO_2 emissions allowing system (2005) – Grants for ecological buildings (2001) – CO_2 emissions trading in the electric sector (2001, 2005) – CO_2 emissions allowance system (2005) – Energy labeling of new cars to inform consumers of energy consumption and CO_2 emissions (2000) – Finance Act of 2009 imposed energy-saving targets for state-owned establishments – 1996 Agreement of industrial energy efficacy – 1998/2005 Agreement on green public transport purchasing by the government

(continued)

Table 2 (continued)

Country	Tax	Duty, fee, and charge	Subsidy[a]
Estonia	– Motor vehicle tax (1995) – Fuel excise tax on consumption of natural gas, coal, and electricity (1993, 2010) – Heavy good vehicle tax for the use of lorries and trailers (2003)	– Air pollution non-compliance fee—charge for excess pollution emissions (1993, 2009) – Air pollution charge for CO_2, SO_2, and heavy metals emissions (1991)	– Environmental investments centers grant scheme (1990)
Finland	– Motor vehicle diesel tax for the use of lorries and trailers, and passenger vans and cars (2004) – Vehicle sticker tax (2004) – Car tax for first time registration of a motor vehicle (1958, 2009) – Amendment of car tax and annual vehicle tax regimes applies to all registered vehicles whether commercial or private vehicles (2008)	– Parking fine for unnecessary idling (1992) – Strategic stockpile fee (2008) – Charge on exceeding GHG emissions limits from 2008–2012 (2004)	– SME soft loans for pollution control investment (1992) – Energy audit program scheme (1992) – Energy grants for residential buildings (1993) – Government decision on energy efficiency measures (2010) – Building code (1976, 2008) – Energy labeling passenger cars (2010)

(continued)

Table 2 (continued)

Country	Tax	Duty, fee, and charge	Subsidy[a]
France	– Promotion of electric and natural gas-powered vehicles tax—to promote the use of natural gas in transport vehicles as a clean fuel alternative (1999) – Annual tax on motor vehicles with large CO_2 emission outputs (245 g/km) (2010) – Tax on company cars – General tax on polluting activities (1999) – Mineral oil tax – Tax on natural gas (1986) – Tax on vehicle access for trucks and tractors that use public traffic routes (1968) – Bonus tax for vehicles with high emission outputs (2008)	– Charge on production of petrol refineries	– Tax rebate on environmentally friendly housing equipment (2001, 2005) – Renewable energy feed-in tariffs: biogas and mechanization, onshore wind power, offshore wind power, photovoltaic, and geothermal (2006, 2010) – Renewable energy feed-in tariff: biomass (2009) – Financing for energy efficiency investments (2002) – Government crediting and loan guarantee for energy efficiency and renewable energy investment (2001) for SMEs to invest in energy efficiency and renewable through renewable energy feed-in tariff: hydropower (2001) – Finance law 2009 for sustainable energy provisions

(continued)

Table 2 (continued)

Country	Tax	Duty, fee, and charge	Subsidy[a]
Germany	– New vehicle car tax system (2009) – Motor vehicle tax based on weight and emission output	– Duty on electric usage buses, railways, and manufacturing (1999) – Duty on mineral oil usage such as petrol and diesel – Extra fee on electric bill for energy usage – Fee for lorries or trucks that use motorways (2005, 2009)	– Preferential Loan Program offered by the Reconstruction Loan Corporation (1999) – CO_2 Building Restructuring Program (2001) – Housing Modernization Program (2005) – Clean Truck Procurement Subsidies (2007) – Special fund for energy efficiency in SMEs (2008) – Policy for wind power to produce electricity (2008) – Mandatory fuel efficacy labeling on passenger cars (2004) – Operators of renewable energy plants not operating under the German feed-in tariffs program sold electricity at a premium on the market—i.e., green power (1996)

(continued)

Table 2 (continued)

Country	Tax	Duty, fee, and charge	Subsidy[a]
Greece	– Tax on motor vehicle purchase – Mineral oil tax – Tax on motor vehicle usage		– Development incentives for renewable energy sources within multisectors policy (2004) – Law 2941/2001: multiple renewable energy framework (2001) – Law 2244/94: multiple renewable energy resources within electricity sector (1994) – Incentives for investment in combined heat and power, multiple renewable energy sources for heating and cooling in domestic sector, and industrial process sector (1990) – National Allocation Plan 2008–12, GHG emissions trading plan that allocates 69.1 Mt CO_2 equivalent of emissions per year to energy and industrial facilities (2008) – Law 1559/85: regulation of alternative forms of energy and specific issues of power production regulatory instruments (1985)

(continued)

Table 2 (continued)

Country	Tax	Duty, fee, and charge	Subsidy[a]
Hungary	– Air pollution tax on exceeding air quality regulations (2009) – Production tax of lubricating oils (2006, 2008) – Tax on motor vehicle and trailer use – Vehicle tax (1991) – Tax on foreign registered vehicles (1991)		– Climate Friendly Home Program (2009) – Lightbulb Change Program (2009) – Preferential loans for energy efficiency investments of SMEs (2000) – 2006 National Energy Savings plan—replaced energy portion of the Széchenyi Plan – Feed-in Tariffs Electricity Act (2006) – ETS for CO_2 (2005, 2008) – Kyoto ETS (2008) – Efficiency labeling of household appliances (2002)
Ireland	– Mineral oil tax (1999, 2009) – Motor vehicle tax based on engine size and CO_2 emissions (2009) -Vehicle registration tax based on CO_2 emissions (1993, 2008)	– Duty on other sorts of oil (1999, 2010)	– New House Grants (2001) – Low-Carbon Home Program (2008) – Energy efficient tax incentives for businesses (2008) – Renewable Energy and Energy Efficiency Partnership (2002) – Ireland National Allocation Plan 2008–12, GHG ETS with allowances available for purchase (2008)

(continued)

Table 2 (continued)

Country	Tax	Duty, fee, and charge	Subsidy[a]
Italy	– Additional regional tax on natural gas consumption (1977, 2010) – Additional tax on electricity for towns and provinces (1988, 2007) – Regional oil and petrol tax (1990, 2008) – Tax on electrical energy at state level for industrial and private consumption (2000)	– Excise duty on oil and petrol consumption (1998, 2010) – Excise duty on consumption of energy from any source (1993, 2009) – Registration fee for vehicle use annually (1999, 2010)	– Motor vehicle scrapping subsides (2007, 2009) – Personal income tax allowance for public transport (2008, 2009) – Tax credit for biomass heating systems – Feed-in tariff for solar thermodynamic energy (2008) – Funding for energy efficiency, renewable energy, and bike-sharing (2010) – Smart-grid development incentives (2010) – Car sharing subsidies (2001) – Tax allowances for high-efficacy fridges and freezers (2007) – Tax allowances for electric motors (2007) – EU ETS (2007) – Finance Act 2008: renewable energy provisions for the Green Certificates System in relation to electricity (2008, 2009)

(continued)

Table 2 (continued)

Country	Tax	Duty, fee, and charge	Subsidy[a]
Latvia	– Fuel excise tax on consumption of natural gas, coal, and electricity	– Old vehicle duty (1999–2001) – Air pollution non-compliance fees (2001) – Fee for annual motor vehicle usage (1994, 2001) – Air emissions charge (1995)	
Lithuania	– Turnover taxation of vehicles that exceed 100,000 LTL – Fuel excise tax for the consumption of natural gas, coal, and electricity (1994, 2002) -Heavy duty vehicle tax for the use of lorries and trailers (1995–2002)	– Motor vehicle duty for the importation of an old vehicle (1993, 1997) – Air emissions non-compliance fee (1991, 1999) – Air pollution charge for the usage of motor oils (1991, 1999) – Air pollution charge for stationary sources (1991, 1999)	– Energy efficiency and housing project (1996) – Tax exemptions for biofuels (VAT and excise duty) (2000)

(continued)

Table 2 (continued)

Country	Tax	Duty, fee, and charge	Subsidy[a]
Luxemburg	– Tax on electric distribution – Tax on electric production – Annual vehicle tax – Mineral oil tax – Taxing on heating fuels – Energy efficient partner tax	– Road usage fees	– Feed-in tariffs for renewable energy and cogeneration (1994), superseded by Law 14 in 2005 – Grants for energy efficiency and renewable energy investments which provided companies investing in infrastructure, buildings, land, and installations which were "eco-friendly" (2005) – Luxembourg National Allocation Plan 2008–12—GHG ETS (2005) – Energy performance of residential buildings education and outreach: regulatory instrument stating that new buildings and existing buildings that are undergoing significant renovation must meet new energy performance requirements (2008) – Energy efficient labeling (1999) – Climate financial aid program for energy savings and renewable energy in housing (2008) – CO_2 emissions reduction plan (2008) policy framework
Malta		– Annual vessel registration of small ships – Annual motor vehicle licenses – Electricity charges based on usage-fee for oil distribution licensing	

(continued)

Table 2 (continued)

Country	Tax	Duty, fee, and charge	Subsidy[a]
Netherlands	– Coal tax (1992) – Energy tax (1996) – Mineral tax (1990, 1996) – Petrol tax (1990, 1996)	– Nutrient surplus charge (2005)	– Energy labeling scheme (2015)
Poland	– Excise taxation on energy products (1990, 2007)	– Charge on air pollution (1990, 2010)	– Rationalization of heat consumption in household sector (1998, 2001) – EU ETS for CO_2 (2005, 2008) – GHG ETS that is for greening of financial resources, ensuring that funds are used for climate protection (2009)
Portugal	– First time motor vehicle registration tax (2007) – Circulation tax for the use of vehicles for transportation of goods (2007) – Excise tax of passenger vehicles and motorcycles (2007) – Municipal tax on vehicles – Taxation on petroleum and energy products – Truck tax – Taxation of less-effective lightbulbs (2008)		– Energy efficient fund to promote citizens and businesses to promote energy efficient projects (2010) – Implementation of the CHP Directive (2010) – Energy efficient requirements for appliances (2002)

(continued)

Table 2 (continued)

Country	Tax	Duty, fee, and charge	Subsidy[a]
Romania	– Annual vehicle tax for usage (1992, 2002) – Fuel excise tax on consumption of natural gas, electricity, and coal based on vehicle pollutions (1998, 2006) – Vehicle import duty tax (1993, 2001)	– Air emissions non-compliance fee (2000/2005) – Fee for registration of vehicles exceeding standard dimensions (1996) – Air emissions charge for CO_2 (2000, 2003) – Fuel road charge for use of fuel (1996)	
Slovak Republic	– Taxation on air pollution from medium and large businesses (1992, 2003) – Excise duty on coal (2008) – Excise duty on electricity (2008) – Excise duty on natural gas (2008) – Excise on mineral oils (1993, 2004) – Road tax usage for al motor vehicles (1993, 2004) – Tax permits on entering historical towns with a motor vehicle (1993, 2004)	– Air pollution charge from small businesses (1967, 2003)	– Act on energy and amendments (2005) – Decree on the regulation of network industries (2007)
Slovenia	– CO_2 tax (1997, 2008) – Energy efficient tax on consumption of natural gas, coal, and electricity (2010) – Fuel excise tax (1999, 2009) – Vehicle excise tax based on value of vehicles (1999, 2006)		– Emission allowances permit trade system (2004, 2009)

(continued)

Table 2 (continued)

Country	Tax	Duty, fee, and charge	Subsidy[a]
Spain	– Taxation on air pollution in Andalusia (2004) – Taxation on environmental damages caused by emissions to the air in Aragon (2006) – Taxation on petroleum fuels in the Canary Islands (1987, 2000) – Taxation on activities that cause environmental harm in Castille la Man (2001) – Taxation on electricity (1998) – Taxation on mineral oils – Taxation on motor vehicle usage – Taxation on retail sales of specific mineral oils – First registration taxation based on CO_2 emissions (2008)		– Grants for energy efficiency in buildings (2008) – Feed-in tariffs for electricity from renewable energy sources (2007, 2009) – Feed-in tariffs for small-scale cogeneration and renewable electricity production (1999–2001) – Creation of a credit line for investments in renewable energy and improving efficiency projects (2002) – Building energy certification (2008)

(continued)

Table 2 (continued)

Country	Tax	Duty, fee, and charge	Subsidy[a]
Sweden	– Energy and CO_2 tax on fuels minus petroleum (1991, 2010) – Energy and CO_2 tax on petrol (2010) – Energy taxation on electricity consumptions (2010) – Motor vehicle tax – Taxation on nuclear power (1983, 2008)	– Road usage charge for lorries (1998, 2001)	– Tax reduction for fossil fuels used for heat production in CHP plants (2004) – Fuel consumption and CO_2 labels for new cars (2002) – Building energy performance certificates (2008) – GHG ETS (2005) – Energy declaration of buildings act—incentive for investments in lower-energy buildings (2006) – Government subsidies for local energy efficiency measures (2010)
UK	– Energy consumption tax – Vehicle insurance tax company – Car tax reform (2004) – Reduced VAT for energy saving materials (2000) – Non-fossil fuel obligation levy (2001) – Climate change levy (2001)	– Duty on hydrocarbon oils (2001) – Vehicle excise duty on fuel type and CO_2 emissions vehicle band (2009) – Congestion pricing fee	– Feed-in tariffs for renewable electricity (2010) – Energy efficient loans for SMEs (2007) – Bioenergy infrastructure scheme (2003) – Pay as you save piolets (2006) – Scottish Government Central Heating Program (2001) – Exception from climate change levy for energy incentive (2004) – Enhanced capital allowance (2001)

[a]SME = small and medium enterprise

the first trading period and were extremely simplistic. On June 30, 2006, the NAPs were accepted. Compared to the first period cap, emissions were 13% lower and 6% lower than 2005. The third trading period (2013–2020) was the first trading period where European regulators set the rules for allocating and creating a single EU-cap. Changes for the third phase are as follows: allowances are not allowed to be allocated for free to electric generators and "the level of auctioning of allowances for non-exposed industry will be raised in a linear manner: 70% by 2020 and reaching 100% by 2027" [29]. In order to achieve GHG reductions, target gaps for this trading period, the carbon cap will need to be tightened by 1.7% per annum with 60% of all allowances being auctioned compared to 3% during the two previous trading periods.

Other important European Union efforts included energy and mineral oil taxes in which a long-standing tradition has been used to raise tax revenues. Since the 1950s, several European Union member states have imposed energy taxes on the consumption of coal, natural gas, other oils, and electricity. During the 1990s, many members started imposing taxes on fuel and taxes on the production of CO_2 emissions from the sectors of energy and mineral oil. Fuel is classified in three different ways: type (i.e., petrol, diesel, LPG, etc.), energy, and usage—whether used for heating, propellent, commercial, or industrial purposes [29]. Transportation taxes within the European Union focalize on vehicle-related taxes which mostly are based on weight, purpose, age of the vehicle, type and fuel usage, insurance of vehicle, road usage, and emissions. Emission and air pollution from multiple member states have imposed fees or charges for regulating both. Finland was among the first countries to implement a tax on CO_2 emissions in 1990—followed by Sweden, Denmark, Germany, the Netherlands, Slovenia, Poland, and the UK. Aside from CO_2 emission fees and taxes, air pollutions taxes and fees have been implemented throughout the European Union. By 1997, the Czech Republic and the Slovak Republic had changed their air pollution fees to include large, medium, and small sourcing agents. Since 1997, Latvia and Italy have implemented taxing policies on air pollution. Since 2001, Bulgaria, Romania, and Latvia implemented non-compliance fees for sources that do not comply with air pollution regulations. Since 1998, the Netherlands have provided subsidies for CO_2 reduction that apply to sources when they cut the transport CO_2 emissions, receiving carbon credits. "In Hungary, an air load tax and air pollution tax were levied in 2001 and 2004" [29]. Spain has implemented various taxes on air pollution, including taxes on air emissions in 1996 and providing tax deduction for environmental instruments and separate emissions and pollutants taxes on activities that cause environmental harm years later. These taxes were modified in 2006 to include an additional tax on air pollution and on environmental damages caused by CO_2 and SO_2 emissions [29].

In terms of tax deductions and provisions, i.e., to increase energy efficiency, special tax provisions are widely used throughout the European Union. These provisions include: subsides, tax credits, grants, tax deductions, and tax exemptions [39]. The Netherlands created the Green Fund Scheme in 1995 which provided environmental tax credits to investors and loans to environmental projects through green banks. Tax credits play a vital role in increasing the production and sale of renewables since they reduce production costs relative to the cost of producing substitutes. Like the Netherlands, Italy also introduced a biomass heating tax credit in 2001 in which subsides for the support of eco-friendly activities were followed up in 2004, while in 2007, the implementation of subsidies for energy efficiency was launched. An example of an exemption to taxes was implemented in Lithuania for biofuels, and from 2003 to 2008, the UK introduced a bioenergy infrastructure scheme to provide grants that would stimulate the use of "small-scale biomass supplier fuel for heating and electricity generation" [29]. Moreover, in 2009, France created a law that produced a consumption tax reduction on a variety of biofuels.

In Germany and France, an interest in promoting renewable energy and energy efficiency in the renovation and reconstruction of public and private buildings provided interest free loans. In France, Grenelle de l'Environnement was an environmental policy put into effect that aimed to cut energy consumption of existing buildings by 38% by 2020 and to reduce existing public buildings' energy consumption by 40% and GHG by 50%. The government established agreements with banks and construction sectors to provide zero interest loans to allow building owners to improve their buildings to meet the standards of their environmental policy and its goals. In Germany, a green tax for renewable feed-in tariffs promoted a strategy to develop "renewable energy technologies such as wind or geothermal-generated electricity" [29]. It offered manufacturers long-term renewable energy purchase contracts at a fair price, typically equal to the cost of production. This allowed investment returns to be stable and calculated. Enterprises or households could participate in the feed-in program which promoted the security of the "domestic energy supply, promoted technological innovation, fair market conditions for renewable technologies, and creation of jobs" [29]. However, it should be noted that innovated technology and the economic scale are supposed to drop production cost for renewable energy sources to become competitive with their fossil fuel-based counterparts. Germany, France, Denmark, the UK, Italy, Spain, and Bulgaria have all introduced feed-in tariffs in their economic structures. Since 2010, the UK has used small-scale and low-carbon electricity produced from a variety of renewable energy technologies such as hydropower, bioenergy, and wind. Since 1998, Germany has had electrical production created from wind and hydro-stations. France has implemented numerous sectors of renewable energy feed-in tariffs since 2001. Current types of feed-in tariffs include "biogas and mechanization, wind power, photovoltaic, hydro, solar, and geothermal" [29]. Up to 2011, there were a number of regulations and policies related to decreasing emissions in the European Union. Member states varied in their approach, but a general consensus indicated a unified position to decreasing emission levels in conjunction with a greening of its energy supplies.

4 American Environmental Regulations and Policies on Emission Reduction

A background on the US regulatory and policy-making protocol indicates environmental policy is geared toward stabilizing environmentally friendly protection and conservation of natural resources while promoting affordable energy, economic development, and job creation [49]. However, in order to achieve these goals of environmental regulations and policies, the US has a lengthy and mundane law creation process that must go through three approving agents and one regulatory body for it to be implemented. Once this process is completed and approved by all parties, it will then be implemented at federal, state, and local levels. This process is unique compared to other countries throughout the world. The process begins once Congress proposes a bill, i.e., a document that outlines a potential law. Both parties within Congress must pass the bill for the next step to begin. Once the bill is approved by both parties in Congress, it is then sent to the President of the US where he has the option to approve or veto (i.e., reject) the bill. If the President approves the bill, it becomes a law known as an "act" or "statute." The approved act is then sent to the House of Representatives where it becomes standardized text and published in the US Code making it an official law [49]. Once the law is official, it will then be given to the proper regulatory agency authorized by Congress to create the regulations for the law. In the case of environmental regulations in the US, the Environmental Protection Agency (EPA) is the approved authority by Congress. Once EPA receives the regulation, it must go through the agencies system to become an official Code of Federal Regulations (CFR), the official record of regulations. Once the regulation becomes a CFR, it will be reviewed annually and revised as needed. EPA then publishes the regulation and the various states must create policies to comply [49]. Table 3 illustrates a breakdown of notable US environmental regulations and policies up to 2011—with key referenced research from EPA [49] and Vicario [50]. The state and local levels are limited to key examples. Limitations in the findings were restricted to states with major environmental policies toward GHGs.

Within the US environmental regulations, its policy-based frameworks, various taxes, subsidy programs, acts and laws, duties, fees, charges, and ETS all exist at the federal and state levels, and sometimes filter to the local levels. A detailed look at how the California ETS and the Regional Greenhouse Gas Initiative (RGGI) compare in terms of a cap-and-trade system—similar to the EU ETS—will exemplify this. The California ETS, otherwise known as Law A.B. 32, is California's solution to global climate change. The goal of this law is to have California's GHG emissions at 1990 levels by 2020 [42]. The ETS system is regulated by the California Air Resource Board and covers up to 80% of GHG emissions sectors within the state. These sectors include electricity generation (i.e., in-state and imports), transportation, industrial, commercial, residential, agriculture, and forestry as well as other non-specified sectors. The system works by allocating allowances and then auctioning them off based on calendar year vintages. Free allotments can be given to certain industrial facilities and are based on transition assistance and leakage preventive. What is unique about this cap-and-trade system is allowances can be banked but only up to a capped quantity. RGGI, another ETS within the US, is a cap-and-trade scheme based in the New England states, New York, and Maryland. RGGI includes the specific focus of reducing CO_2 emissions in the power plant sector [51]. Revenue from this system is supposed to help each state improve energy efficiency by modernizing electric grids and purchasing renewable energy systems from wind and solar power [52]. Since its start in 2009, the program holds quarterly boards where members can buy allowances at auctions for a seal-bid at a uniform price. Aside from auctions, allowances can be traded in secondary markets [53]. Until 2011, the foundation of these environmental regulations and policies have had a long-standing history of leadership. However, it has also had various stages in which its bureaucratic elements have halted advancement in favor of economic development and skeptics.

5 Conclusion and Comparative Findings

To express how effective policy frameworks can decrease emission levels while providing economic sustainability—the European Union and the US green energy regulations and policies indicate strong similarities with some distinct differences. The chapter described both environmental policies with a focused attention on CO_2 emission reduction as well as background information on how frameworks have

Table 3 Notable US environmental regulations and policies, adapted from EPA [49] and Vicario [50]

	Federal	State	Local
Act	– Clean Air Act 1963—designed to regulate air pollution at a national level, modified with the 1990 Amendment – Energy Policy Act 2005—provided tax incentives as well as loan guarantees for various types of energy productions – Synthetic Fuels Corporation established in 1980 and was funded by the government to develop and construction of synthetic fuel manufacturing plants – Energy Independence and Security Act—encouraged biofuel development, phased out non-energy efficient lightbulbs, and increased fuel economy requirements – Energy Improvement and Extension Act of 2008—gave a tax credit to hybrid vehicles owners	– Examples of State Programs for Clean Air Act are: New York: Clean Air NY, Georgia: The Clean Air Campaign, and Texas: Drive Clean Across Texas Campaign	– County sales tax on purchases of goods and services – Additional road usage toll – Permits for bringing motor vehicles with weight greater than 5 tons – Additional registration fees for motor vehicles

(continued)

Table 3 (continued)

	Federal	State	Local
Tax, fee, charge, and duty	– Commercial building tax deductions – Federal tax on local parking fees – Vehicle sales tax – Diesel fuel tax – Heavy vehicle use tax – Tariff on imported oils – Vehicle registration fee – Truck tire tax – Truck/trailer sales tax – Motor fuel tax -automobile tire tax – Vehicle miles traveled fee – Carbon tax/cap-and-trade – Freight waybill tax – Freight ton miles tax – Driver's license surcharge fee – Freight ton-based tax – Vehicle inspection and traffic citation surcharge – Petroleum franchise tax – Mineral severance tax – Federal tax on local transit fares	– Motor fuel tax (California 2002, Iowa 1950, etc.) – One-time motor vehicle registration fee (Iowa 2008, etc.) – Motor vehicle rental excise tax (Kansas) – Motor vehicle purchase tax (Louisiana, etc.) – Natural gas franchise tax (Louisiana 1950) – Annual air emissions licenses fee (Maine) – Annual waste discharge license fee (Maine 1997) – Energy generation tax (Maryland) – Tire recycling fee (Maryland 1950) – Annual motor vehicle excise tax (Massachusetts 1950) – Aircraft fuel tax (Massachusetts) – Airport parking tax (Michigan 1987) – Gas tax (Mississippi) – Battery fees (Missouri 2005) – Petroleum inspection fee (Missouri 1950) – Coal severance (Montana 1950) – Mineral tax (Nevada 1950) – Electricity consumption tax (New Hampshire 2001) – Conservation tax on coal (New Mexico) – Highway use tax (New York 1950) – Coal conservation facilities privilege tax (North Dakota 1950) – Annual vehicle registration fee (California) – Motor vehicle tax (Oklahoma)	
ETS		– California ETS – Regional greenhouse gas Initiative	

(continued)

Table 3 (continued)

	Federal	State	Local
Subsides and incentives	– Federal tax credit for improvement of existing homes for energy efficiency – Federal and state tax credit for consumer energy efficiency	– California emerging renewables program—rebate for small-scale renewable electricity generating technology – Clean alternative fuels credit (Connecticut 2008) – Kansas biomass to energy credit – Tax credit for pollution reducing boilers (Maine) – Alabama tax deduction program for wood-burning heating system – Arizona tax credit for solar and wind power	

implemented successful policy-rated programs. To determine if frameworks play a role in CO_2 reduction, data pertaining to CO_2 emissions within all industries are assessed. Data retrieved from the OECD statistical database showed that emission levels, for the timeframe 2005–2011, have fallen short of the Kyoto Protocol targets. Both parties agreed to the Kyoto Protocol, although the US never ratified the regulations. Illustrative findings for CO_2 emission data from the European Union (Fig. 1) and the US (Fig. 2) both show unreached Protocol target levels.

Trends between both the European Union and the US show decreasing CO_2 emissions from 2005 to 2009 with an increase from 2009 to 2010 and then a decrease again from 2010 to 2011. A possible reason for this could be the global 2008 economic crisis and recession. The European Union never had CO_2 emissions over 8000 tons within this timeframe while the US never went below 16,500 tons. The biggest decrease for both entities was from 2008 to 2009. The US decreased its CO_2 emissions from approximately 18.5 tons to 17.0 tons while the European Union decreased from

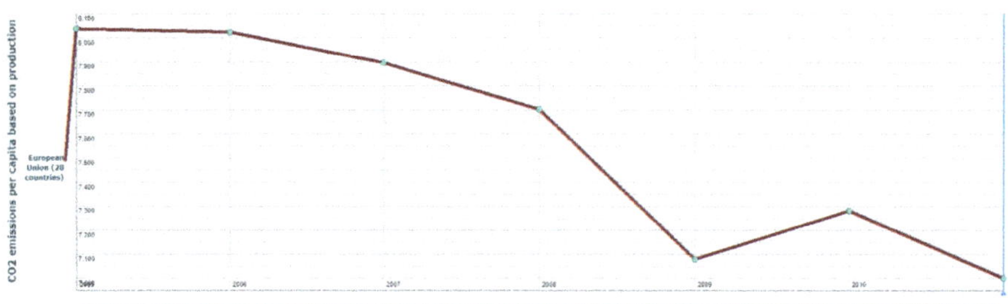

Fig. 1 European Union CO_2 emissions data per capita based on production from 2005-2011, sourced from OECD [32]

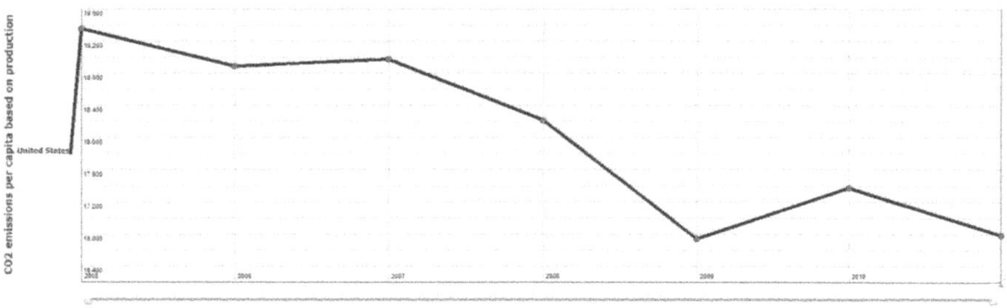

Fig. 2 US CO_2 emissions data per capita based on production from 2005-2011, sourced from OECD [32]

approximately 7.7 tons to 7.1 tons. The US had a slight increase in CO_2 emissions from 2006 to 2007 by approximately 0.3 tons. It can be deduced from the data that the European Union is more successful at decreasing CO_2 emissions than the US. This is evident by the two graphs that express the highest amount of CO_2 tonnage in which the Europeans are half that of the Americans' lowest CO_2 tonnage. Nonetheless, a direct relationship between the two economies is significant as they both follow the same general trend of CO_2 emission decreases and increases within the same time period. Further investigation would be useful to pinpoint exactly why the European Union is more successful at decreasing CO_2 emissions than the US or why there is a direct relationship between the two economies. Based on background information within this chapter, an educated guess can be formulated in attempt to answer these questions.

Key diplomatic affairs between the European Union and the US date as far back as 1953 [54]. Bilateral trade between both parties is seen as prestigious by many countries since both play key roles in the global economy, international political relations, and trade as well as having the largest economies and military forces deployed around the world [54]. These dominating powers have major influences on global events and set the standard for other countries to achieve, especially in terms of development. The consensual accords between the European Union and the US in various economic and political factors entail how there is a direct relationship between the two. Although the US did not sign the Kyoto Protocol, they have implemented their own regulations and laws to decrease environmental pollutants during the same timeframe as the European Union. The decrease by both parties could express how both regulatory frameworks were successful at decreasing emission levels—with the European side showing overall better success. Data show both parties had a spike from 2009 to 2010. This spike, related to the economic crisis at the time, effected gas and oil prices by dramatically causing both to fall. The price of gas dropped from USD 150 a barrel in July 2008 to USD 40 in February 2009 [55]. The average price of oil in 2009 was USD 61.7 compared to USD 96.94 in 2008 [56]. The cheap oil prices might have allowed for more residential citizens within the European Union and the US to consume more gas and oil than in previous years leading to higher levels of GHGs. Moreover, climate can affect the consumption of energy usage, i.e., from fossil fuel consumption. The European Environmental Agency (EEA) stated that winter weather within Europe was extremely cold in 2010 compared to previous years. The energy usage in the European Union that had an increase in GHG emissions, over this timeframe, was public heat and electricity production as well as an "increase stemmed from manufacturing industries and construction ([i.e.,] including iron and steel process emissions)" [57]. This indicates that weather and seasonal patterns play a critical role in energy consumption and can therefore express a possible reason for the increase in GHGs in 2010. The oil drop from the 2008 recession expressed how low oil prices effected both Europeans and Americans and, potentially, due in part by its citizenry's rates of energy consumption. A positive correlation between seasonal weather patterns and energy consumption is evident by EEA data, pointing toward a conclusion that both events affected the increase in GHG emissions in period ending in 2010.

Green taxes have been implemented to combat environmental pollution in order to prevent or slow ecological changes. Pollution increases the chances of environmental fluctuations that can be large-scale and unforeseen. The rationale for climatological solutions should be carefully studied and considered. To help combat the increase in emissions and polluting agents, the European Union and the US created cap-and-trade systems to help in the reduction of GHGs. Both parties implemented various taxes and regulations on energy consumption and CO_2 emissions. After the global 2008 recession, examination of the two bilateral trade and political and economic agreements expressed a direct relationship between GHGs policy making and the amount of CO2 emissions being produced. Linkages between ecological change, pollution, and the green regulations and policies play a vital role in our current global paradigm. This chapter does an excellent job in recognizing the gap that exists between developed and underdeveloped countries when it came to expressing the major contributors in environmental pollution. It positively expresses how the developed countries in the European Union and the US are taking ownership of their negative ecological consequences and creating regulations to reduce GHGs. To date, the successful implementation of ETS in the European Union and the US can be used as standards or guidelines for other countries aiming to implement green energy regulations and policies. This historical-environmental research can serve as a starting point to better understanding where and why certain green energy policy objectives have been implemented, e.g., the United Nations Sustainable Development Goals, as well as emerging green deal policies that will significantly alter the way energy is sourced and used.

References

1. Guo Y, Xia X, Zhang S, Zhang D (2018) Environmental regulation, government R&D funding and green technology innovation: Evidence from China provincial data. Sustainability 10:1–21. https://doi.org/10.3390/su10040940

2. de Serres A, Llewellyn J, Llewellyn P (2011) The political economy of climate change mitigation policies: how to build a constituency to address global warming? OECD, Economics Department Working Papers, No. 887, Paris
3. Fan X, Li X, Yin J (2019) Impact of environmental tax on green development: a nonlinear dynamical system analysis. PLoS One 14:e0221264. https://doi.org/10.1371/journal.pone.0221264.t002
4. Sun C, Tong Y, Zou W (2018) The evolution and a temporal-spatial difference analysis of green development in China. Sustain Cities Soc 41:52–61. https://doi.org/10.1016/j.scs.2018.05.006
5. Feng C, Wang M, Liu GC, Huang JB (2017) Green development performance and its influencing factors: a global perspective. J Clean Prod 144:323–333. https://doi.org/10.1016/j.jclepro.2017.01.005
6. Barbier E (2010) How is the global green new deal going? Nature 464:832–833. https://doi.org/10.1038/464832a
7. Mao Y, Hong S, Fuqiang Y (2011) The true cost of coal
8. Raskin PD, Electris C, Rosen RA (2010) The century ahead: Searching for sustainability. Sustainability 2:2626–2651. https://doi.org/10.3390/su2082626
9. Woolf D, Amonette JE, Street-Perrott FA et al (2010) Sustainable biochar to mitigate global climate change. Nat Commun 1:56. https://doi.org/10.1038/ncomms1053
10. Diffenbaugh NS, Burke M (2019) Global warming has increased global economic inequality. Proc Natl Acad Sci USA 116:9808–9813. https://doi.org/10.1073/pnas.1816020116
11. Frieler K, Meinshausen M, Golly A et al (2013) Limiting global warming to 2C is unlikely to save most coral reefs. Nat Clim Chang 3:165–170. https://doi.org/10.1038/nclimate1674
12. Château J, Dellink R, Lanzi E (2014) An Overview of the OECD ENV-Linkages model: version 3. OECD, Environment Working Papers, No. 65, Paris
13. UNESCO (2009) The United Nations world water development report 3: water in a changing world. In: Earthscan (ed) World Water Assessment Programme, p 349
14. NASA (2017) Climate change: vital signs of the planet. Natl. Aeronaut. Sp. Adm. https://climate.nasa.gov/evidence/. Accessed 11 Mar 2021
15. FitzRoy FR, Papyrakis E (2016) An introduction to climate change economics and policy, 2nd edn. Taylor and Francis, London
16. Buzan JR, Oleson K, Huber M (2015) Implementation and comparison of a suite of heat stress metrics within the Community Land Model version 4.5. Geosci Model Dev 8:151–170. https://doi.org/10.5194/gmd-8-151-2015
17. Field CB, Barros V, Stocker TF, et al (2012) Managing the risks of extreme events and disasters to advance climate change adaptation: Special report of the intergovernmental panel on climate change
18. Knabb R, Brown D, Rhome J (2011) Tropical cyclone report. moz-extension://b25df011-8780-4be7-9be7-6714e4992816/enhanced-reader.html?openApp&pdf=https%3A%2F%2Fwww.nhc.noaa.gov%2Fdata%2Ftcr%2FAL122005_Katrina.pdf. Accessed 17 Feb 2021
19. Plyer A (2016) Facts for features: katrina impact. The data center. Data Center. https://www.datacenterresearch.org/data-resources/katrina/facts-for-impact/. Accessed 17 Feb 2021
20. Patt A, Glantz MH (2001) Currents of change: impacts of El Nino and La Nina on climate and society. Int J Afr Hist Stud 34:173. https://doi.org/10.2307/3097309
21. Sahu N, Singh RB, Kumar P et al (2013) La Niña impacts on austral summer extremely high-streamflow events of the Paranaíba River in Brazil. Adv Meteorol. https://doi.org/10.1155/2013/461693
22. Capotondi A (2015) Extreme La Niña events to increase. Nat Clim Chang 5:100–101. https://doi.org/10.1038/nclimate2509
23. Cai W, Wang G, Santoso A et al (2015) Increased frequency of extreme La Niña events under greenhouse warming. Nat Clim Chang 5:132–137. https://doi.org/10.1038/nclimate2492
24. Bell GD, Halpert MS, Kousky VE et al (1999) Climate assessment for 1998. Bull Am Meteorol Soc 80:1040–1040. https://doi.org/10.1175/1520-0477(1999)080%3c1040:caf%3e2.0.co;2

25. Bove MC, Elsner JB, Landsea CW et al (1998) Effect of El Niño on U.S. landfalling hurricanes revisited. Bull Am Meteorol Soc 79:2477–2482. https://doi.org/10.1175/1520-0477(1998)079%3c2477:EOENOO%3e2.0.CO;2
26. Ropelewski CF, Halpert MS (1987) Global and regional scale precipitation patterns associated with the El Niño/southern oscillation. Mon Weather Rev 115:1606–1626. https://doi.org/10.1175/1520-0493(1987)115%3c1606:garspp%3e2.0.co;2
27. NASA (2019) Facts—climate change: vital signs of the planet. Natl. Aeronaut. Sp. Adm. https://climate.nasa.gov/evidence/. Accessed 17 Feb 2021
28. Saboori B, Sapri M, Bin Baba M (2014) Economic growth, energy consumption and CO_2 emissions in OECD (Organization for Economic Co-operation and Development)'s transport sector: a fully modified bi-directional relationship approach. Energy 66:150–161. https://doi.org/10.1016/j.energy.2013.12.048
29. ILO (2011) Towards a greener economy: the social dimensions. International Labour Organization, Geneva
30. Bygrave S, Ellis J (2003) Strategies to reduce greenhouse gas emissions from road transport: analytical methods strategies to reduce greenhouse gas emissions from road transport. OECD, Paris
31. Eichner T, Pethig R (2007) Efficient CO2 emissions control with national emissions taxes and international emissions trading. CESifo Work Pap Ser
32. OECD (2019) Taxing energy use 2019. OECD, Paris
33. Timilsina GR (2018) Where is the carbon tax after thirty years of research? World Bank, Washington, DC
34. Timilsina RR, Kotani K, Kamijo Y (2017) Sustainability of common pool resources. PLoS One 12:e0170981. https://doi.org/10.1371/journal.pone.0170981
35. Horn J, Escobedo FJ, Hinkle R et al (2015) The role of composition, invasives, and maintenance emissions on urban forest carbon stocks. Environ Manage 55:431–432
36. Zimmermannová J, Hájek M, Rozenský L (2017) Carbon taxation in the European countries. In: Theoretical and practical aspects of public finance. University of Economics, Prague
37. Andersen MS (2015) An introductory note on carbon taxation in Europe a Vermont briefing. Aarhus University and European Environment Agency, Aarhus
38. Starbatty N (2010) Emissions trading schemes. IASB, London
39. Caciagli V (2018) Emission trading schemes and carbon markets in the NDCs: their contribution to the Paris agreement. In: Alves F, Filho WL, Azeiteiro U (eds) Theory and practice of climate adaptation. Springer, Berlin, pp 539–571
40. Lecocq F, Capoor K (2003) State and trends of the carbon market 2003. The World Bank
41. Brunnermeier MK, Cheridito P, Brunnermeier MK, Cheridito P (2019) Measuring and allocating systemic risk. Risks 7:46. https://doi.org/10.3390/risks7020046
42. C2ES (2018) California cap and trade. Cent. Clim. Energy Solut. https://www.c2es.org/content/california-cap-and-trade. Accessed 17 Feb 2021
43. Cameron P (2000) From principles to practice: the Kyoto protocol. J Energy Nat Resour Law 18:1–18. https://doi.org/10.1080/02646811.2000.11433184
44. IETA (2021) IETA website. Int. Emiss. Trading Assoc. https://www.ieta.org/. Accessed 13 Mar 2021
45. OECD (2021) OECD iLibrary. https://www.oecd-ilibrary.org/. Accessed 30 Jan 2021
46. Kawabe Y, Wang Q, Yamashita M (2012) The truth about the reduction of greenhouse gas emissions stated in the Kyoto protocol: environmental problems used as political and economic strategies by European countries. World Environ 4:206–212
47. IPCC (2007) Climate change 2007 : an assessment of the intergovernmental panel on climate change. Cambridge University Press
48. IPCC (2013) AR5 climate change 2013: the physical science basis—IPCC. Working Group I of the Intergovernmental Panel on Climate Change, Geneva
49. EPA (2019) The basics of the regulatory process. Environmental Protection Agency, Washington, DC

50. Vicario L (2015) The United States environmental policy: an in-depth review for the ENVI Committee. Brussels
51. Stevenson DT (2016) A review of the regional green gas initiative. Cato Inst. https://www.cato.org/working-paper/review-regional-green-gas-initiative. Accessed 17 Feb 2021
52. IER (2012) Is the regional greenhouse gas initiative having an impact? Inst. Energy Res. https://www.instituteforenergyresearch.org/uncategorized/regional-greenhouse-gas-initiative-impact/. Accessed 17 Feb 2021
53. Harrison F, Maniloff P (2015) Beneficial leakage: the effect of the regional greenhouse gas initiative on aggregate emissions. Cent Environ Resour Econ Policy WP2015-06
54. European Parliament (2021) EU-US relations. Eur. Parliam. Liaison Off. Washingt. https://www.europarl.europa.eu/unitedstates/en/eu-us-relations?fbclid=IwAR0NmCEcKdWp_CEuKpYXHL1Q0vCMeB7t8a9F9RPEOBucxnfwv0bKg2Giq9c. Accessed 19 Mar 2021
55. ECB (2012) The development of prices and costs during the 2008–09 recession. Mon Bulletim April:71–85
56. FRED Economic Data (2021) Crude oil prices: Brent-Europe. FRED Econ. Data. https://fred.stlouisfed.org/series/ACOILBRENTEU. Accessed 19 Mar 2021
57. EEA (2012) Annual European Union greenhouse gas inventory 1990–2010 and inventory report 2012. Eur. Environ. Agency. https://www.eea.europa.eu/publications/european-union-greenhouse-gas-inventory-2012. Accessed 19 Mar 2021

Shelterbelt Planning in Agriculture: Application from Bulgaria

Veselin M. Shahanov and Giuseppe T. Cirella

Abstract The first attempt to build wind protection belts on the territory of Bulgaria dates back to the mid-twentieth century. At present, plantings are mature and perform to a great extent their function. However, in order to continue their positive role on agricultural landscapes they should be improved, current planned requirements reviewed, and factors that have the greatest impact on their sustainability analyzed. Unstable climate processes are one of the key parameters that change growth conditions, as a result, setting new requirements for these living technical facilities is necessary. The purpose of this chapter is to explore the past experience and create guidelines for planning shelterbelts in accordance with climate change and other modern trends. An examination of the proposed areas examines the construction of wind protection belts, types of green system-based structures, and optimal species composition. Agricultural landscape science structures the arguments for this chapter and aids in piecing together the recommendations.

Keywords Agricultural landscape · Land use · Wind protection belts · Climate change · Dobrudja

1 Introduction

Historically, people have had a deeper understanding of nature both in a terrestrial and cosmic sense. Through emancipating our surroundings, human beings gradually have lost their ability to see objective interrelationships and adapt to external knowledge [1]. Human activities are not accompanied by a total understanding of their actions, but by the inherent idea between right and wrong. As fields of knowledge differentiate, there is a greater need for experimentation, i.e., through trial-and-error,

V. M. Shahanov (✉)
Landscape Architecture Department, University of Forestry, Sofia, Bulgaria
e-mail: veshaha@abv.bg

G. T. Cirella
Faculty of Economics, University of Gdansk, Sopot, Poland
e-mail: gt.cirella@ug.edu.pl

G. T. Cirella (ed.), *Human Settlements*, Advances in 21st Century Human Settlements,
https://doi.org/10.1007/978-981-16-4031-5_8

to best achieve a desired outcome. Such experimentation is very applicable to the environment and land use research. Economic and political interests that supersede the environment, planned activities that are inconsistent with basic natural conditions, and inappropriate environmental measures all result in consequences human beings need to adjust. In agriculture, one of these problems is deteriorating growth conditions. Although not a complete solution, the use of protection belts, i.e., made of plants to protect agricultural landscapes, has a number of advantages, including the improvement of the microclimate, partially returning the landscape to a more natural state, and supporting, and restoring wildlife.

The use of wind protection belts (i.e., shelterbelts or windbreaks) has been applied in various places for quite a few centuries. Their integrated use in agriculture has been documented since at least the fifteenth century, when the Scottish Parliament required construction of protection belts for agricultural crops (Brandle et al. [2], referred to Droze [3]). In the twentieth century, there is a strong impetus for the application of wind protection plantings within the agricultural landscape both to the West and East. Two large-scale projects are known. In the United States, Franklin Roosevelt issued the creation of the Windbreak Belt Project, also known as the Prairie States Forestry Project. It was developed between 1935 and 1942, after which, its focus moved more toward soil conservation [2, 4, 5]. In 1948, a similar idea was developed in the Soviet Union due to water and wind erosion [6, 7] problems; namely, the so-called Stalin's Nature Transformation Plan [8, 9] in which protective forest plantings were of great importance. The construction of wind protection belts in Bulgaria dates back to the mid-twentieth century. Under the influence of the Russian experience, with similar climatic and soil conditions, an inherent steppe nature zone, Dobrudja implemented several thousand hectares of shelterbelts (Fig. 1).

Fig. 1 Wind protection belts system (i.e., fragment), General Toshevo Municipality, Dobrudja, Bulgaria. *Source* Google Maps

The aim of this chapter is to create guidelines for wind protection belt planning within agricultural landscapes with case research from Bulgaria. The chapter will outline current trends in the field of agricultural landscaping as well as constructive application of protective barriers. A breakdown of the chapter is as following: (1) historical overview of creating forest protection belts in Bulgaria, (2) vulnerable landscape components related to climate change and resulting problems in agricultural territories throughout the country, (3) some new trends via the European Union's land use planning and policies in the field of agriculture (i.e., specific to sustainable development), (4) main features of protection belts and their practical benefit, and (5) specific solutions for their construction.

2 Overview of Agriculture in Bulgaria and History of Windbreak Construction

Alongside the historical record, the dense and impassable Bulgarian forest is well-documented as far back as the chroniclers of the Crusades—mentioning the obstacles of developing agriculture and cattle breeding over Bulgarian lands on both sides of the Balkan Mountains [10]. This captive information alludes to the lush, diverse landscape where along the forested areas, it surrounded the fertile arable lands. The development of the agricultural landscape in Bulgaria, over the past few centuries, has led to changes in the natural territory's complexity. The anthropogenic influences found in some agricultural areas has led to instability of the environment which has become entirely dependent on human activities [11]. The fragmentation of the landscape has led to the emergence of new problems that require adequate landscape planning measures. In 2018, according to the Ministry of Agriculture, Food, and Forestry [12], 47% was appointed for agricultural land use. From within this area, the arable land (e.g., crop rotation fields, temporary meadows with cereals and legumes, and fallow and greenhouses) accounted for 66.3% or 3.46 million ha. This translates into land that is cultivated and that needs improvement, i.e., in terms of growth conditions, is approximately a third of the whole country's landmass. The 2019 Annual Report on the State and Development of Agriculture shows a decline in crop yields in some cereals (i.e., in 2018 compared to the previous 2017 yields), one reason being, unfavorable weather conditions such as floods, frost, hail, and drought. Dating back to the 2017 Report, it is found again that the yields and quality of agricultural produce are heavily dependent on climate conditions. For example, as a result of the drought in the summer of 2016, the harvested areas of maize was reduced by 18.4% in which the dry and hot weather from September to October resulted in the speeding up of the development of late-season crops. Little rain was also compensated by watering. In 2016, unfortunately irrigated areas were only 818,000 ha of which only 541,779 ha or 15.6% of the arable land was eligible for irrigation and the percentage of real irrigated areas during that year was below 1%. As such, plowing the land and reducing permanent crop coverage lead to a disruption of ecological stability, resulting in processes that have led to unnatural changes and instability in the environment. To better control these processes, the implementation of effective windbreak belts is considered. At present, there are a number of examples of tree vegetation belts in agricultural landscaping—both positive [13–20] and negative [21–24].

Afforestation, applied via narrow strips, has significant benefits, i.e., increasing carbon stock, restoring soil fertility, protecting against wind and water erosion, and increasing agricultural yields [25, 26]. The benefits of windbreaks are multilateral, since research on the subject is mainly related to their positive impact on agricultural activity. Fewer studies focus on non-agricultural benefits, i.e., increasing biodiversity, increasing the esthetic qualities of the landscape, providing recreational opportunities, and harvesting timber, fruits and other products [27, 28]. Along with positive qualities, it should be noted that the presence of tree vegetation around arable land can lead to competition between the protecting plants and crops, including: shading, emission of substances, and absorption of soil moisture [2, 18]. As highly favorable agricultural landscapes, planar territories with fertile soils located in the steppe areas of southeastern Europe, i.e., where grass communities dominate, the planting of woody vegetation is not an easy task. According to Isachenko [9], the forest plantation in the steppe region is sustainable and lasting when it does not displace the steppe grassland. This is a long-term activity; it includes the correct selection of plant species and production of plant material (i.e., collection of seeds from appropriate locations, adaptation of young plants to new conditions in their cultivation, etc.). Wind protection belts and their benefits are related to the intense development of soil science.

In the first half of the twentieth century, a strong impulse for this development was the deterioration of the Chernozems (i.e., Mollisols in the United States soil taxonomy) in Eastern Europe [25, 29]. Although the experience of the United States and Canada has been explored, the creation of windbreaks in Bulgaria is mainly influenced by the practice in the former Soviet Union since it shares similar climatic conditions and geography [27, 30]—the Stalin Plan included a large-scale construction of shelterbelts (Fig. 2). The wind protection belts construction in Bulgaria are accompanied by several research books and a dissertation, developed and defended in 1957 by G. Georgiev on the creation of forest protection belts in Dobrudja. After a lengthy career, the author recommends the construction of windbreaks in other non-irrigated areas throughout Bulgaria [30].

Despite surveys showing many economic and environmental benefits of windbreaks, their construction is not common practice [5, 15, 31]. To a large extent, this also applies to the territory of Bulgaria—a system of wind protection belts that are built mainly in the northeastern part of the country (Fig. 3). According to the Executive Forestry Agency [32] in 2014, northern Bulgaria had more than 9000 ha of forest protection belts. It is also note stating, other sources (e.g., Lyubenova et al. [33] and Radev [34]) state the area of shelterbelts in Dobrudja in 1957 was 9000 ha and presently this has been reduced to 6500 ha which protects (i.e., blocks) nearly

Fig. 2 Stalin's Nature transformation plan. Adopted from Chendev et al. [25]

Fig. 3 Wind protection belt in Dobrudja, Bulgaria. Photograph taken by V. M. Shahanov on 20 July 2018

300,000 ha of arable land. Hence, at present, the agricultural territories with built up windbreaks approximate 8–9%, in which the belts themselves occupy about 2.2% of the protected areas.

3 Agricultural Landscape and Climate Change

The construction of wind protection tree belts is a long-term activity. They work well and fully when the vegetation reaches a certain age and is in good health. For these reasons, their planning should be based not only on past experience. It is necessary to take into account current and possible future processes and phenomena related to environmental, social, and economic factors [35, 36]. Substantial influence over recent decades has, for instance, been effected by climate change and environmental pollution [37–45]. Although part of the arable land in Bulgaria in moderate latitudes is steppe, i.e., the prevailing natural vegetation is grassy and climatically has appropriate soils for the development of tree and shrub plantations [46, 47], other parts of the country should also be considered. The most important environmental factors for shelterbelt construction are climate and soils, namely, their fragile interrelationship can generate serious environmental backlash if improperly applied. Environmentally, properly shelterbelts favor growth not only for agricultural crops but also for forest plantations and ornamental plants [48, 49].

In connection with the problems that some meteorological phenomena frame, models are developed to identify the adverse consequences. Soil erosion is the most serious cause of soil degradation [50–54]. Rousseva and Stefanova [55] suggest models for assessing (1) rainfall erosivity and (2) soil erodibility for the whole territory of Bulgaria. Southwestern and southern Bulgaria have low degrees of rainfall erosivity. The rest of the country, i.e., 65% of the territory, falls between the fourth and sixth grade on a six-degree scaling. In term of soil erodibility, a large part of the country's landmass, i.e., 61.5%, falls between the fourth and sixth grade on the scaling. However, parts of Dobrudja and the Thracian Lowlands are exceptions. As such, these two indicators' high values are highly correlated with water erosion and indicate the need for additional measures to increase environmental sustainability. One important aspect of these measures for agricultural landscapes are windbreaks.

In terms of climate change, this is a global concern. At the national level, it affects all economic sectors. A number of strategic documents have been developed by the Ministry of Environment and Water and follow the National Action Plan on Climate Change, National Climate Change Risk and Vulnerability Assessment for the Sectors of the Bulgarian Economy, Financial Disaster Risk Management and Insurance Options for Climate Change Adaptation in Bulgaria, and The Climate Change Mitigation Act—enforced on 11 March 2014. As a result of the objectives set in these documents and the introduction of relevant measures, greenhouse gas emissions are expected to be reduced. Several measures in the National action plan on climate change are related to increasing forest cover in agricultural landscapes, they include: land use, land use change, and the forestry sector which is directly related to afforestation of abandoned agricultural land, eroded and erosion-threatened areas, and restoration and maintenance of shelterbelts. Current legislation, however, is not considering the construction of new windbreaks—stating them as not necessary—and focalizes on restorative features and improving current projects. At the same time, a measure introducing water-saving and energy-saving irrigation technologies is included in the agriculture sector. Although they are an inseparable part of some agricultural practices, irrigation systems do not have the complex benefits inherent in long-term sustainable shelterbelt designs.

Climate change will directly affect agricultural yields, which is the essential argument for conducting mitigation action. Changes may also have a positive effect on yields (i.e., a prolonged vegetation period and increased rainfall), but the overall impact of destabilizing the environment, associated with extreme temperature change and climatic events (e.g., natural disasters) can easily outweigh the benefits [40, 56, 57]. The clearly expressed moderate-continental climate in northern Bulgaria, characterized by prolonged droughts and strong winds, creates prerequisites for compromising agricultural activity. This is particularly important for a region with fertile soils, where agriculture is the main land use type. The analysis of meteorological elements shows that it is precisely for the northern and northeastern parts of the country where the greatest need for construction of windbreaks exists [46]. The established shelterbelts provide an opportunity to correct the impact of small amounts of rainfall (i.e., 530–650 mm) and strong winds in the area. Trends in temperature increases have been identified over the last decades along the Danube coast [58–63]. According to Nikolova et al. [63], the biggest temperature difference is in the summer, i.e., 1–1.2 °C, and lowest is in the winter, i.e., 0.5 °C—dating back from 1931 to 2013. However, in transition seasons it was found that there is hardly any difference in seasonal temperature. As such, the most significant climate change impact on agricultural activity is the reduction of precipitation [64]. Koleva-Lizama [64] state that all climate models for the coming decades show rainfall reduction in spring which will adversely affect soil moisture and spring crops grown in non-irrigated areas. Other authors also believe that reduction in yields will be the result of worsening conditions in non-irrigated agriculture [57, 65]. This is particularly true for southeastern Europe, where rainfall will decrease and restructuring of irrigated agriculture should be closely looked at [65–71]. In-line with this notion, protection belts will also help, especially if irrigation systems have open channels.

Climate change also raises the issue of windbreak sustainability. In some models, developed for managing forest landscapes, significant theories focalize on sustainability and environmental factors [72–84]. Adaptation of forests to climate change is needed [47]. Following an analysis of the program of measures for climate change adaptation of forests in Bulgaria and reduction of negative impact adopted in 2011, research concludes that the climate in Bulgaria will become warmer and dryer, especially in the second half of the twenty-first century, and that rainfall reduction during the warm half-year will negatively affect ecosystems, including forests. Figure 4 illustrates the measures identifying the vulnerability zones and adaptation measures for forests. Optimistic, realistic, and pessimistic scenarios for the present climate and possible changes in the coming decades of this century have been developed.

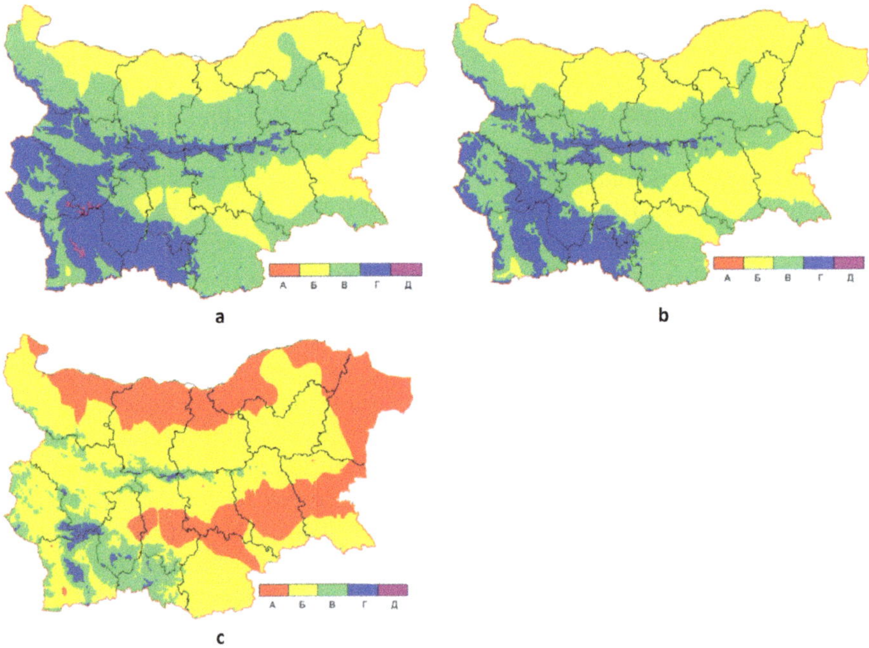

Fig. 4 Vulnerability zones of forest ecosystems in Bulgaria, adopted from Marinova et al. [47]: **a** contemporary climate from 1961–1990, **b** 2080—optimistic scenario, **c** 2080—pessimistic scenario. *Legend* zone A = very high vulnerability, zone Б = high vulnerability, zone B = medium vulnerability, zone Г = low vulnerability, zone Д = very low vulnerability

The optimistic scenario for 2080 is largely identical to the current situation when it comes to agricultural territories. If they fall into zone B, i.e., a high degree of vulnerability, it would mean that at present, in these territories the conditions would become no longer optimal for forestry activities. Under the pessimistic scenario, a large percentage of agricultural areas located in the lowest parts of the country fall into zone A which would place them under a very high degree of vulnerability.

4 Windbreak Characteristics

For a descriptive review of windbreaks, knowledge-based on the topic is taken from the Training Manual for Applied Agroforestry Practices—2018 Edition [85] in combination with the Handbook for Agroforestry Planning and Design [85]. Wight and Straight [31] describe the following shelterbelt characteristics: height, density, orientation, length, width, continuity, and cross-sectional shape. The windbreak height (*H*) has the most significant influence on the length of the wind protected area. Wind reduction is 2*H* to 5*H* from the windward and 30*H* from the leeward barrier side. The windbreak length determines the actual area of protected zone. For maximum efficiency the ratio length:height should be at least 10:1. This prevents further turbulent movements along the periphery of the protected area. The density of plants strongly influences wind speed. It is defined as a percentage of an impermeable barrier in three degrees: dense = 60–80% (i.e., minimum wind speed, but also small area), moderately dense = 40–60% (i.e., less reduced speed but more protected area), and permeable = less than 40% (i.e., suitable for evenly distributed snowfall). Therefore, the denser the barrier, the more the wind speed is reduced, but the protected area is smaller. Determining the correct density depends on what the barriers will be used for. For example, low density belts are suitable for uniform distribution of snow cover and are not suitable for protection against wind erosion. The windbreak orientation must comply with the prevailing strong winds. Since the direction is not constant and fluctuates, the belts should be oriented perpendicularly to the problem winds. The windbreak width is related to achieving appropriate density. It depends on number of rows, planting distances, and species composition. Greater width can contribute to additional benefits, i.e., biodiversity, product production, and more. The windbreak continuity is related to its effectiveness. Openings or breaks along the belt act as a funnel, concentrating the airflow and increasing its speed above normal levels for open field. In the cross section, the belt may be rectangular, trapezoidal, or triangular in shape. To a certain extent, the speed is affected by these forms, but it depends more on windbreak density.

5 Windbreaks Planning

The planning and design of wind protection belts must comply with the legal framework, purpose, and status of territories as well as their ownership, forest plans and programs, specific site conditions, and necessary functions of the plantations. According to Art. 2, para. 1. of the Forestry Act, these belts are forests. According to their functions, windbreaks belong to protective afforestation (i.e., Art. 9 of Ordinance No. 2 of 7 February, 2013 on the conditions and order for afforestation of forest territories and agricultural lands used for creation of special, protective, and economic forests and forests in protected areas, inventory of created crops, their reporting, and registration). The Ordinance defines requirements for development of technological plans for afforestation, exploratory work, determination of type of habitat, selection of tree, and shrub species. The selection of species considers the specific functions of vegetation, e.g., for sheltered forest belts the species must be fast growing, long-life, wind resistant, snow proof, and dry resistant. The creation of the pan-European ecological network Natura 2000 requires some changes in the use of vegetation within protected areas. Under Regulation (i.e., Ordinance No. 2 of 7 February 2013), afforestation of non-afforested areas within protected territories and Natura 2000 protected areas will need to be considered—respective of their management plans. For example, in artificial restoration of existing forest crops in Natura 2000 protected areas, the same tree species and cultivars should be adopted.

In terms of windbreak efficacy, if the break is greater when a complete network of belts is built it should take into account the belt system barrier level, i.e., primary and secondary [46, 47]. Specifically, primary belts are oriented perpendicular to the direction of the prevailing strong winds, i.e., on an average of 500 m and secondary ones perpendicular to the them, i.e., at a distance of 1000 m. Other research states that belt systems include three types of plantations—main, intermediate, and additional [86]. In this setup, the first type coincides with the dual barrier system, with the addition that distance between belts varies between 300 and 500 m. Intermediate belts support the function of the main belts located parallel to them in order to maintain the elevated air flow. The additional barriers, similar to the secondary belts, are located perpendicular to the main ones, at a distance of 800–1200 m in order to protect against winds from other than the general direction. By connecting all forest plantations, the technical benefits can add ecological value and protected natural areas adjacent to the agricultural landscape can be interlinked. This can offer additional economic motivation for creation of so-called green infrastructure [87, 88]. It provides much more benefits for people, wildlife, and the economy [89]. Its improvement is also in line with one of the European Union's new objectives for biodiversity conservation in 2020 [90] and 2030 [91]. The Green Infrastructure Bulletin [92] outlined some basic arguments. Green infrastructure supports the state of the natural environment, e.g., by mitigating the impact of climate change, which is more cost-effective than expensive artificially made technological solutions. A more integrated approach to land management is the best way to build green infrastructure. This can be done by spatial planning on a wider scale and covering a larger area, e.g., at the regional or municipal level.

The direction for increasing sustainability and integrity of territories is another important factor into properly planning and setting up windbreaks. One of the most important principles in large-scale planning is the development of multifunctional territories [93]. Except protected areas, in other places where anthropogenic impact is significant, the function integrity principle needs to be applied. This is particularly relevant for the agricultural landscape, where farming restricts land use for other purposes, thus changing not only the environment but also the way of life. Taking these general issues into account, specific windbreak designs should structure themselves consistent with the desired effect as well as what it intends on defending, e.g., crops or orchards, soils, and livestock farms. In Wight and Straight [31] the following sequence for the design process is laid out: (1) determination of purpose of wind protection barriers, (2) analysis and evaluation of existing site conditions, (3) specification of windbreaks characteristics and choice of plant species, and (4) choice of planting technology and definition of maintenance activities.

6 Conclusion

In relation with the priceless provisions of forests regulating water, energy, and the carbon cycle, Ellison et al. [20] propose a change in paradigm and a shift from a carbon-oriented model to a model that puts hydrological and cooling effect of forests first. In-line with this, long-term and by the implementation of necessary measures for increasing woods and building up windbreaks, it can be possible the avoid some of the more pessimistic scenarios. The main problems of agricultural territories in Bulgaria are the climatic fluxes in decreased rainfall and increased temperature. Moreover, the frequency of extreme meteorological phenomena—i.e., storms with intense rainfall leading to floods by accelerating erosion processes and activating landslides—is increasing [67, 94]. The relief features in turn reduce strong winds in plain areas that are predominantly for agricultural use. Taking into account the complexity of the problems, the negative consequences due to irrigated agriculture as well as all non-agricultural benefits of forest protection belts becomes clear, i.e., windbreaks are a better alternative compared to constructed irrigation systems. As such, it is possible to redirect planned means of constructing irrigation systems in harmony with shelterbelt systems—a practice that would bring more benefit as well as minimize land use problems. Alongside the restoration of existing belts, it is possible of providing a further measure of creating new plantings in agricultural areas where climate variables are expected, specifically, not to limit the development of cultivated crops.

In order to minimize competition between crops and protecting plantings, the choice of tree species must take account of past research and potential climatic variability. It was found, e.g., that the most suitable tree species for Dobrudja are *Quercus robur* L., *Quercus cerris* L., *Quercus petraea* (Matt.) Liebl., *Frainus excelsior* L. and *Fraxinus oxycarpa* Willd., *Gleditchia triacanthos* L., *Juglans regia* L., *Ulmus pumila* L., and *Robinia pseudoacacia* L. [30]. From the species that have historically proven to be appropriate, the species that would resist and best perform against changing conditions would need to be more resource-friendly, especially in terms of water consumption. Georgiev [30] explains that *Quercus* is the most economical in terms of soil moisture use, followed by *Fraxinus oxycarpa* Willd., *Acer negundo* L., *Gleditchia triacanthos* L. and *Robinia pseudoacacia* L. When setting up a windbreak system, it is first necessary to clarify the main function of the barriers. They can work in two main directions: (1) reduction of soil erosion and protection of agricultural crops and (2) uniform distribution of snow cover. Although in both cases all other benefits are present, the specific function requires differences in density and location of plantations. In the first case, it is required a density of 40–60% [31], while in case of wind-sensitive crops, the distance between belts needs to be reduced to six to ten times their height. This means that with 10–15 m of belt height, there should be at a distance of 100–150 m instead of 300–500 m. Conversely, in sloping terrains, in territories with extreme precipitation events, or in the presence of soils with light mechanical composition susceptible to wind erosion, a significant ecological effect will require such barriers to be properly located with a higher density. Also, when there is a lack of precipitation, uniform distribution of snowfall is a necessary option. This is achieved precisely by more permeable belts, with a density of up to 25–30% [31]. In this case, the shelterbelts should be located at a greater distance, equal to 15–20 times their height—all functional considerations for an application in Bulgaria and for the proper design of viable agricultural land use.

References

1. Steiner R (1909) Die Geheimwissenschaft Im Umriss. Jessica Duber, Orleans
2. Brandle J, Hodges L, Tyndall J, Sudmeyer R (2009) Windbreak practices. north american agroforestry: an integrated science and practice, 2nd edn. American Society of Agronomy, New York, pp 75–104
3. Droze WH (1977) Trees, prairies and people: a history of tree plantings in the plains states. Texas Woman's University Press, Denton, TX
4. Dahl J (1940) Progress and development of the prairie states forestry project. J For 38:301–306. https://doi.org/10.1093/jof/38.4.301
5. Brandle J, Hodges L, Zhou X (2004) Windbreaks in North American agricultural systems. Agron Hortic Fac Publ 389:65–78
6. Kangaspuro M, Smith J (2006) Modernisation in Russia since 1900. SKS Finnish Literature Society, Helsinki
7. Popov VP (1992) Golod I gosudarstvennia politika. Otchestvennaya Istor 6:36–60
8. Brain S (2010) The Great Stalin plan for the transformation of nature. Environ Hist Durh N C 15:670–700. https://doi.org/10.1093/envhis/emq091
9. Isachenko H (1950) Forestry properties of main and secondary wood species for creation of state protective belts. Zemizdat, Sofia
10. Yordanov K (2003) The image of Bulgarians and Bulgarian lands in the chronicles of the first, second and third crusades. http://www.modvsvivendi.org/bg/biblioteka/bg_and_crusades.htm. Accessed 10 Sept 2018
11. Rangelov V, Karatoteva D (2017) Analysis of the green system in landscapes of the Sitovo Municipality. Int Acad J Web Sch 1:6–8
12. Government of Bulgaria (2019) 2019 Annual report on the state and development of agriculture. Ministry of Agriculture, Food, and Forestry, Sofia
13. Mertia RS, Upadhya AK (1993) Performance of tree species suitable for shelterbelt plantation in Thar desert. Curr Agric 17:109–111
14. Zobel JM, Ek AR, Edgar CB (2021) Assessing the impact of 41 years of forest management on native wildlife habitat in Minnesota, USA. J For 119(2):164–176. https://doi.org/10.1093/jofore/fvaa050
15. Heisler GM, Dewalle DR (1988) 2. Effects of windbreak structure on wind flow. Agric Ecosyst Environ 22–23:41–69. https://doi.org/10.1016/0167-8809(88)90007-2
16. Ritchie KA (1988) 24. Shelterbelt plantings in semi-arid areas. Agric Ecosyst Environ 22–23:425–440. https://doi.org/10.1016/0167-8809(88)90037-0
17. Sturrock JW (1988) Shelter: its management and promotion. Agric Ecosyst Environ 22–23:1–13. https://doi.org/10.1016/0167-8809(88)90004-7
18. McNaughton KG (1988) 1. Effects of windbreaks on turbulent transport and microclimate. Agric Ecosyst Environ 22–23:17–39. https://doi.org/10.1016/0167-8809(88)90006-0
19. Schroeder WR (1988) 25. Planting and establishment of shelterbelts in humid severe-winter regions. Agric Ecosyst Environ 22–23:441–463. https://doi.org/10.1016/0167-8809(88)90038-2
20. Ellison D, Morris CE, Locatelli B et al (2017) Trees, forests and water: Cool insights for a hot world. Glob Environ Chang 43:51–61. https://doi.org/10.1016/j.gloenvcha.2017.01.002
21. Gregory NG (1995) The role of shelterbelts in protecting livestock: a review. New Zeal J Agric Res 38:423–450. https://doi.org/10.1080/00288233.1995.9513146
22. Valduga MO, Zenni RD, Vitule JRS (2016) Ecological impacts of non-native tree species plantations are broad and heterogeneous: a review of Brazilian research. An Acad Bras Cienc 88:1675–1688
23. Reddy V, Reddy PR, Reddy DV (2016) Impact due to sunspot activity on climate change: some salient results article in. Int J Earth Sci Eng
24. Solheim JE, Stordahl K, Humlum O (2012) The long sunspot cycle 23 predicts a significant temperature decrease in cycle 24. J Atmos Solar-Terrestrial Phys 80:267–284. https://doi.org/10.1016/j.jastp.2012.02.008

25. Chendev Y, Sauer T, Ramirez G, Burras C (2015) History of east European chernozem soil degradation. Protection and restoration by tree windbreaks in the Russian steppe. Sustainability 7:705–724
26. Alam H, Khattak JZK, Ppoyil SBT et al (2017) Landscaping with native plants in the UAE: a review. Emirates J Food Agric 29:729–741. https://doi.org/10.9755/ejfa.2017.v29.i10.319
27. Georgiev G, Koserkov Al, Panaiotov P (1961) Additional uses of wind protection belts. State Publ, Varna
28. Kudryashov P, Erusalimsky V, Knyazeva L (1985) Management of state forest shelter belts. Agropromizdat, Sofia
29. Kögel-Knabner I, Amelung W (2013) Dynamics, chemistry, and preservation of organic matter in soils. In: Holland HD, Turekian KK (eds) Treatise on geochemistry, 2nd edn. Elsevier Science, Berlin, pp 157–215
30. Georgiev G (1960) Forest protection belts in Bulgaria. State Publ, Varna
31. Wight B, Straight R (2019) Windbreaks. Training manual for applied agroforestry practices, 2018th edn. University of Missouri, Columbia, Center for Agroforestry, pp 92–114
32. Executive Forests Agency (2021) Executive forests agency documents. In: Minist Agric Foods. http://www.iag.bg/lang/2/index. Accessed 15 Feb 2021
33. Lyubenova M, Petrov J, Radev R et al (2017) Forest ecosystem services and payment schemes (case study). St. Kliment Ohridski University Press, Sofia
34. Radev R (2018) The system of protective forest belts in Dobrudja: Importance, condition, prospects and management problems. PESFOR-Forest for Water, COST Action CA15206, Albena
35. Gimenez C, Sierra V, Rodon J (2012) Sustainable operations: Their impact on the triple bottom line. Int J Prod Econ 140:149–159. https://doi.org/10.1016/J.IJPE.2012.01.035
36. Cirella GT, Mwangi SW, Paczoski A, Abebe ST (2020) Human-nature relations: The unwanted filibuster. In: Cirella GT (ed) Sustainabile human-nature relations: environmental scholarship, economic evaluation, urban strategies. Springer Singapore, Singapore
37. Cao L, Eby M, Ridgwell A et al (2008) The importance of ocean transport in the fate of anthropogenic CO_2. Biogeosciences Discuss 5:4521–4557. https://doi.org/10.5194/bgd-5-4521-2008
38. Walters CE, Meslé MMI, Hall IM (2018) Modelling the global spread of diseases: a review of current practice and capability. Epidemics 25:1–8
39. UNESCO (2009) The United Nations World water development report 3: water in a changing world. In: Earthscan (ed) World Water Assessment Programme, p 349
40. Spasova Z (2013) Climate change: a threat to human health. Ministry of Health, Sofia
41. Wilson E, Piper J (2010) Spatial planning and climate change. Routledge
42. Mahmood R, Jia S, Zhu W (2019) Analysis of climate variability, trends, and prediction in the most active parts of the Lake Chad basin, Africa. Sci Rep 9:6317. https://doi.org/10.1038/s41598-019-42811-9
43. Miah MG, Ahmed M, Afroz T (2013) Climate change and adaptation: evidence from a forest-dependent community in Bangladesh. J For Livelihood 11:94–108. https://doi.org/10.3126/jfl.v11i1.8615
44. Cramer W, Bondeau A, Schaphoff S et al (2004) Tropical forests and the global carbon cycle: impacts of atmospheric carbon dioxide, climate change and rate of deforestation. Philos Trans R Soc London Ser B Biol Sci 359:331–343. https://doi.org/10.1098/rstb.2003.1428
45. Jansson JK, Hofmockel KS (2020) Soil microbiomes and climate change. Nat Rev Microbiol 18(1):35–46
46. Marinov I, Genova F (2008) Ecological prerequisites for creation of forest protection belts in northern Bulgaria. For Sci 99–107
47. Marinova V, Trichkov L, Vassilev N (2011) Hot spots—forests and climate change. Program of measures for adaptation of forests in the Republic of Bulgaria and reduction of negative impact of climate change on them. Manag Sustain Dev 1:240–250
48. Shahanova M (2016) Creating compositions of perennial flowers in changing urban environment. Manag Sustain Dev 83–87

49. Shahanova M, Kabatliyska Z (2018) Introduction of new ornamental floral species in the context of sustainable development—experience of cultivation of a native and non-native ecotype of Bituminaria (L.) Stirt. Manag Sustain Dev 5:1–6
50. Pimentel D (2006) Soil erosion: a food and environmental threat. Environ Dev Sustain 8:119–137. https://doi.org/10.1007/s10668-005-1262-8
51. Jie C, Jing-zhang C, Man-zhi T, Zi-tong G (2002) Soil degradation: a global problem endangering sustainable development. J Geogr Sci 12:243–252. https://doi.org/10.1007/bf0283 7480
52. Gruver (2013) Prediction, prevention and remediation of soil degradation by water erosion. Nat Educ Knowl 4:2
53. Issaka S, Ashraf MA (2017) Impact of soil erosion and degradation on water quality: a review. Geol Ecol Landscapes 1:1–11. https://doi.org/10.1080/24749508.2017.1301053
54. Shi ZH, Fang NF, Wu FZ et al (2012) Soil erosion processes and sediment sorting associated with transport mechanisms on steep slopes. J Hydrol 454–455:123–130. https://doi.org/10.1016/j.jhydrol.2012.06.004
55. Rousseva S, Stefanova V (2006) Assessment and mapping of soil erodibility and rainfall erosivity in Bulgaria. https://www.researchgate.net/publication/251813968_Assessment_and_Mapping_of_Soil_Erodibility_and_Rainfall_Erosivity_in_Bulgaria. Accessed 10 Sept 2018
56. Alexandrov V, Simeonov P, Kazandzhiev V, Korchev G, Yotova A (2010) Climate change. Sofia
57. Alexandrov V (2014) Climate change in Bulgaria: past, present, and future. http://catrisk.insmarket.eu/docs/Veselin-Alexandrov.pdf. Accessed 10 Sept 2018
58. Adams FG, Milgram G, Green EW, Mansfield C (1968) Undeveloped land prices during urbanization: a micro-empirical study over time. Rev Econ Stat 50:248. https://doi.org/10.2307/192 6200
59. Stagl J, Hattermann F (2016) Impacts of climate change on riverine ecosystems: alterations of ecologically relevant flow dynamics in the Danube River and its major tributaries. Water 8:566. https://doi.org/10.3390/w8120566
60. Ionita M, Badaluta CA, Scholz P, Chelcea S (2018) Vanishing river ice cover in the lower part of the Danube basin-signs of a changing climate. Sci Rep 8:1–12. https://doi.org/10.1038/s41 598-018-26357-w
61. Webb BW, Nobilis F (1995) Long term water temperature trends in Austrian rivers. Hydrol Sci J 40:83–96. https://doi.org/10.1080/02626669509491392
62. Pekarova P, Halmova D, Miklanek P et al (2008) Is the water temperature of the Danube River at Bratislava, Slovakia, rising? J Hydrometeorol 9:1115–1122. https://doi.org/10.1175/2008JH M948.1
63. Nikolova N, Assenova N, Daskalova V, Hristov G, Dyakova N (2015) Multi-year changes in air temperature on Bulgarian coast of Danube river. In: Yearbook of Sofia University "St. Kliment Ohridski." Sofia University "St. Kliment Ohridski," Sofia, pp 71–84
64. Koleva-Lizama I (2014) Analysis and assessment of risk and vulnerability in the field of agriculture. In: Analysis and risk and vulnerability assessment of bulgarian economy sectors from climate change, special part. Ministry of Environment and Water, Sofia, pp 7–27
65. Alexandrov V (2014) Climate change and adaptation measures. http://institutfrancais.bg/media/29/fc/4aa8674ed5598794a39e3df8b3b9/vesselin-alexandrov.pdf. Accessed 10 Sept 2018
66. Jacob D, Kotova L, Teichmann C et al (2018) Climate impacts in Europe under +1.5 °C global warming. Earth's Futur 6:264–285. https://doi.org/10.1002/2017EF000710
67. Beillouin D, Schauberger B, Bastos A et al (2020) Impact of extreme weather conditions on European crop production in 2018: random forest—yield anomalies. Philos Trans R Soc B Biol Sci 375:20190510. https://doi.org/10.1098/rstb.2019.0510
68. Pasqui M, Di Giuseppe E (2019) Climate change, future warming, and adaptation in Europe. Anim Front 9:6–11. https://doi.org/10.1093/af/vfy036
69. Ciscar JC, Iglesias A, Feyen L et al (2011) Physical and economic consequences of climate change in Europe. Proc Natl Acad Sci U S A 108:2678–2683. https://doi.org/10.1073/pnas.1011612108

70. Hosseinzadehtalaei P, Tabari H, Willems P (2019) Regionalization of anthropogenically forced changes in 3 hourly extreme precipitation over Europe. Environ Res Lett 14:124031. https://doi.org/10.1088/1748-9326/ab5638
71. Behrens A, Georgiev A, Carraro M (2010) Future impacts of climate change across Europe. Centre for European Policy Studies, Brussels
72. Gunn EA (2007) Models for strategic forest management. Handbook of operations research in natural resources. Springer, New York, pp 317–341
73. Elith J, Leathwick JR (2009) Species distribution models: ecological explanation and prediction across space and time. Annu Rev Ecol Evol Syst 40:677–697. https://doi.org/10.1146/annurev.ecolsys.110308.120159
74. He HS (2008) Forest landscape models: definitions, characterization, and classification. For Ecol Manage 254:484–498. https://doi.org/10.1016/j.foreco.2007.08.022
75. Twery MJ, Weiskittel AR (2013) Forest-management modelling. In: Wainwright J, Mulligan M (eds) Environmental modelling: finding simplicity in complexity, 2nd edn. John Wiley and Sons, Oxford, pp 379–398
76. Dragozova-Ivanova E, Paligorov I, Ivanov I, Kovacheva S (2016) Management modelling for forest landscapes. Econ Bus 8:90–97
77. Doxford SW, Freckleton RP (2012) Changes in the large-scale distribution of plants: extinction, colonisation and the effects of climate. J Ecol 100:519–529. https://doi.org/10.1111/j.1365-2745.2011.01902.x
78. Corlett RT, Westcott DA (2013) Will plant movements keep up with climate change? Trends Ecol Evol 28:482–488
79. He HS, Gustafson EJ, Lischke H (2017) Modeling forest landscapes in a changing climate: theory and application. Landsc Ecol 32:1299–1305. https://doi.org/10.1007/s10980-017-0529-4
80. Sano M, Miyamoto A, Furuya N, Kogi K (2009) Using landscape metrics and topographic analysis to examine forest management in a mixed forest, Hokkaido, Japan: guidelines for management interventions and evaluation of cover changes. For Ecol Manage 257:1208–1218. https://doi.org/10.1016/j.foreco.2008.10.005
81. Martinez del Castillo E, García-Martin A, Longares Aladrén LA, de Luis M (2015) Evaluation of forest cover change using remote sensing techniques and landscape metrics in Moncayo Natural Park (Spain). Appl Geogr 62:247–255. https://doi.org/10.1016/j.apgeog.2015.05.002
82. Sharifi A, Amini J, Tateishi R (2016) Estimation of forest biomass using multivariate relevance vector regression. Photogramm Eng Remote Sensing 82:41–49. https://doi.org/10.14358/PERS.83.1.41
83. Pretzsch H, Grote R, Reineking B et al (2008) Models for forest ecosystem management: a European perspective. Ann Bot 101:1065–1087. https://doi.org/10.1093/aob/mcm246
84. Baskent EZ, Jordan GA (1996) Designing forest management to control spatial structure of landscapes. Landsc Urban Plan 34:55–74. https://doi.org/10.1016/0169-2046(95)00200-6
85. Gold M, Hemmelgarn H, Ormsby-Mori G, Todd C (2019) Training manual for applied agroforestry practices, 2018th edn. University of Missouri, Columbia, Center for Agroforestry
86. Troeva V, Tzolova G (1997) Landscape planning. University of Architecture, Civil Engineering and Geodesy Publ., Sofia
87. Forest Research (2010) Benefits of green infrastructure. Report by Forest Research. Farnham
88. Cirella GT, Tao L (2008) Measuring sustainability: an application using the index of sustainable functionality in South East Queensland, Australia. Int J Interdiscip Soc Sci 3:231–240
89. Mihova K, Tzolova G (2009) Green Infrastructure—an approach to sustainable landscape. For Ideas 2:219–229
90. Commission E (2011) The EU biodiversity strategy to 2020. Publications Office of the European Union, Luxembourg
91. European Commission (2020) Biodiversity strategy for 2030. https://ec.europa.eu/environment/strategy/biodiversity-strategy-2030_en. Accessed 15 Feb 2021
92. Commission E (2010) Green infrastructure. EU Bulletin, Brussels

93. Evrev P (1999) Territorial and spatial planning of recreation and tourism. “St. Kliment Ohridski” University Press, Sofia
94. Skougaard Kaspersen P, Høegh Ravn N, Arnbjerg-Nielsen K et al (2017) Comparison of the impacts of urban development and climate change on exposing European cities to pluvial flooding. Hydrol Earth Syst Sci. https://doi.org/10.5194/hess-21-4131-2017

Synchronizing Agricultural Trade Regulations: Case Study from Subang Regency

Mustika S. Purwanegara, Nurrani Kusumawati, Rini H. Ekawati, Herry Hudrasyah, and Giuseppe T. Cirella

Abstract Government agricultural policy in Indonesia in terms of product trade has not reflected perfectly the principles of fairness, i.e., fair play, for the poor. Powerless farmers are strongly affected by inadequate laws, regulations, and bureaucracy. This condition might be complicated by the enactment of decentralization in the country, i.e., stipulated in Article 10 and 11 of Law No. 22 of 1999. Decentralization empowers local governments the ability to regulate and make policies that would be more pro-welfare to farmers as well as to local trading of agricultural commodities at the provincial and regency, i.e., city, level. The national government only facilitates and coordinates national policies which leaves local authorities the power to control and guide local development. In this chapter, secondary data obtained from government sources alongside direct interviews of stakeholders examine how pineapple performs as a cash crop in different parts of Indonesia. The research attempts to synchronize national-provincial-regency level difficulties by considering Indonesia's legislation in trading agricultural commodities. This chapter suggests the need for better national policies on price stabilization, growth stimulus, legal protection, and prosperity-based strategies for increasing pineapple farmers' welfare.

Keywords Pineapple supply chain · Decentralization · Food crops trading · Commodities · Legislation · Indonesia

---

M. S. Purwanegara (✉) · N. Kusumawati · R. H. Ekawati · H. Hudrasyah
School of Business and Management, Bandung Institute of Technology, Bandung, Indonesia
e-mail: mustika@sbm-itb.ac.id

N. Kusumawati
e-mail: nurrani.k@sbm-itb.ac.id

R. H. Ekawati
e-mail: rini.hanjani@sbm-itb.ac.id

H. Hudrasyah
e-mail: herryhud@sbm-itb.ac.id

G. T. Cirella
Faculty of Economics, University of Gdansk, Sopot, Poland
e-mail: gt.cirella@ug.edu.pl

G. T. Cirella (ed.), *Human Settlements*, Advances in 21st Century Human Settlements,
https://doi.org/10.1007/978-981-16-4031-5_9

1 Introduction

Indonesia is a tropical country with a climate that favors more than 38,000 species of plants of which 55% are endemic to its territory—including 450 species of fruits [1]. Among a large number of types of fruit in Indonesia, there is some exotic, tropical fruits that are widely grown (e.g., durian, mangosteen, papaya, mango, and star fruit). One of the most popular fruits from tropical countries is pineapple (Ananas Comosus). Pineapple is a fruit crop of shrubs that can grow in various seasons. Generally, pineapple plantations are grown from 800 m above sea level in areas with rainfall of 1000–1500 mm per year and with temperatures ranging from 23 to 32 °C [2]. The global production of pineapples amounted to 25.8 million metric tons in 2019. Costa Rica became the biggest producer of pineapples, producing approximately 2931 thousand metric tons every year. With its tropical environment, Costa Rica is the perfect environment for promoting pineapple growth. The country diligently invested in developing that sector of its economy at the beginning of 2000 and since has increased production by almost 300%. Second place is Brazil, where pineapple is thought to have originated. Production has reached approximately 2695 thousand metric tons annually. Del Monte, one of the largest fruits producing conglomerates in the world, has huge pineapple plantations and factories throughout Brazil. Third on the list is the Philippines, with an annual production rate of 2612 thousand metric tons each year. The majority of the harvested fruit in the Philippines is exported and accounts for 17% of the global supply [3]. In 2016, Indonesia became the eighth largest pineapple producing country in the world. Many of the pineapple producing countries have advanced technologies to increase production. The largest producing countries grow MD2 and Queen type pineapples as their flagship product [4]. Still, many countries still produce smooth cayenne pineapple including India, Sri Lanka, Malaysia, and Thailand [5].

As one of the major pineapple producing countries around the world, Indonesia exported 9586 tons in 2017 and 11,247 tons in 2018. Its export target markets are the United Arab Emirates, Japan, Hong Kong, Singapore, Oman, Canada, Kuwait, and Korea [6]. Pineapple production in Indonesia is shown in Fig. 1. Lampung became the most significant provincial supplier (i.e., 32.8%), followed by North Sumatra (i.e., Sumatera Utara with 12.8%), West Java (i.e., Jawa Barat with 10.4%), and then other provinces. In 2011, Indonesia's pineapple production grew from 1.5 million tons to 1.7 million ton in 2018 representing a productivity increase from 124.9 tons/ha in 2011 and 117.71 tons/ha in 2018.

Subang is the primary producer of pineapple in West Java province (i.e., 96.9%), followed by Bogor (i.e., 2.2%) and others (Fig. 2). Subang's farmers can produce 40,000 pineapples per hectare [6]. Subang is located in the north of West Java with an altitude between 0 and 1500 m above sea level and annual rainfall between 2000 and 4000mm. Air temperature in Subang regency ranges from 21 to 31 °C with humidity levels from 78 to 84 °C—making it a very suitable place to grow pineapple. Key pineapple plantations are located in the districts of Jalan Cagak, Ciater, Cijambe, and Kaso Malang. Based on the shape of the leaves and fruits, Subang pineapple plant is

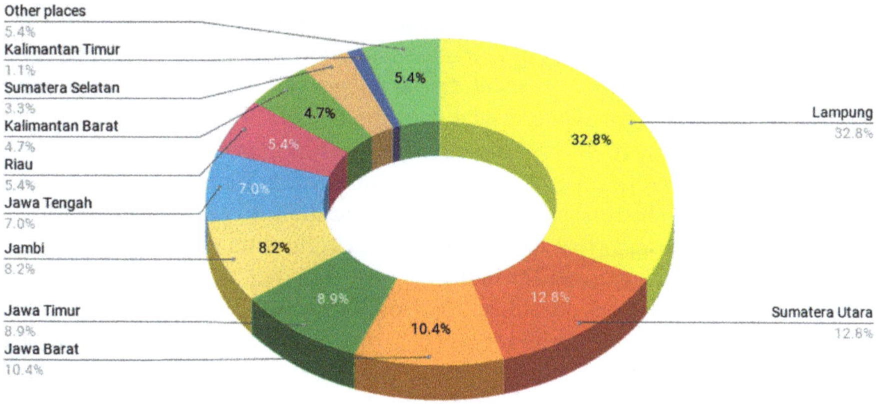

Fig. 1 Pineapple production by province, 2011–2018, adapted from the Ministry of Agriculture [7]

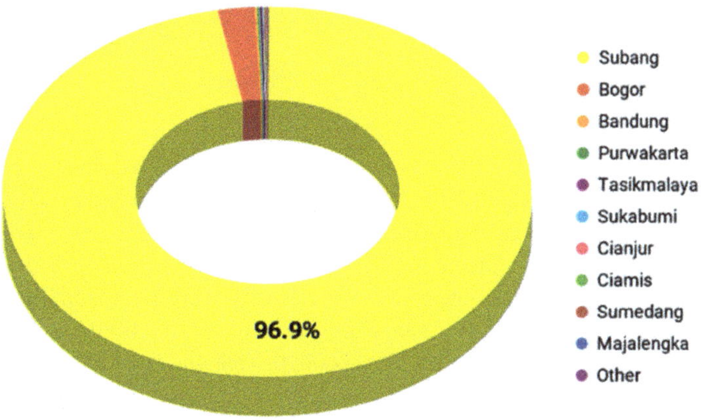

Fig. 2 Pineapple production in West Java, 2011–2018. Adapted from the Ministry of Agriculture [7]

classified as smooth cayenne. Smooth cayenne pineapple has the characteristics of delicate leaves, bubbly fruit, small fruit crown, watery, strong aroma, and taste tends to be sweet. This variety has a large size and weight between 1.5 and 5 kg (i.e., an average of 2.3 kg) [8] and is classified accordingly, i.e., 2.5–3 kg = grade A, 2–2.5 kg = grade B, 1.5–2 kg = grade C, and 1–1.5 kg = grade D. Pineapple which is too young, too mature, too small, bruised, or defective is categorized as out-of-grade [9]. Subang regency has a much higher potential compared with its real production, i.e., its production capacity is only 45.45%. As such, the district of Jalan Cagak is the highest producer of fresh pineapple in the regency. Production of top pineapple must be supported by adequate production facilities, stable and professional farmer institutions, and government policies that favor pineapple farmers [6].

Pineapple is usually marketed in the form of fresh fruit that sells, from its destination to the factory or traditional market, usually via a middleman. There is an increasing trend in pineapple prices at the producer level in Indonesia. The price of pineapple in West Java is the cheapest in the country. Pineapple produced in Subang is the cheapest (IDR 1391/kg, i.e., USD 0.10/kg) in Indonesia compared to Lampung (IDR 2797/kg, i.e., USD 0.20/kg) and North Sumatra (IDR 2516/kg, i.e., USD 0.18/kg). This discrepancy bares the question of why Subang sells its pineapple so cheaply and what additional barriers control its market compared to the two largest producing provinces in the country. Based on this development, this chapter will address several questions. (1) How is the supply chain designed and what is the position of the Subang pineapple farmer in the chain? (2) What are the interventions from the government at the national-provincial-regency level and are regulations or programs put in place to resolving this problem? How can the synchronization of the regulations, policies, and programs of the national-provincial-regency chain of command solve the trade system issues for agricultural commodities as well as assist in achieving a level playing field bottom-up?

2 Decentralization in Indonesia

Nur [10] grouped decentralization into political, administrative, fiscal, and market —i.e., economic. Rondinelli et al. [11] define decentralization from an administrative perspective as the transfer of accountability in terms of planning, management, and allocation of resources from the central government and its agencies to field units of government agencies, subordinate units or levels of government, semi-autonomous public authorities or corporations, area-wide, regional, or functional authorities, and non-governmental organizations (NGOs)—both private and voluntary. Decentralization is a corrective action to handle poor performance brought by excessive central control. Four typologies are considered, noting they can be used at the same time: deconcentration, delegation, devolution, and privatization [12, 13]. Deconcentration is taking apart some workload, administrative authority, and responsibility from centrally located government to subordinate lower levels of government. Delegation is handing over managerial responsibility for the specific function to organizations outside the central government which can carry it out, but the principal authority is still with sovereign authority. Devolution is strengthening local government financially and legally by giving autonomy and independent legal status. Privatization is releasing responsibility for functions to other organizations or private enterprises. In reality, a decentralization strategy is most typically a mixture of the first three forms, depending on the specific objectives of the strategy [14–17]. However, devolution offers the most potential for obtaining governance and economic benefits by way of transferring responsibilities to local units of government by way of accountability, problem-solving, and citizen participation [18–27].

Indonesia is the biggest archipelago state in the world; it has more than 17,000 islands consisting of 300 ethnicities divided into 34 provinces, 416 regencies, and

98 cities. It would be very ineffective if an archipelago state like Indonesia adopted a fully centralized system of government. In the 1945 Constitution of the Republic of Indonesia, Article 1 Para. 3, stipulates that the government administration is ruled by law, meaning that all activities performed must refer to the applicable law. The legal power of law follows the hierarchy mentioned in the 2004 Act of the Establishment of Legislation, No. 10, Article 7 Para. 1, as follows: (1) 1945 Constitution of the Republic of Indonesia, (2) act and government regulation in lieu of the law, (3) government regulation, (4) president regulation, and (5) local regulation. As a result, with decentralization every local regulation from the province down to the regency in Indonesia that make policy—is in accord and does not go against the constitution—based on the Pancasila. Decentralization, in this case, is equivalent to a right, i.e., for an authority to take responsibility to govern and manage itself locally, while still upholding compatibility with national law. In Indonesia, local authorities cover all authorities in the field of government, with exception to foreign policy, defense, security, justice, monetary and national fiscal, and religion as regulated in the provisions of Article 10 Para. 3 of the 2004 Act No. 32. With this in mind, mandatory affairs which become regional issues are regulated in the provisions of Article 13 and 14 which have been further regulated by government regulation No. 38/2007 concerning the division of government affairs between national, provincial, and regency governments. In the framework of the implementation of local government, the government has also stipulated government regulation No. 41/2007 in which regional device organization should be taken into account. Legislation that governs the application of decentralization in Indonesia is constitutionally based on Article 18 Para. 1–7, Article 18A Para. 1–2, and Article 18B Para. 1–2. Decentralization in Indonesia is applied through the 1999 Act of Local Government No. 23., 2014 amendment No. 23, and the 2015 amendment No. 9. Decentralization has the goal of increasing public service so it can better and advance democratic life in Indonesia. In the name of harmonization, a hierarchy principle is applied through the annulment of local regulations that may go against the constitution or public importance by disturbing harmony in society, public service, public order, or any discriminatory issue. Local regulation is structured as follows: the provincial house of representative member is assigned as the provincial governor, the regency house of representative member is assigned over the regent, the city house of representative member is assigned as the mayor, and the village representative is assigned the village leader who consults with villagers or others of the same level. Further legislation about the rules of at the village level is governed at the regency regulatory level.

The growing interest of governments in developing countries in decentralizing the provision, financing, and maintenance of local services and infrastructure will require a more comprehensive and integrated framework for policy analysis than is offered by either neoclassical economic theories or public administration and finance theories [11, 13]. An integrated political economy framework that draws from the most useful aspects of each set of approaches can provide policy analysts with a more powerful set of concepts and methods for designing and implementing decentralization policies more effectively [11]. Research indicates that successful implementation of decentralization policies depends heavily on political factors who have commitments and supports, especially from the top-down, i.e., starting with national leaders [11]. Organizational factors conducive to decentralization include the appropriate allocation of planning and administrative functions among levels of government and local organizations with each set of functions suited to the decision-making capabilities of each level of organization [11]. "Behavioral and psychological conditions supporting decentralization include appropriate attitudes and behavior of central and lower-level government officials toward the decentralization and a willingness on their part to share authority with citizens and accept their participation in public decision-making" [11]. In Indonesia, institutional change is not reflected by the way social and economic power is concentrated that being, many of the old faces continue to dominate politics and business, while the new are drawn to follow predatory practices of the former [28]. Although the liberal democratization in Indonesia is successful in making local politics institutionally more democratic, it still suffers a lot of deficiencies and limitations especially in terms of accommodating the participation of local people [29]. Perhaps the most concrete impact of decentralization is the strengthening of regional political elites. With the introduction of elections and greater resource power, local level politics has led to a newly empowered class of local political elites. i.e., while more independent of national and provincial elites, local elites are now beholden to the "vicissitudes of district-level money politics in order to win and maintain office" [30]. In effect, the case for studying decentralization in Indonesia will now be interlinked with the synchronization of agricultural trade regulations—at all levels.

3 Data Collection

To best accumulate case study data from Subang regency's current pineapple farming state-of-the-art, administrative data from several public service institutions in Indonesia were utilized. Moreover, secondary data also came from sources such as World Bank Data [31, 32], various levels of government data and policy (e.g., Statistics Indonesia, Statistics Subang, and Statistics West Java), and academic journal articles. Interviews with the government officials of Subang regency working in the Department of Agriculture, Department of Cooperative and Small-Scale Businesses, Department of Trade and Regional Planning, and farmers from Jalan Cagak district were administered. At length, the data collection focused on three pineapple farmer associations (Table 1) as actors in the pineapple supply chain to acquire value-added data at the actor-level (Fig. 3).

Table 1 Profiles of farmer groups

Financial profile[a]	Farmer group 1	Farmer group 2	Farmer group 3
Revenue (IDR)	2,067,000,000	4,457,400,000	5,100,000,000
Expenses (IDR)	1,403,970,000	2,344,657,000	2,326,283,200
Net income (IDR)	663,030,000	2,112,743,000	2,773,716,800
Harvest area (ha)	79.5	64.6	50
Harvest time (months)	7	17	14
Average income/ha	8,340,000	32,705,000	55,474,336
Group income (IDR/month)	94,718,571	124,279,000	198,122,629
Number of farmers	25	22	25
Individual income (IDR/farmer/month)	3,788,742	4,971,160	7,924,905

[a]USD 1 = 14,769 IDR (Indonesian Rupiah), XE.com 4 September 2020
Source Farmer interviews

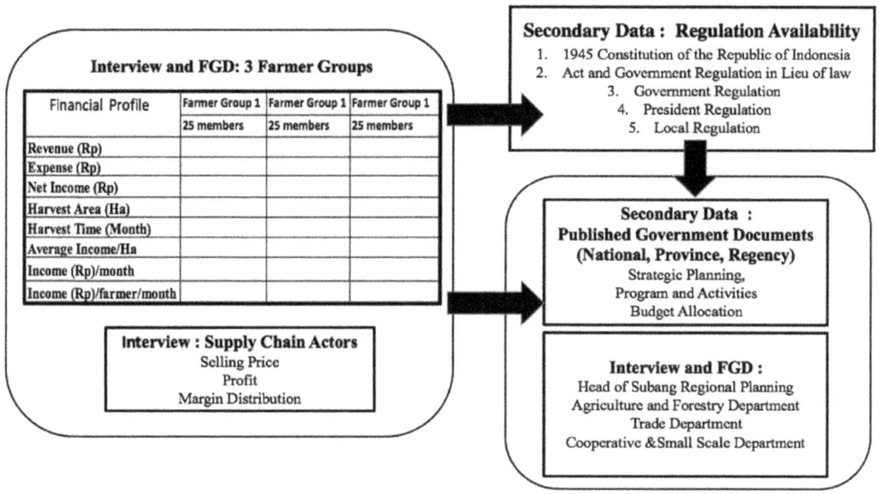

Fig. 3 Data collection process; Rp = Indonesian Rupiah (IDR)

4 Case Study Results

4.1 Supply Chain Analysis

Within the supply chain, the selling price of pineapple at the farmer level is the lowest. Farmers supply pineapple to farmer groups at a price of IDR 2000/kg (i.e., USD 0.14/kg). Farmer groups, as the party collecting pineapple from farmers, sell their pineapple to food processing industries and middlemen (Fig. 4).

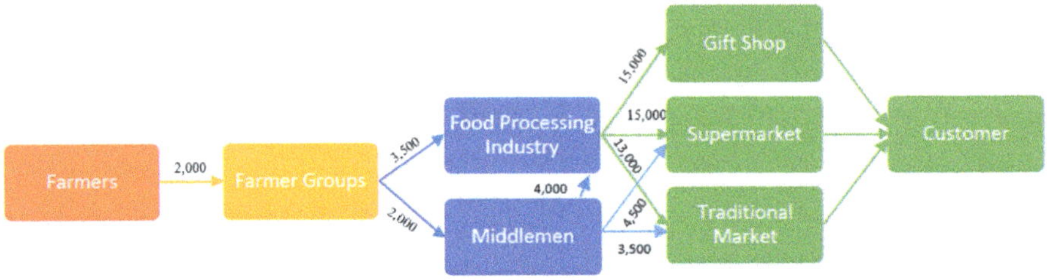

Fig. 4 Supply chain and price of pineapple in Jalan Cagak district, IDR, 2020

The quality that is usually required by food processing industries is pineapple of grade A or B with the average product selling price of IDR 3500/kg (i.e., USD 0.25/kg), while grade C or D goes to the middlemen for IDR 2000/kg (i.e., 0.14/kg). Farmer groups do not take any profit. The profits go directly to the farmers who supplied the pineapple. Middlemen will collect products from several farmer groups and sell them to their customers. Middlemen sell pineapple to the food processing industry for IDR 4000/kg (i.e., USD 0.28/kg), supermarkets for IDR 4500/kg (i.e., USD 0.32/kg), and traditional markets for IDR 3500/kg (i.e., USD 0.25/kg). Food processing industries will market its products to gift shops and supermarkets with an average price of processed foods IDR 15,000/pack (i.e., USD 1.07/pack) and traditional markets for IDR 13,000/pack (i.e., USD 0.93/pack). The selling price to traditional markets is smaller due to varying target segments which have lower purchasing power. Processed pineapple products produced by the food processing industry are snacks that can be consumed daily and as gifts. Farmer groups must know their financial position among the actors of the supply chain to find out whether their position in the chain is monetarily sound. Based on the book, Making Value Chains Work Better for the Poor: A Toolbook for Practitioners of Value Chain Analysis [33], there are several ways to calculate the financial position of actors in the supply chain. One of them is to calculate the value-added margin and profits as shown in Table 2. Figure 5 illustrates how sales to the food processing industry are much more profitable than to the middleman even though the demanded quality is higher. Alternative sales of medium quality fresh pineapple are made to middlemen, but they distributed at a lower price.

Table 3 elucidates the percentage of the overall value-added to the supply chain. The most significant percentage added cost occurred in the processors stage, i.e., 81.8%. This is due to a lot of additional costs to make a processed product such as additional raw material and operational costs such as salaries for workers, electricity, and information technology and communication costs (e.g., internet and telephone). The farmer, as a major supplier of pineapple, requires much cost during the planting period until the harvest. This makes the presentation added cost to farmers 17.9%. While the smallest added cost percentage is held by middlemen and the farmer group, which are 0.2% and 0.1%, respectively. This is due to the fact that not much cost is incurred in adding to the overall supply, i.e., the cost incurred includes the

Table 2 Formulae for calculating marketing margin

Supply chain actor	Costs			Revenues	Profits		Margins
	Unit total costs	Added unit cost	Added cost %	Unit price	Unit profit	Total profits %	Unit margin
Farmer	A	–	A/f	G	G − a	(g − a)/(k − f)	G
Assembler	G	B	B/f	H	H − b − g	(h − b − g)/(k − f)	H − g
Processor	H + c	C	C/f	I	I − c − h	(i − c − h)/(k − f)	I − h
Trader	I + d	D	D/f	J	J − d − i	(j − d − i)/(k − f)	J − i
Retailer	J + e	E	E/f	K	K − e − j	(k − e − j)/(k − f)	K − j
Total		F = a + b + c + d + e	100		K − f	100	K

Source Food and Agricultural Organization of the United Nations [33]

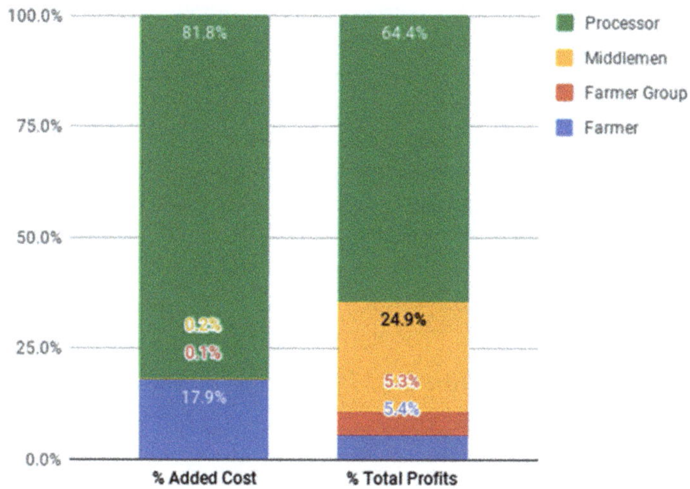

Fig. 5 Distribution of profit in the supply chain

cost of transportation of the product such as the purchase of gasoline and vehicle maintenance, telephone costs, and labor costs (e.g., drivers and pineapple sorters).

Regarding profit, the processors and retailers again get the highest percentage of the highest profit compared to the other four supply chain actors. With a total of 64.4%, this demonstrates that if the product is sold with value-added then it will become much more profitable. As a result, innovative management in the agricultural food chain is necessary to protect poor local farmers from acts of unfair marketing

Table 3 Supply chain margin

Value chain actor[a]	Costs			Revenues	Profits		Margins
	Unit total costs (IDR)	Added unit cost (IDR)	Added cost %	Unit price (IDR)	Unit profit (IDR)	Total profits %	Unit margin (IDR)
Farmer	1678	1678	17.9	2000	322	5.4	2000
Farmer group	1687	9	0.1	2000	313	5.3	322
Middlemen	2021	21	0.2	3500	1479	24.9	1500
Processor and retailer	11,174	7674	81.8	15,000	3826	64.4	11,500
Total		9383	100		5939	100	15,322

[a]USD 1 = 14,769 IDR (Indonesian Rupiah), XE.com 4 September 2020

practices [34, 35]. The second position is occupied by the middlemen with 24.9% of the cut. Middlemen do business with the principle of taking out as minimal and low as possible to getting as much as possible to make a return. While this is an ethical business principle, it can harm others, since the percentage given to farmer groups and farmers is the least profitable. Farmers become supply chain actors who suffer the most due to their high added costs. Evidently, farmers have been identified as the weakest party in the supply chain of pineapple. Institutional strengthening of farmers should be somewhat better supported by the government. As such, government policy is still based on pineapple production and lacks institutional elements and marketing expertise. Nonetheless, on the institutional side, it is necessary to strengthen farming-based associations and farmer groups' roles and participation to be independent and not dependent on overly exuberant government projects [36, 37]—allowing for the market economy to flourish.

4.2 Indonesian Regulation and Policies of Food Crops and Horticulture

In Indonesia, a total of 21 acts are linked with agriculture. These acts are associated with edible crops and horticulture, farmer empowerment from unhealthy competition, and land ownership sharing and co-cultivation. At the same time, there is only one government regulation in the lieu of law that discusses local government authority. There are 40 government regulations that assure the functioning of the farming business via the use of pesticide, fertilizer, seed, and protection of the crops and its variety—plus food security, its quality and nutrient, and production availability of area, land, and tools facilitation. There are 22 presidential regulations in Indonesia that mainly abide to governing institutions, organizations, middle-term development, and only one that is linked with farming, i.e., the application of subsidized fertilizer as a regulated item. Presidential decisions and instructions governing farming issues concern primarily with need for sugar and rice with explanatory rules for business, its monetary access, capital investment, and credit. Marketing of agricultural products and its price setting are not found in the constitution, acts, or regulation of government. While bilateral trade continues to grow, Indonesia has increasingly adopted trade-restrictive measures in an attempt to fulfill food self-sufficiency goals, i.e., in-line with the United Nations Sustainable Development Goals that call for consuming local food more sustainably [38–44]. As a result, restrictions on agricultural imports have already caused shortages of some commodities and have exacerbated already high prices such as corn and rice [44]. These findings support Indonesia's decentralization effort that states the authority of operational management is given more to the local government through the instruments of local governance regulation. The legal basis of this authority can be seen in Article 10 and 11 that describe local governments as the authority that should have operational control over agricultural product marketing while the national government is only to facilitate and coordinate national regulation applied at the regency level. Hence, the national government no longer regulates the trade organization and technical operation within regencies. As such, stable policy is important for sustained interest of the private sector in agriculture business and target reforms should move toward enhancing integration of regional agricultural markets, developing grades and standards for inputs and outputs, evolving food safety standards and regulations and their implementation mechanisms, warehousing laws, and regulating foreign investment in domestic markets [45].

Each province has their right to formulate their own regulation. However, the regulation and policy put in place still must abide by the regulations and policies already set at higher levels of government. Three major governmental ministries may affect the pineapple industry, they include: (1) Ministry of Food, Crop, and Horticulture, (2) Ministry of Trade and Industry, and (3) Ministry of Unions. The position of ministerial regulations is higher than provincial ones—since ministries act as an auxiliary body to the president who runs general policy guidelines that have been determined and are the scope of the national agenda, each ministry plays a vital role toward the pineapple industry. The Ministry of Food, Crop, and Horticulture focuses on controlling the process of the production sector in terms of the actual produce, the Ministry of Trade and Industry controls the processing and consumption of the outcomes of the agriculture nationwide, and the Ministry of Union handles and offers service for welfare of the stakeholder within the industry. Focusing on government regulation and policies in relation to pineapple productivity, the key impact from the Ministry of Food, Crop, and Horticulture's regulation and policy directly effects pineapple pricing and stability. Analyzed data from this ministry indicates impacts from are top-down—i.e., they abide by the national strategy and directly affect West Java's produce including pineapple. Specifically, four key top-down strategic issues affect and should be taken into account when considering the pineapple farmer groups in Jalan Cagak, including:

1. Law of the Republic of Indonesia No. 19 of 2013 concerning farmer protection and empowerment. This law regulates the obligations of the central and regional government to facilitate and encourage farmers to become agricultural insurance participants who can protect farmers from the loss of harvest due to natural disasters, attacks on organisms, effects of climate change, and other types of risks. Farmer protection and empowerment aim to realize the sovereignty and independence of farmers in order to improve the level of welfare, quality, and betterment of life of farmers, protect farmers from crop failures and price risks, and provide agricultural infrastructure and facilities needed to develop farming and develop financing institutions that are agriculture-friendly, serving the interests of the farmers.
2. Human resources are one of the keys to success in agricultural development. In order to improve the quality of human resources, one of the efforts made is through education and training. This refers to the Minister of Agriculture Regulation No. 75/Permentan/Ot. 140/12/2012, concerning guidelines for organizing education and training and horticultural human resource competency certification. The regulation explains that in order to obtain good horticultural human resources and meet the expected competency standards, education, and training must be held, and competency certification must be systematic and enforced. Education and training in Jalan Cagak district were carried out to improve the quality of the farmers by the Agriculture Department of Horticulture in the Subang regency along with the Agriculture, Fisheries, and Forestry Extension Agency of Jalan Cagak.
3. Regulation of the Minister of Agriculture No. 273/Kpts/OT.160/4/2007 concerning guidelines for farmer institutional development. The government combines farmer groups that already exist with farmer groups called Gapoktan. This merger is in order to develop farmer groups that can expand the reach of business, increase agricultural production, facilitate the marketing of agricultural products, increase business capital, and increase the welfare of its members. Several things should be used as material for evaluation of these policies, one of which is the weak access of farmer groups and an association of farmer groups in terms of banking access. Agricultural business needs to be managed professionally so that the business assistance process, especially the development of agricultural business, meet the needs for proper implementation. This is based on many farmer group-based businesses that have gone bankrupt due to a lack of capital and poor business management.
4. Regulation of the Minister of Trade and Industry of the Republic of Indonesia No. 24/M-DAG/PER/6/2008 concerning the provisions on the export of bananas and pineapple to Japan in the framework of the Indonesia Japan-Economic Partnership Agreement. With this collaboration, the opportunity to export pineapple to Japan is open. The demand for pineapple from Japan is very high, from a report written by FruiTrop [46] (i.e., a specialized magazine in world trade of fruits and vegetables), the average consumption of pineapple in 2010–2016 reached 163,975 tons per year. This can be used by pineapple farmer groups in Jalan Cagak to increase their sales to Japan. However, in conducting commodity

sales abroad, several standards are implemented to define the quality requirements for pineapple export. HS Code 080430 is a number code in classifying fresh pineapple and dried pineapple products. The use of the harmonized system code is to provide an official international system for the provision of codes, explanations, and classifications of goods for trading purposes.

Based on these four points, the regulations originating from the central government should be oriented toward benefitting pineapple farmers advancement as well as their businesses. Focusing on fresh food, their marketing channels, and supply chains have been described in different regions showing how farmers supply their fresh fruit and vegetables to consumers through various types of intermediaries. A number of policy gaps still exist to protect the interests of the farmers in maximizing their benefits, especially in the areas of participation between public and private sectors [45].

Benchmarking from Thailand, farmers still need national government policy intervention. Various programs have been implemented to improve the fresh fruit and vegetables market through five major strategies including price stabilization and subsidy schemes, production efficiency improvement, domestic market development, and export promotion. Supermarkets also play a role in connecting farmers to their markets through direct procurement from farmers and cultivation improvement at farms [45]. To enable smallholder farmers to capture the benefits of modern value chains, the governments should encourage producer organizations to act as aggregators of farm produce for marketing, and for sourcing and delivering inputs, information and financial and non-financial services to farmers. This may be achieved through government agencies or NGOs. Perhaps, Indonesia's central government could strategize from Thailand, India, China, and other countries regarding national intervention and develop similar policies that are more supportive to the welfare of farmers [45]. Unfortunately, from the government level of Subang regency, there are still no policies contained in regional regulations. Even though policies made by the provincial government still significantly impact the sustainability of the pineapple farmer groups in Jalan Cagak district—more should be done.

4.3 Strategic Planning and Program Development

Strategic planning is needed to make sure organizations are kept on track when doing business. Well-organized strategic planning will aid the pineapple industry's ability to contribute to the economic development in Indonesia [45]. Strategy planning provides the organization road map. It supplies a plan for systematic operational decision-making as well as assists in developing competitive advantages and competencies stakeholder-wide [47]. The Ministry of Agriculture states Indonesia's strategic planning will require self-reliance, proper and sound development, and a prosperous agricultural community if increasing food security and agricultural competitiveness is to be achieved [48].

4.3.1 National Government

Changes in action or political policy in a country can have a significant impact on the business sector of that country. Hence, if the political situation is supportive, then business, in general, will run smoothly [49–51]. The government system in Indonesia is involved or intervenes in business. This can be seen in the law and policies issued by the government to support business [52, 53]. During the reign of Jokowi-Jusuf Kalla, there was a draft of nine priority agendas called the Nawa Cita program, with one of its focal points relating to agriculture with the government promising to accelerate the release of 9 million ha of land for farmers, i.e., as written in the 2015 National Medium-term Development Plan (RPJMN). The division of land would be given per farmer group, not per individual. However, due to limited land in Java, the area of land distribution was limited—unlike outside Java where the distribution of land reaches more than 50 ha per farmer group. To follow up on RPJMN's agriculture initiatives, the provincial government helped create the 2013–2018 Regional Medium-term Development Plan (RPJMD). An amended strategy specific to agriculture to maintain and replace the area of paddy fields that change the function of land from agriculture to non-agriculture, with the direction of the policy of printing new rice fields to achieve sustainable agricultural land. Presently, RPJMD has been renewed and extended for an additional five-year plan.

Strategic planning of the regulation and policy for the period of 2015–2019 was a continuation and advancement from the last period of realization. To achieve the desired target that may have come up in the next five-year plan, an evaluation of previous agricultural development, Indonesia-wide, was performed, including: conversion of estate crops, limited land for new crops, land degradation concerns, limited land owned by farmers, unidentified land ownership, damaged irrigation systems, availability of high-quality seed and plantation fertilizer, lack of agricultural tools and machinery, lack of organizational skill and human resources, ineffective and directed government regulations, and funding problems for farmers. The Ministry of Agriculture [7] has set seven main strategies as solving for these problems—sector-wide. The activities needed derived from the strategy are illustrated in Table 4.

4.3.2 West Java Provincial Government

The provincial government strategic plan for agriculture published by West Java's Department of Food, Crop, and Horticulture and synchronized with RPJMD as well as the Strategic Plan of Cities and Districts of All West Java, laid out support for their 2029 agenda, namely the West Java Provincial Strategic Living Environment Study. The study identified issues that affect West Java's food, crop, and horticulture service needed in developing a vibrant agricultural sector via a number of planned five-year blocks. In addition, a number of interviews, with highly ranked officials in the department, point out that farmers' key concerns remain bountiful, including: land conversion; low quality of human resources in the agriculture business sector; much of the

Table 4 National strategic plan for food, crops and horticulture

Strategy	Activity
Increase the availability and use of land	Land audit and update the database
	Effectively implement Indonesia's constitution
	Undertake land protection efforts
	Optimize the using of available land
	Help farmers with the administration of their land
	Increase the quality of the land
	Optimize the existing water resources
Agriculture facilities and infrastructure upgrading	Create and fix the needed infrastructure
	Coordination with related ministries to implement the strategy
	Allocate grants for agriculture facilities
	Strengthening the organizational system
	Strengthening the role of farmer groups
Develop the logistic system of seedling and seeds	Rearrangement of the organizational system of seedling and seeds
	Protect and use national genetic resources
	Strengthening the supervisory roles of the seedling and seed system
	Empowering local seedling and seed producers
	Increase private parties and their role to develop the industry
	Create the seedling and seed industry
	Coordination with seedling and seed importers
	Provide plantation resources
Strengthen farmer groups	Improve the quality and quantity of farmer groups
	Provide technical guidance and assistance for farmers
	Extend group type
	Strengthen the funding system for farmer groups
Develop and strengthen the funding system	Fix credit program to help farmers
	Grow and develop channeling agents for formal financial institutions
	Increase the function of agriculture instruction as well as instructors
	Develop cooperation patterns between farmers and local entrepreneurs
	Grow agribusiness micro financial institutions
	Provide funding for farmers
	Support local investment
	Support the development of agriculture-based banks

(continued)

Table 4 (continued)

Strategy	Activity
Develop and strengthen bioenergy and bio-industry	Arrange the development map of bioenergy and bio-industry
	Strengthen the production commodity supply
	Develop simple processing industry
	Support industry to apply zero waste management
	Support the development of domestic advanced processing
	Support investment on agriculture production result
Strengthen the market channel for agriculture products	Arrange the marketing channel map for strategic commodities
	Strengthen the market service information system
	Facilitate the warehouse receipt system
	Support agriculture product export market
	Conduct the positive campaign for agriculture distinct commodities
	Strengthen the international trading negotiation for agricultural commodities
	Accompaniment to the implementation of a standard quality
	Open new target market(s) outside existing market

plantation commodities already being too old and damaged with limited resources, budget, infrastructure, and irrigation; agricultural commodity losses; using inefficient traditional processing and lack of technology; final products being underdeveloped and not marketed or promoted well; role of farmer groups as not optimal; food, crops, and horticulture diseases; commodity production, productivity, and quality as still low; high dependence of farmers to government grants; and eco-friendly technology usage as limited. Based on these identified problems, the department defines several important programs to assist with the sector's transformation, including: increasing agricultural production (i.e., food, crop, and horticulture), activating and encouraging the agricultural resource empowerment program, implementing the prevention programs for plant disease, augmenting related processing and marketing programs, and starting a renewed reduction of food, crop, and horticulture losses program. These programs in concert are an excellent avenue to bettering the agriculture sector West Java-wide.

4.3.3 Subang Regency Government

The strategic plan published by Subang's Department of Food Crop offers a descriptive outline of its middle-term development plan consist with the regency's vision and mission statement, its policy direction, and strategy in putting together the Subang Development Plan. The plan includes three key points that are worth mentioning: (1)

government regulation No. 8/2008 puts together stages to facilitate tutorials, restructuring, monitoring, and evaluating its execution; (2) home secretary regulation, No. 54/2010 focalizes on the execution of government regulation No. 8/2008; and (3) other legislation that governs and mandates the arrangement of the plan to engage with the local task force for five-year blocks and act as an instrument for planning and measuring work performance of Subang's Department of Food Crops. Key outcome and goals that will need to be followed, include: production and productivity of horticulture and food crops, quality, and continuity of food crops and horticulture product in terms of global competition, prevention and action against infestation of pests, quality and quantity in terms of infrastructure (i.e., facilities, space, and watering systems), application of technology (i.e., environmentally and economically oriented), capital access for farmers, human resource capabilities in farming, and farming organization and structure. Program initiatives will need to include: increasing production, productivity, and quality of agriculture products, provision and development of agricultural infrastructure and facilities, extension of education and agricultural training, strengthening farmer groups, reducing agricultural commodities losses, increasing farmers selling margin, marketing Subang's rice, marketing Subang's distinctive agricultural commodities, supporting new product development and value-added processed food, and preventing plant disease. These programs will greatly benefit and should be considered essential to developing a vibrant agricultural sector Subang-wide.

5 Budgetary Allocation of Indonesia's Agriculture Sector

A significant amount of funding is needed in running agricultural development. The success of the executive in facilitating agricultural development and management processes needs to start from the Indonesian state budget then local, i.e., provincial and regency, budgets. Funding needed to run the coordination, supervision, scouting process, and evaluation of all planned programs and activities. However, the source of the funds should not be solely from a top-down approach, but should be supported from other funding sources such as private and state-owned enterprises, regionally owned enterprises, foreign and domestic direct investment, banks (e.g., skin credit and commercial credit), and from the citizenry directly involved—including farmers. With the use of the incremental capital output ratio (ICOR) [54–56], it can be applied to reach the specific investment needed for agriculture development to be fulfilled for the period 2021–2026. Utilizing Marissa et al.'s [54] ICOR approach for South Sumatera and Jambi province, a nationwide estimate would be approximately IDR 500 trillion (USD 35 billion) annually. Most of the fund (i.e., 85–90%) would need to be from the private sector and banks, since historically the government would only be able to facilitate about 10–15% of the needed amount. As such, about 30% of the annual state budget has gone to directly developing the agriculture sector, while the other 70% has been in support of agricultural program and activity implementation. A more prominent role in agriculture development is needed from the authority at the provincial and regency level. For example, in 2017 Subang's Department of Food Crops received two tranches of IDR 870 million (USD 62,000) and IDR 130 million (USD 9200) from the state budget which it stated was not sufficient. Fieldwork research disclosed evidence that financial assistance from the national government can be partly sporadic since it currently focuses on food security (i.e., mostly rice) and the allocation of funds distributed to other food commodities is less prioritized.

In terms of Subang's case, the budgetary programs from the national, provincial, and regency governments have been very supportive of the pineapple business in Jalan Cagak district. Assistance in the form of additional land is still not been fully realized as pineapple farmers district-wide usually acquire land via inheritance, buy it by themselves, or rent vacant land. Top-down budgets are managed by the Department of Agriculture in which distribution is usually in the form of fertilizer and seed subsidies as well as the provision of training to increase farmer knowledge and skills in agricultural science. Seventeen programs currently being looked at are illustrated in Table 5. Budget for the food crops and horticulture marketing programs to the domestic and foreign market are minimal, i.e., only 4% of the total annual budget, while in Subang regency its even lower, i.e., 1%. There is no domestic market structure arrangement effort visible, relative to price establishment policy and agriculture product distribution and storage facilitation at the regency level. As a result, in this case, the decentralization principle is conducted and the head of Subang regency together with its house representatives have the authority to make the local regulations. Hence, Subang regency, with its minimal budget, still—to the best of its ability—develops programs that are needed to increase the prosperity of farmers by increasing commodity added value and facilitating the commerce of its district's agriculture commodities.

6 Conclusion

Indonesian legislation in commerce and agriculture commodities, and especially in the food crops and horticulture sectors, is nonexistent. The only evidence of it can be found in its national strategic planning on agriculture commerce and how commerce functions, i.e., its executed via marketing management, buying, assembling, selling, distributing, transportation, storage, standardizing, financing, risk-taking, and market information. Moreover, legislation within the scope of the Ministry of Trade and Industry has almost no regulations for regulated the trade systemization of agricultural commodities within Indonesia other than refined sugar (i.e., Law 40/M-DAG/PER/6/2017). There are only eight pieces of the trade law concerning prices, all of which are regulations aimed at protecting consumers, namely the price policy of rice (i.e., Law 7/M-DAG/PER/8/2017), farmer's selling and consumer's buying reference price (i.e., Law 27/M-DAG/PER/5/2017), and maximum export prices for tariff-affected commodities such as agricultural commodities, forestry, and mining—subject to export duty. It would seem that participation as a member of the World

Table 5 Budget allocation for food, crops, and horticulture at national, West Java provincial, and Subang regency level

No	Program	Budget allocation (%)		
		National	West Java province	Subang regency
1	Management support and technical duties	5.2		
2	Increasing people accountability in the Ministry of Agriculture	0.35		
3	Increasing productivity, quality of agriculture products	27.8	89	96
4	Increasing productivity of eco-friendly horticulture	5.25		
5	Increasing productivity of sustainable plantation crops	11.37		
6	Food fulfillment program of livestock origin and agribusiness	4.73		
7	Value-added programs, quality, marketing, and investment	4	1	<1
8	Development of agricultural infrastructure and facilities	25.52		3
9	Technological innovation of sustainable bio-industry	7.92		
10	Extension education and agricultural training programs	8.08	1.60	
11	Strengthening farmer associations			<1
12	Reducing agricultural commodity losses		6.20	<1
13	Increasing farmers selling margin			<1
14	Marketing Subang's rice			<1
15	Marketing distinctive agricultural commodities			<1
16	Supporting product development and value processed food			<1
17	Preventing programs for plant disease		2.2	

Trade Organization has confined Indonesia to paying a high price for membership. As such, Indonesian farmers still need facilitation from the national level to better distribute their produce internationally, assist with domestic prize stabilization, aid growth stimulus, advance economic development, enforce legal protection, and increase prosperity via strategic safeguards.

Based on the Jalan Cagak district case research, farmer groups should create value-added in order to increase the selling price of their pineapple. Farmers have become supply chain actors who suffer due to the high added cost which offers a very low profit margin. The bargaining power of farmers is still the main trigger point in which sales to the food processing industry is far more profitable than sales to the middleman. As a result, farmer groups will benefit more if they sell directly to the food processing industry. In the long run, it would be most profitable if farmer groups created a value-added system by which they processed their own crops, but this is unrealistic at this time. This lack of financial gain has left farmers in a very weak position—establishing an unfair playing field between farmers and other agricultural stakeholders as well as between farmers from other provinces. The system of decentralization in Indonesia has supported autonomous-based competition—something that continues to plague farmers who amount to 40% of the population's productive age. Concepts of creating policies to reduce the effect of economic liberalization should move to improve farmers' economic growth and establish a synchronicity between the national-provincial-regency levels for best practices in the trading of the agricultural commodities domestic and international market.

References

1. Ministry of Environment (2018) Keaneka Ragaman Hayati. Republic of Indonesia, Jakarta
2. Lathifah SN (2013) Pembuatan Nata De Phina dari Limbah Bonggol Nanas Menggunakan Sumber Nitrogen Ekstrak Kacang Hijau, Bandung (ID). Universitas Pendidikan Indonesia, Jakarta, Indonesia
3. WorldAtlas (2020) Top pineapple producing countries. In: WorldAtlas. https://www.worldatlas.com/articles/top-pineapple-producing-countries.html. Accessed 23 Mar 2021
4. FreshPlaza (2020) Global fresh produce and banana news. In: FreshPlaza News. https://www.freshplaza.com/article/184712/OVERVIEW-GLOBAL-PINEAPPLE-MARKET/. Accessed 23 Mar 2021
5. TFNet (2016) Pineapple—common varieties. In: TFNet—international tropical fruits network. https://www.itfnet.org/v1/2016/05/pineapple-common-varieties/. Accessed 23 Mar 2021
6. AgroIndonesia (2018) Produksi dan Ekspor Nanas Terus Meningkat. http://agroindonesia.co.id/2018/12/produksi-dan-ekspor-nanas-terus-meningkat/. Accessed 23 Mar 2021
7. Ministry of Agriculture (2018) Outlook: Komoditas Pertanian Sub Holtikutura Pineapple. Pusat Data dan Sistem Informasi Pertanian Sekretariat Jenderal Kemen-terian Pertanian. Jakarta
8. Santoso HB (1998) Sari Buah Nanas. Kanisius, Yogyakarta
9. Redaksi AgroMedia Pustaka (2009) Buku Pintar: Budi Daya Tanaman Buah Unggul Indonesia. Agro Media, Jakarta
10. Nur S (2013) Decentralization and development in public policy implementation perspective: case study in Indonesia. IOSR J Humanit Soc Sci 17:27–33
11. Rondinelli DA, McCullough JS, Johnson RW (1989) Analysing decentralization policies in developing Countries: a political-economy framework. Dev Change 20:57–87. https://doi.org/10.1111/j.1467-7660.1989.tb00340.x
12. Poniatowicz M (2015) Determinanty autonomii dochodowej samorządu terytorialnego w Polsce. Nauk o Finans 1:11–30. https://doi.org/10.15611/nof.2015.1.01
13. Rondinelli DA (1981) Government decentralization in comparative perspective. Int Rev Adm Sci 47:133–145. https://doi.org/10.1177/002085238004700205
14. Kifli FW, Umami A (2017) Contribution of economic sector and agricultural development to support food security in North Sumatra Province. J Comput Theor Nanosci 23:6428–6431. https://doi.org/10.1166/asl.2017.9643

15. Weingast BR (2014) Second generation fiscal federalism: political aspects of decentralization and economic development. World Dev 53:14–25. https://doi.org/10.1016/j.worlddev.2013.01.003
16. Enikolopov R, Zhuravskaya E (2007) Decentralization and political institutions. J Public Econ 91:2261–2290. https://doi.org/10.1016/j.jpubeco.2007.02.006
17. Filippetti A, Sacchi A (2016) Decentralization and economic growth reconsidered: the role of regional authority. Environ Plan C Gov Policy 34:1793–1824. https://doi.org/10.1177/0263774X16642230
18. Chen Y, Pereira I, Patel PC (2020) Decentralized governance of digital platforms. J Manage XX No. X:014920632091675. https://doi.org/10.1177/0149206320916755
19. Schneider A (2003) Decentralization: conceptualization and measurement. Stud Comp Int Dev 38:32–56. https://doi.org/10.1007/BF02686198
20. Dick-Sagoe C (2020) Decentralization for improving the provision of public services in developing countries: a critical review. Cogent Econ Financ 8:1804036. https://doi.org/10.1080/23322039.2020.1804036
21. Schneider N (2019) Decentralization: an incomplete ambition. J Cult Econ 12:265–285. https://doi.org/10.1080/17530350.2019.1589553
22. Frug J (1993) Decentering decentralization. Univ Chicago Law Rev 60:253. https://doi.org/10.2307/1600075
23. Budiyanto MN (2006) Reformasi Administrasi Pemerintahan Lokal dalam Pelayanan Publik di Indonesia. J Demokr 5:1–22
24. Supriyadi A, Wang T, Chu S et al (2018) Urban area and population change in a new autonomous City in Indonesia: Evidence Banjar Municipality. In: Cirella GT (ed) Conference proceeding of the 2nd international conference on sustainability, human geography and environment, Krakow, Poland, 28 Nov–2 Dec 2018. PCS Publishing, Co., Gdansk University Press, Sopot, Poland, pp 87–97
25. World Bank (2001) What, why, and where. In: World Bank. http://www1.worldbank.org/publicsector/decentralization/what.htm. Accessed 30 Oct 2019
26. Mills ES (1968) Studies in the structure of the urban economy. Resources for the Future, Washington
27. Miller MA (2013) Decentralizing Indonesian City spaces as new "Centers." Int J Urban Reg Res 37:834–848. https://doi.org/10.1111/j.1468-2427.2013.01209.x
28. Hadiz VR, Robison R (2013) The political economy of oligarchy and the reorganization of power in Indonesia. Indonesia 2013:35–57. https://doi.org/10.5728/indonesia.96.0033
29. Mariana D, Husin LH (2017) Democracy, local election, and political dynasty in Indonesian politics. J Wacana Polit 2:88–97. https://doi.org/10.24198/jwp.v2i2.13998
30. Ostwald K, Tajima Y, Samphantharak K (2016) Indonesia's decentralization experiment: motivations, successes, and unintended consequences. Southeast Asian Econ 33:139–156. https://doi.org/10.1355/ae33-2b
31. World Bank (2019) World Bank Open Data. In: World Bank. https://data.worldbank.org/. Accessed 27 Jul 2019
32. World Bank (2020) Data catalog. https://datacatalog.worldbank.org/. Accessed 30 Dec 2020
33. FAO (2008) Making value chains work better for the poor: a toolbook for practitioners of value chain analysis. Food and Agricultural Organization of the United Nations, Rome
34. Caiazza R, Volpe T, Audretsch D (2014) Innovation in agro-food chain: policies, actors and activities. J Enterprising Communities 8:180–187. https://doi.org/10.1108/JEC-06-2014-0009
35. Oliveira PM, Natário MM (2016) Territorial innovation systems and strategies of collective efficiency: the case of Tagus Valley agro-food complex. Eur J Innov Manag 19:362–382. https://doi.org/10.1108/EJIM-07-2014-0072
36. Singagerda F, Marantika A (2016) Indonesia pineapple supply chain management and the improvement of global market and competitiveness. In: Fifth international conference on entrepreneurship and business management. Tarumanagara University, Tainan

37. FAO (2017) Country fact sheet on food and agriculture policy trends—Indonesia. In: Food Agriculture Organanization. United Nations. moz-extension://b25df011-8780-4be7-9be7-6714e4992816/enhanced-reader.html?openApp&pdf, http://www.fao.org/3/a-i7696e.pdf. Accessed 23 Mar 2021
38. Karp DG (1996) Values and their effect on pro-environmental behavior. Environ Behav 28:111–133. https://doi.org/10.1177/0013916596281006
39. Kangun N, Carlson L, Grove SJ (1991) Environmental advertising claims: a preliminary investigation. J Public Policy Mark 10:47–58. https://doi.org/10.1177/074391569101000203
40. Arquitt SP, Cornwell TB (2007) Micro-macro linking using system dynamics modeling: An examination of eco-labeling effects for farmed shrimp. J Macromarketing, pp 243–255
41. Cornwell TB, Drennan J (2004) Cross-cultural consumer/consumption research: dealing with issues emerging from globalization and fragmentation. J Macromarketing 24:108–121. https://doi.org/10.1177/0276146704269301
42. Do Valle PO, Reis E, Menezes J, Rebelo E (2004) Behavioral determinants of household recycling participation: The Portuguese case. Environ Behav 36:505–540. https://doi.org/10.1177/0013916503260892
43. Griskevicius V, Cantú SM, Van Vugt M (2012) The evolutionary bases for sustainable behavior: implications for marketing, policy, and social entrepreneurship. J Public Policy Mark 31:115–128
44. Testa F, Russo MV, Cornwell TB et al (2018) Social sustainability as buying local: effects of soft policy, meso-level actors, and social influences on purchase intentions. J Public Policy Mark 37:152–166. https://doi.org/10.1509/jppm.16.215
45. Chen KZ, Joshi PK, Birthal PS, Cheng E (2015) Innovations in financing of agri-food value chains in China and India Lessons and policies for inclusive financing. China Agric Econ Rev 7:616–640. https://doi.org/10.1108/CAER-02-2015-0016
46. FruiTrop (2017) World fresh pineapple market. In: FriuTrop Online. https://www.fruitrop.com/en/Articles-by-subject/Economic-analyses/2016/World-fresh-pineapple-market. Accessed 16 Feb 2021
47. Kaereem MA (2021) Strategic planning for agricultural development . National Institute of Agricultural Extension Management
48. Tahlim S (2021) Strategic plan of the Indonesian Ministry of Agriculture 2020–2024. In: FFTC agricultutal policy platform https://ap.fftc.org.tw/article/1842. Accessed 23 Mar 2021
49. Gemeda BS, Abebe BG, Paczoski A et al (2019) What motivates speculators to speculate? Entropy 22:59. https://doi.org/10.3390/e22010059
50. Gallagher KP (1997) Economic globalisation and the environment. Organisation for Economic Cooperation and Development, Paris
51. Henderson J, Dicken P, Hess M et al (2002) Global production networks and the analysis of economic development. Rev Int Polit Econ 9:436–464. https://doi.org/10.1080/09692290210150842
52. Nurhemi SG (2015) Regional financial autonomy. Bull Monet Econ Bank 18:171–192
53. Khusaini M (2015) A shift-share analysis on regional competitiveness—a case of Banyuwangi District, East Java, Indonesia. Procedia Soc Behav Sci 211:738–744. https://doi.org/10.1016/j.sbspro.2015.11.097
54. Marissa F, Yulianita A, Fitriyah A (2019) The Effect of Investment Efficiency toward Economic Growth in South Sumatera and Jambi Province. J Ekon Pembang 17:71–80. https://doi.org/10.29259/jep.v17i2.9399
55. Taguchi H, Limpanonda Lowhachai S, Taguchi H, Limpanonda Lowhachai S (2014) A revisit to the incremental capital-output ratio: the case of Asian economies and Thailand. Int J Econ Policy Emerg Econ 7:35–54
56. Vanek J, Studenmund AH (1968) Towards a better understanding of the incremental capital-output ratio. Q J Econ 82:452–464. https://doi.org/10.2307/1879517

Policy Development and Future Scenarios

Environmental Safety in the Sustainable Development Goals: Public Survey

Andrea Farkas, Janos Mika, and Giuseppe T. Cirella

Abstract This chapter looks at three aspects of environmental safety: energy preference, air pollution, and climate change. It is divided into two parts: (1) global outlook and (2) case research. First, the United Nations Sustainable Development Goals, established in 2015 with a fifteen-year objective of implementing them, are broken down and analyzed. Direct and indirect relations are established. Second, empirical case research from Hungary is grouped and stratified sampling is reported upon. Results indicate that renewable energy sources were shown to receive the highest score in terms of energy, while nuclear and fossil fuel power scored the lowest. The worst type of air pollution was emitted by enterprises and everyday CO_2 emission in combination with communal waste. In terms of Hungary's climate change impacts, water-related and agricultural aspects were ranked the highest. Additional energy concerns showed cooling systems as highly consumptive. In terms of environmental safety, tests indicated women were more sensitive to the environment than men. Moreover, people with higher education levels evaluated the risks lower than those who just finished technical college. Evaluation of air pollution monotonically increased with settlement size.

Keywords Sustainability policy · Public opinion · Energy preference · Air pollution · Climate change · Hungary

---

A. Farkas
National University of Public Service, Budapest, Hungary
e-mail: andrea@andreafarkas.hu

J. Mika (✉)
Eszterhazy Karoly University, Eger, Hungary
e-mail: mika.janos@uni-eszterhazy.hu

G. T. Cirella
Faculty of Economics, University of Gdansk, Sopot, Poland
e-mail: gt.cirella@ug.edu.pl

G. T. Cirella (ed.), *Human Settlements*, Advances in 21st Century Human Settlements,
https://doi.org/10.1007/978-981-16-4031-5_10

1 Global Outlook: Energy, Air Pollution, and Climate Change in the Sustainable Development Goals

The United Nations accepted the 2030 Agenda for Sustainable Development from 2016–2030, made up of seventeen Sustainable Development Goals (SDGs) [1] with 169 detailed associated targets. SDGs broadly span topics on environmental, social, and economic aspects of sustainability. The SDGs document is utilized in terms of its global outlook on energy, air pollution, and climate change [1]. The 126 targets contain quantitative objectives, mostly related to 2030. Moreover, a minority of the targets point at organizational needs as preconditions of the objective targets, i.e., 43 of them. As a rule, the quantitative targets are marked by two numbers (e.g., 1.1), while the nonquantitative ones by one number and one letter (e.g., 1.a). The seventeen established SDGs are not ordered in any logical format. The coloring of the logos also do not establish any clear classification. The trial by Mika and Farkas [2] attempted to appropriately group SDGs for a better understanding and breakdown (Table 1). It should be noted that SDG 13, i.e., climate action, is the only environmental problem looked at as a separate goal [1]. All other issues, e.g., the reduction of biodiversity and ozone depletion, are considered in other goals as an aftereffect or consequence in terms of vulnerability to the environment or human health. As a result, it could be considered as an individual group with one single goal, but it is recommended that the goal be included in the efficient and sustainable economy group. Another remark is that this goal refers to the Paris Agreement [3], which deals with several aspects of climate change not specifically mentioned in the goal per se.

SDGs reflect the important 5Ps of humanity: people, planet, prosperity, peace, and partnership [1]. These concepts, however, are not well-represented in the document. The first two groups of the classification, basic human needs and equality and justice, deal, in most part, with people. The following two groups, efficient and sustainable economy and protecting vulnerable landscapes, fit in with prosperity and planet, while peace and partnership relate to the smallest group, cooperation toward common goals.

An in-depth look at how the SDGs' 169 targets relate to the primary aspects of environmental safety is analyzed and denoted by way of direct and indirect causation. Utilizing this premise, a focalized examination into energy, air pollution, and climate change indicated that several SDGs are significantly related. From Tables 2, 3, 4 and 5, the use of the symbol " + " denotes a direct relatedness to one or more aspects of environmental safety, whereas "(+)" symbolizes those indirectly related. In order to distinguish between direct and indirect relatedness, not only was the official text of targets used but also an in-depth use of professional sources, e.g., all improvements related to weather extremes indicate an indirect relation since they occur independent from the climate changing. Moreover, in many cases, energy-relatedness or consequences in air pollution load is not mentioned, but the relatedness is reported. This is most likely the case for the long-term protection of various landscapes where climate change influences the process. Table 2 displays the targets

Table 1 Grouping of the seventeen SDGs

Group	Goal
Basic human needs	SDG 2. End hunger,achieve food security andimproved nutrition, and promotesustainable agriculture SDG 3. Ensure healthy lives and promote well-being for all at all ages SDG 6. Ensure availability andsustainable managementof water and sanitation for all SDG 7. Ensure access to affordable, reliable, sustainable, andmodern energyfor all
Equality and justice	SDG 1.End povertyin all its forms everywhere SDG 4. Ensure inclusive and equitablequality education and promote lifelong learning opportunities for all SDG 5. Achieve gender equality and empower all women and girls intheir social role SDG 10. Reduce in equality within and among countries
Efficient andsustainable economy	SDG 8. Promote sustained, inclusive, andsustainable economic growth, full and productive employment, and decent work for all SDG 9. Build resilient infrastructure, promote inclusive andsustainable industrialization, andfoster innovation SDG 12. Ensure sustainable consumptionand production patterns [a]SDG 13. Take urgent action tocombat climate change andits impacts
Protecting vulnerable landscapes	SDG 11. Make citiesand human settlements inclusive, safe, resilient, and sustainable SDG 14. Conserve and sustainably use the oceans, seas, and marine resources for sustainable development SDG 15. Protect, restore, and promote sustainable use ofter restrial ecosystems, sustainably manage forests,combat desertification, and halt and reverse land degradation and halt biodiversity loss
Cooperation toward common goals	SDG 16. Promote peaceful and inclusive societies for sustainable development, provide access to justice for all, and build effective, accountable, and inclusive institutions at all levels SDG 17. Strengthen the means of implementation and revitalizethe global partnership for sustainable development

[a]United NationsFramework Conventionon Climate Changeis the primary international, intergovernmental forum for negotiatingthe globalresponse to climate change [4]

related to the aspects of energy, air pollution, and climate change in the basic human needs group. Four SDGs are represented by six targets relating to one or more of the three aspects. The total number of related aspects is eleven, from which five targets are associated with climate change. Only three direct relationships were established, with eight indirect ones.

Table 2 SDGs targeting energy, air pollution, and climate change in terms of basic human needs

Basic human needs	Energy	Air pollution	Climate change
2.4 By 2030, ensure sustainable food production systems and implement resilient agricultural practices that increase productivity and production, that help maintain ecosystems, that strengthen capacity for adaptation to climate change, extreme weather, drought, flooding, and other disasters, and that progressively improve land and soil quality			(+)
3.9 By 2030, substantially reduce the number of deaths and illnesses from hazardous chemicals and air, water, and soil pollution and contamination	(+)	+	
6.4 By 2030, substantially increase water-use efficiency across all sectors and ensure sustainable withdrawals and supply of freshwater to address water scarcity and substantially reduce the number of people suffering from water scarcity			(+)
6.6 By 2020, protect and restore water-related ecosystems, including mountains, forests, wetlands, rivers, aquifers, and lakes			(+)
7.2 By 2030, increase substantially the share of renewable energy in the global energy mix	+	(+)	(+)
7.3 By 2030, double the global rate of improvement in energy efficiency	+	(+)	(+)
Targets with direct and indirect relatedness	3	3	5

Table 3 illustrates the targets related to energy, air pollution, and climate change in two groups of SDGs, i.e., the equality and justice and the cooperation toward common goals groups. Altogether six SDGs are represented by five targets related to one or more of the three aspects. The total number of related aspects is ten, of which five targets are associated with climate change and four to air pollution. All established relationships are found to be indirect. This is understandable considering the somewhat abstract social character of the two groups.

Table 4 displays the targets related to energy, air pollution, and climate change in terms of the efficient and sustainable economy group. The four SDGs are represented by eight targets. The total number of related aspects is 20, of which eight targets are related to energy, seven to climate change, and five to air pollution. As a result, all the selected targets are related to energy and climate change, except one. Only six direct relations were established, equally distributed among the three related environmental aspects, together with 14 indirect ties to energy, air pollution, and climate change.

Table 5 shows the targets related to the three aspects in terms of the protecting vulnerable landscapes group. The three SDGs are represented by ten targets. The total number of related aspects is 15, of which nine targets are related to climate change. Only three direct relations were established together with 12 indirect ones.

Table 3 SDGs targeting energy, air pollution, and climate change in terms of equality and justice as well as cooperation toward common goals

Equality and justice, and cooperation toward common goals	Energy	Air pollution	Climate change
1.5 By 2030, build the resilience of the poor and those in vulnerable situations and reduce their exposure and vulnerability to climate-related extreme events and other economic, social, and environmental shocks and disasters		(+)	(+)
4.7 By 2030, ensure that all learners acquire the knowledge and skills needed to promote sustainable development, including, among others, through education for sustainable development and sustainable lifestyles, human rights, gender equality, promotion of a culture of peace, and non-violence, global citizenship, and appreciation of cultural diversity and of culture's contribution to sustainable development	(+)	(+)	(+)
10.7 Facilitate orderly, safe, regular, and responsible migration and mobility of people, including through the implementation of planned and well-managed migration policies			(+)
16.10 Ensure public access to information and protect fundamental freedoms in accordance with national legislation and international agreements		(+)	(+)
17.7 Promote the development, transfer, dissemination, and diffusion of environmentally sound technologies to developing countries on favorable terms, including on concessional and preferential terms, as mutually agreed		(+)	(+)
Targets with direct and indirect relatedness	1	4	5

None of the direct targets in terms of protecting vulnerable landscapes are related to climate change.

A summary of the statistics from Tables 2, 3, 4 and 5 illustrates a global outlook on environmental safety. As a result, there are 29 targets from the 169, which are related to energy, air pollution, and climate change. Climate change is related to 26 targets, whereas energy and air pollution are related to sixteen and fourteen targets, respectively. The number of direct relationships to energy are counted in six targets. The number for air pollution and climate change is four and two targets, respectively. Indirect relationships for climate change are established in 24 targets. Correspondingly, indirect relatedness of energy and air pollution could be established equally in ten targets. We utilize the assessed direct and indirect SDG relationships by looking at public survey data from Hungary specific to environmental safety. SDG relatedness is used to better aid the empirical findings and piece together causation.

Table 4 SDGs targeting energy, air pollution, and climate change in terms of an efficient and sustainable economy

Efficient and sustainable economy	Energy	Air pollution	Climate change
9.2 Promote inclusive and sustainable industrialization and, by 2030, significantly raise industry's share of employment and gross domestic product, in line with national circumstances, and double its share in the least developed countries	+	(+)	(+)
9.4 By 2030, upgrade infrastructure and retrofit industries to make them sustainable, with increased resource-use efficiency and greater adoption of clean and environmentally sound technologies and industrial processes, with all countries taking action in accordance with their respective capabilities	(+)	+	(+)
12.2 By 2030, achieve the sustainable management and efficient use of natural resources	(+)		(+)
12.4 By 2020, achieve the environmentally sound management of chemicals and all wastes throughout their life cycle, in accordance with agreed international frameworks, and significantly reduce their release to air, water, and soil in order to minimize their adverse impacts on human health and the environment	(+)	+	
12.5 By 2030, substantially reduce waste generation through prevention, reduction, recycling, and reuse	(+)	(+)	(+)
13.1 Strengthen resilience and adaptive capacity to climate-related hazards and natural disasters in all countries	(+)		(+)
13.2 Integrate climate change measures into national policies, strategies, and planning	+	(+)	+
13.3 Improve education, awareness-raising, and human and institutional capacity on climate change mitigation, adaptation, impact reduction, and early warning	(+)		+
Targets with direct and indirect relatedness	8	5	7

2 Public Survey on Environmental Safety: Empirical Case Research from Hungary

2.1 Methodology of the Survey

In light of the global outlook on environmental quality, case research from Hungary was conducted to examine localized findings and compare Hungary's state of the art from a global perspective. An internet-based questionnaire was launched in May 2016, and answers were collected until the end of September of that year. The survey had 107 questions, was conducted in Hungarian, and all respondents had an IP address

Table 5 SDGs targeting energy, air pollution, and climate change in terms of protecting vulnerable landscapes

Protecting vulnerable landscapes	Energy	Air pollution	Climate change
11.2 By 2030, provide access to safe, affordable, accessible, and sustainable transport systems for all, improving road safety, notably by expanding public transport, with special attention to the needs of those in vulnerable situations, women, children, persons with disabilities, and older persons	+	(+)	(+)
11.5 By 2030, significantly reduce the number of deaths and the number of people affected and substantially decrease the direct economic losses relative to the global gross domestic product caused by disasters, including water-related disasters, with a focus on protecting the poor and people in vulnerable situations			(+)
11.6 By 2030, reduce the adverse per capita environmental impact of cities, including by paying special attention to air quality and municipal and other waste management	(+)	+	(+)
14.2 By 2020, sustainably manage and protect marine and coastal ecosystems to avoid significant adverse impacts, including by strengthening their resilience, and take action for their restoration in order to achieve healthy and productive oceans			(+)
14.3 Minimize and address the impacts of ocean acidification, including through enhanced scientific cooperation at all levels	+		
15.1 By 2020, ensure the conservation, restoration, and sustainable use of terrestrial and inland freshwater ecosystems and their services, in particular forests, wetlands, mountains, and drylands, in line with obligations under international agreements			(+)
15.2 By 2020, promote the implementation of sustainable management of all types of forests, halt deforestation, restore degraded forests, and substantially increase afforestation and reforestation globally	(+)		(+)
15.3 By 2030, combat desertification, restore degraded land and soil, including land affected by desertification, drought, and floods, and strive to achieve a land degradation-neutral world			(+)
15.4 By 2030, ensure the conservation of mountain ecosystems, including their biodiversity, in order to enhance their capacity to provide benefits that are essential for sustainable development			(+)

(continued)

Table 5 (continued)

Protecting vulnerable landscapes	Energy	Air pollution	Climate change
15.8 By 2020, introduce measures to prevent the introduction and significantly reduce the impact of invasive alien species on land and water ecosystems and control or eradicate the priority species			(+)
Targets with direct and indirect relatedness	4	2	9

Table 6 Applied stratifications of the public survey

Stratification (Number)	Strata	Percentage
Sex (2)	Man: woman	52: 48
Location in the country (3)	West: central: east	26: 52: 22
Settlement size (4)	<5000: 5000–50,000: 50,000–1 million: >1 million	15: 29: 27: 29
Age in years (5)	<30: 30–40: 40–50: 50–60: >60	7: 18: 32: 25: 18
People in one household (5)	1: 2: 3: 4: $\geq$ 5	13: 29: 23: 20: 14
Highest education (6)	$\leq$8 yrs: techn. coll.: sec. school: college: university: doctorate	4: 11: 7: 34: 38: 6

registered in Hungary. Altogether, 831 responses were received. Representativity of the samples was enhanced by using stratified sampling using six factors (Table 6). The average calculation for each stratum was weighted against the country's actual proportions, i.e., published by Hungary's Central Statistical Office in correspondence with 2016, utilizing the following indicators: sex, location in the country, settlement size, age, number of people in one household, and the highest level of education. The survey was made up of three sub-sections: (1) household and lifestyle (i.e., 32 questions), (2) energy preference, air pollution, and climate change (i.e., 37 questions), and environmental safety (i.e., 38 questions).

2.2 *Household and Lifestyle of the Respondents*

Equipment statistics for each household and lifestyle choice are indirectly related to the focal interest of this chapter, i.e., understanding the opinion of the respondent's selected level of environmental safety. Acquiring this information, an assessment on how natural (i.e., basic and resource non-intensive) versus artificial (i.e., complex and resource-intensive) one's life is was calculated. In terms of the financial level and lifestyle, some key characteristics are illustrated in Fig. 1. The findings per household indicated the number of mobile telephones was 1–2, while the number of televisions, laptop computers, vacuum cleaners, and other kitchen appliances was

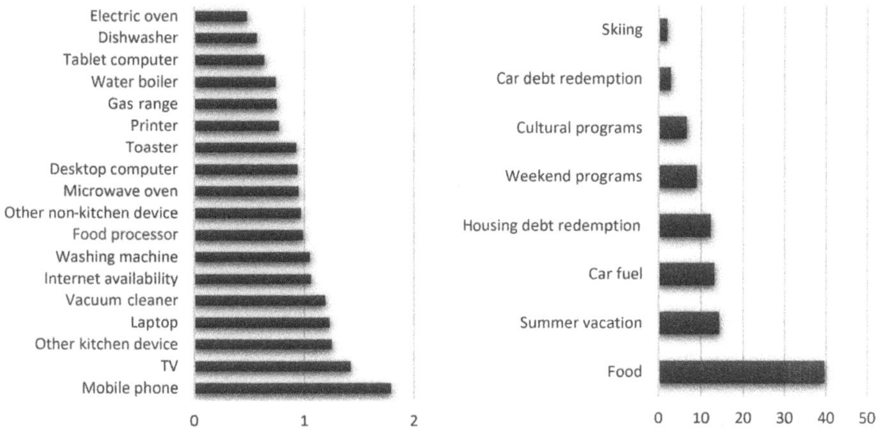

Fig. 1 (Left) Availability of household machines stratified according to location. (Right) Distribution of family costs stratified according to the sex of the respondents

also slightly over 1. Moreover, internet availability, washing machines, food processors, microwave ovens, toasters, desktop computers, and other non-kitchen devices were also found in an average household. A greater number of families were equipped with a gas cooker rather than an electric cooker. The main component of the family budget was spent on food consumption (i.e., 40%), while summer vacation (i.e., 10–20%), car fuel, and housing debt reduction rounded the next three positions. Nearly, 10% of the budget or less was used for weekend and cultural programs, car debts, and winter skiing vacations. Finally, the average sum of the budget share equated to 67.7% (i.e., respondents were not asked to provide 100% of their financial expenses); as a result, averaged shares are proportionally corrected.

2.3 *Results on Energy Preference, Air Pollution, and Climate Change*

In the survey, the respondents were asked to assess various sources of energy by using a Likert scale [5], i.e., 5 = strongly agree or very important, 4 = agree or important, 3 = neutral, 2 = disagree or not important, 1 = strongly disagree or not important at all. Renewable energy sources received the highest evaluation (i.e., almost 5) among the different types of energy used. Moreover, nuclear and fossil energy was found to be the worst (Fig. 2.) The evaluations were driven by environmental consideration and not by actual energy feasibility. Energy consumption and CO_2 emission levels from households were also noted (i.e., over 4) and considered significant. Questions relating to nuclear energy were evaluated in terms of energy dependence outside the nuclear grid (i.e., with no change in CO_2 concentration) and air quality detection. The study found that energy dependence would improve (i.e., decrease) without nuclear

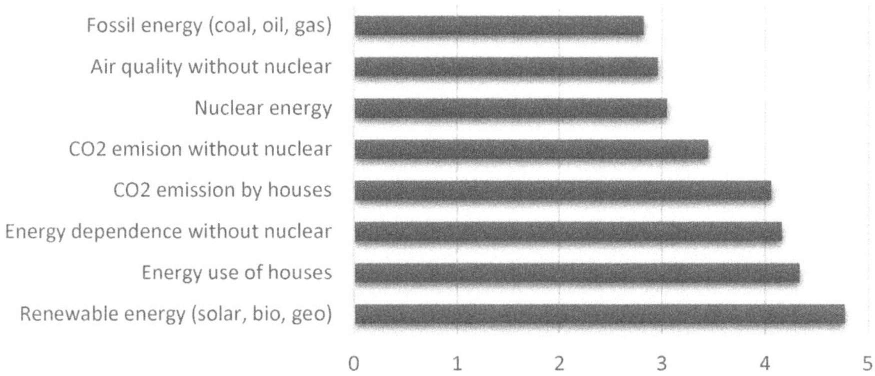

Fig. 2 Evaluation of energy sources according to sex

energy even though a worsening of air quality was reported. This, however, would only be doable if non-fossil energy sources could be applied instead of the nuclear option.

The sources of air pollution were primarily seen as polluting materials emitted by enterprises, everyday CO_2 emissions, and communal (i.e., public) waste (Fig. 3). The importance of urban overpopulation was somewhat less, but still, around 4, climate change (i.e., not a reason, but a consequence) and anachronistic heating systems of flats were also significant. Much less important were the ozone hole (i.e., a result of pollution) and animal ordure.

The key reasons and the likely consequences of climate change are presented in Fig. 4. Highly ranked was CO_2 emissions among possible reasons for climate change. This evaluation was highly correlated to climate change rather than internal processes of the earth [6–9] or solar activity [10–14]. Deforestation is marked somewhat higher than its real danger [15–18], but the evaluation of animal husbandry [19, 20] and agriculture [21–23] were evaluated nearly according to their actual contribution. It

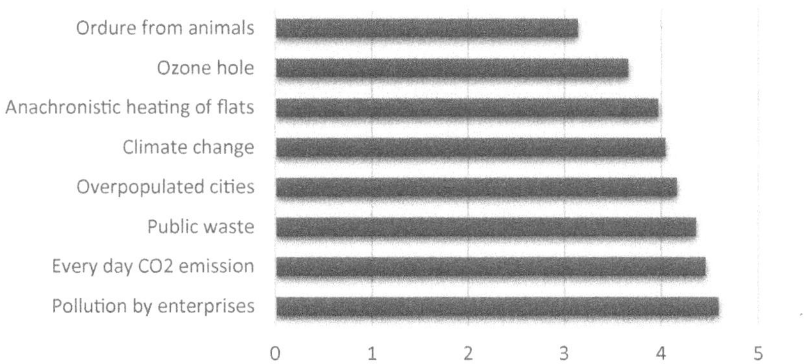

Fig. 3 Evaluation of various forms of air pollution according to education

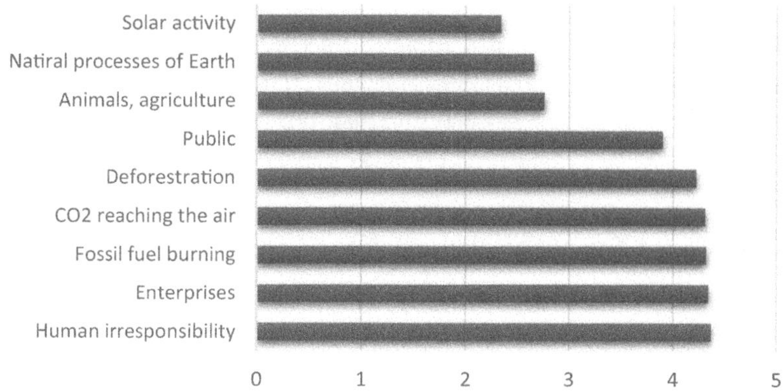

Fig. 4 Evaluation of potential reasons for climate change according to the highest level of education

should also be noted that reasons for climate change are highly correlated to the irresponsibility of humans—i.e., enterprises and the public in general. Between the highest six and the lowest three values, there is over a 1 (and sometimes as much as 2) point difference separating the reasons accepted and rejected by the respondents.

Among the possible impacts of climate change in Hungary, 13 items were selected and asked to be ranked. From the list, presented in Fig. 5, the impacts on water supply were ranked the highest, i.e., reconsidering water management, higher prices for drinking water, and water reservoirs to be built. This ranking is slightly higher than those related to plant production, i.e., some plants can be grown with irrigation only and more expensive field cultivation, while energy demands equated to more energy for air conditioning and annual electricity use increases. However, the difference

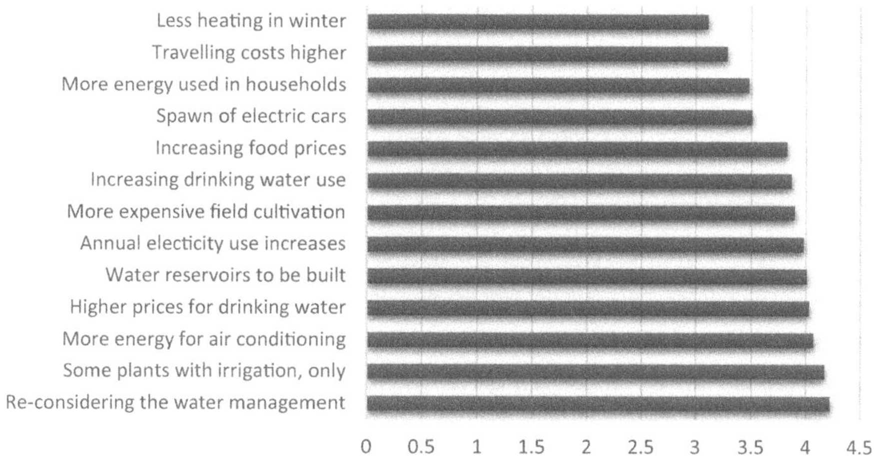

Fig. 5 Evaluation on the impacts of climate change according to location in the country

between the most popular option, reconsidering water management, and the least popular, less heating in winter, is only slightly over 1. This reflects the high sensitivity of the respondents to all potential climate changes. Nonethless, with even the lowest ranked evaluation above 3, it is interesting that positive changes were likened to less energy for heating which makes parts of the findings less believable or under-evaluated.

3 Comparative Results

3.1 Results on Environmental Safety

Three sets of questions were analyzed concerning environmental safety. Two slightly different questions were used to evaluate 12 environmental problems—in the first set, the word safety was not included in the question. Half of these problems were expressed in global terms, while the other half were an evaluation of an individual's everyday experience (Fig. 6). The survey respondents evaluated environmental services as the highest-perceived direct factor (e.g., drinking water supply, waste collection, wastewater treatment, and selective waste collection). In terms of environmental danger, littering and pollution by enterprises were reported as the most unpleasant, but climate change and pollution of cities also remained very high. Somewhat lower scores were given to noise, the ozone hole, acid rain, and overpopulated cities. The evaluation of these general environmental questions showed that, generally, they received 10–20% higher value when just the problem was addressed. The

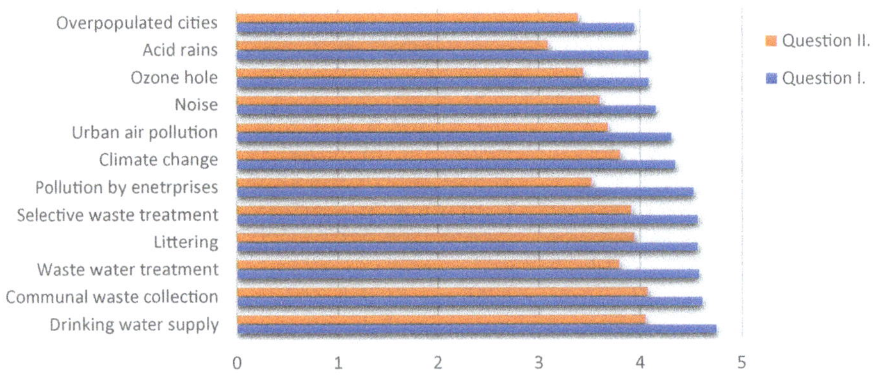

Fig. 6 Importance of selected problems judged in response to two questions according to sex. Question I. How important are the following environmental problems for you? Question II. How much do the following problems affect your life?

mean difference was 0.68, respectfully. Evaluation of responses to the softer question indicated that the respondents were highly sensitive to the exposed questions in which even the lowest evaluation was near or above 4.

In the second set of questions, the word safety was formulated into two questions via slightly different formats (i.e., To what extent does the mean risk or threaten the following? To what extent do your feelings influence safety in the following?). As such, the evaluation was slightly dependent on the question's composition, even if the change did not alter the question's meaning. The 15 problems corresponding to environmental challenges on safety were found to be somewhat lower. When comparing environmental safety, i.e., Fig. 7.

In the third set of questions, comparative threats to the environment with other types of safety risks were examined (Fig. 8). Answering just one question, the respondents considered a lack of clean water, energy, and healthy food as not of high risk,

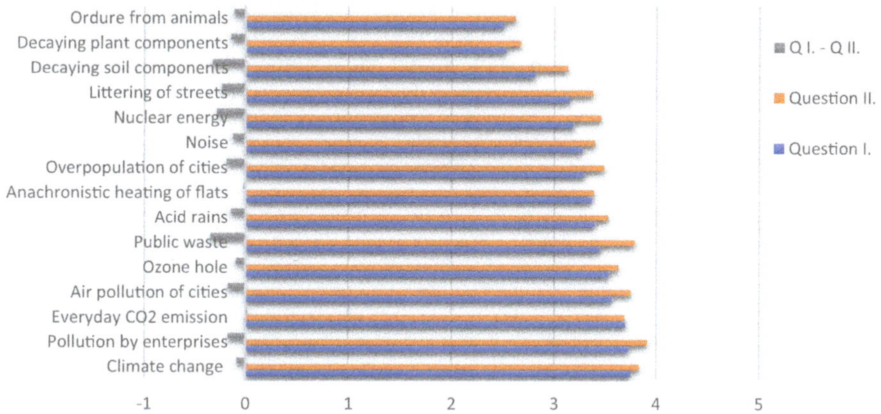

Fig. 7 Evaluation of environmental safety problems. Question I. To what extent does the mean risk or threat the following? Question II. To what extent do your feelings influence safety in the following?

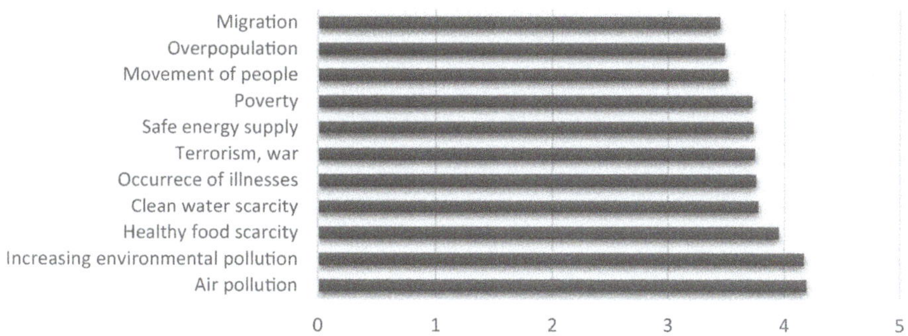

Fig. 8 Evaluation of environmental safety factors according to sex

but air and environmental pollution as important, i.e., scoring a value of over 4. Somewhat surprisingly, direct threats to safety such as healthy food and clean water scarcity, as well, as the occurrence of illnesses, terrorism, and war, together with a safe energy supply and poverty, were evaluated between 3.5 and 4. Most likely, the reason for these medium-level values was that respondents in Hungary have little personal experience or exposure to such problems. Migration (i.e., including mass movement) and overpopulation beared slightly a lower risk in 2016, i.e., at the time of the survey.

3.2 Comparative Differences

Besides some differences among the strata, socioeconomic analyses rarely established any significant differences among the strata averages. Four such cases were presented in Figs. 9 and 10. The first comparison was evident, while the second elucidated a concrete contribution to the social background of public evaluation on environmental challenges. First, comparing the differences among the strata was the increasing number of computer printers in a household with the increasing age of

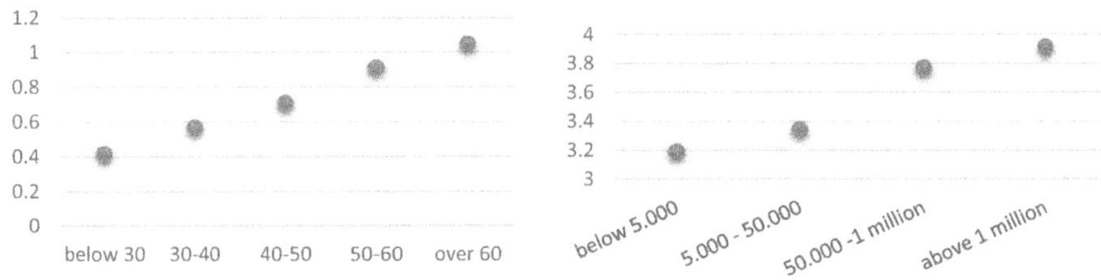

Fig. 9 (Left) Unique difference in frequency of printers in the family versus age. (Right) Unique difference in the evaluation of urban air pollution (i.e., daily air pollution index (PI) rating) according to the settlement size

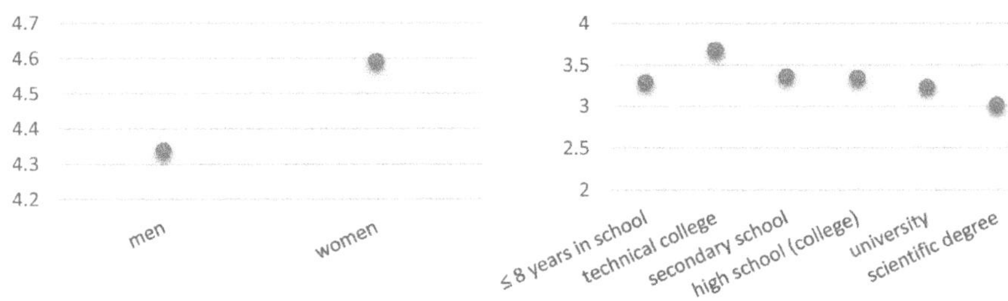

Fig. 10 (Left) The characteristic difference between environment sensitivity of men and women. (Right) Characteristic decrease in the evaluation of the problems by the respondents with higher levels of education. Both examples represent the evaluation of selective waste management

the respondent. The other plausible difference was the evaluation of urban pollution growing with the increasing number of inhabitants in the settlement (Fig. 9).

Second, women are more sensitive to the state of the environment in almost all analyzed aspects, i.e., they give it more importance than men. Figure 10 illustrates just one aspect of the difference, that being, via the evaluation of selective waste management. At the same time, it can be said that those with higher education levels valued the risks less than those who finished just technical college. This behavior is far more general than the evaluation of public waste delivery.

4 Conclusion

In democratic societies, policy and politics are two notions that deserve close attention—the latter being highly ranked by political parties. In order to convince political parties to develop sound policy, it is essential to understand and learn what potential voters think and believe regarding various policy aspects concerning environmental problems and safety. When we consider SDGs and associated targets that directly or indirectly tackle energy preference, air pollution, and climate change, they signal critical global steps to overcome these environmental challenges at the broadest scale of sustainability. The empirical case research from Hungary serves as a factual base for bridging policy and politics by reflecting upon what people think about energy, pollution, and climate as well as on other aspects of environmental safety. Additional information on the status of sustainability and the analyzed environmental issues in Hungary are considered key to sound development.

In Hungary, sustainability, in general, is driven by the national document, National Framework Strategy on Sustainable Development (NFSSD) for 2012–2024, Resolution 18/2013, adopted 25 March 2013. With the Hungarian Parliament's adoption of the Strategy, it specified the sustainable development tasks of the government and the country's municipalities [24]. The strategy's main task for developing a sustainable society is the preservation and enhancement of national resources, which can be categorized as human, environmental, social, and economic. The sustainability indicators [25], published bi-annually, are also presented and evaluated according to NFSSD [24]. The status of SDGs in Hungary are assessed by the Ministry of Foreign Affairs and Trade [26] and the National Council for Sustainable Development [27]. The latter establishes that Hungary continues to face a sustainability crisis, and none of the country's four national resources are thriving. The majority of the relevant indicators are not only below the European Union average but also worse-off in terms of the Visegrad Group, which are similarly positioned countries historically and economically [27]. OECD establishes that Hungary has "made significant progress in decoupling its output growth from the main environmental pressures, largely due to implementing requirements of European Union directives. The recent rebound of economic activity and energy consumption is intensifying pressures on the natural environment. […] Hungary has gradually reduced its reliance on coal and natural gas in favor of low-carbon energy sources. However, fossil fuels still make up about two-thirds of the energy supply. The residential sector is the biggest energy consumer, with 80% of the building stock lacking modern and efficient heating systems. Hungary's total gross greenhouse gas (GHG) emissions have decreased by 35% since 1990. However, emissions have recently started to increase, driven by transport and agriculture. Since 2000, Hungary has substantially reduced emissions of sulfur and nitrogen oxides, ammonia, and non-methane volatile organic compounds. Particulate emissions from transport and residential heating have been increasing, worsening urban air quality. The cost of premature death in Hungary, due to exposure to $PM_{2.5}$ and O_3, has grown to an estimated 9% of gross domestic product, the second-highest value in the OECD" [28]. As such, the European Environmental Agency [29] reports that Hungary performs very well in terms of GHG emissions. Its targeted emission value was 52.8 $MtCO_2eq$ for the 2020s has already matched the year prior at 43.5 $MtCO_2eq$. The country also achieved its renewable energy share targets of 13% in terms of its total primary energy consumption—i.e., matched in 2019 with an estimated 12.6%. Total energy consumption, however, remains above the planned value of 14.4 Mtoe for the 2020s, with 2019 estimates being 19.0 Mtoe. At length, direct and indirect relations can help establish renewable energy sources in Hungary, with provisionary efforts to improve air pollution and mitigate climate change. Public survey research should be complimented with a media-based response to better inform and establish initial steps in educating citizenry awareness as well as foster policy-makers in best practice reciprocity of environmental safety in SDGs.

Acknowledgements This study was supported by the National Research, Development, and Innovation Office (NKFIH), Project No. K 116595.

References

1. United Nations (2015) Sustainable development goals. In: United Nations Resolution. A/RES/70/1 25 Sept. 2015. http://www.un.org/ga/search/view_doc.asp?symbol=A/RES/70/1&Lang=E. Accessed 3 May 2019
2. Mika J, Farkas A (2017) On synergies and conflicts between the Sustainable Development Goals (2016–2030) and renewable energy sources for education of and by sustainability. Probl Educ 21st Century 75:182–193
3. UNFCCC (2015) The Paris agreement. United Nations Framework Convention Climate Change. https://unfccc.int/process-and-meetings/the-paris-agreement/the-paris-agreement. Accessed 14 Feb 2021
4. UNFCCC (2021) United Nations: climate change. United Nations Framework Convention Climate Change. https://unfccc.int/. Accessed 14 Feb 2021
5. Joshi A, Kale S, Chandel S, Pal D (2015) Likert scale: explored and explained. Br J Appl Sci Technol 7:396–403. https://doi.org/10.9734/bjast/2015/14975
6. IPCC (2013) AR5 climate change 2013: the physical science basis—IPCC. Working Group I of the Intergovernmental Panel on Climate Change, Geneva
7. IPCC (2017) AR6 climate change 2021: the physical science basis—IPCC. Intergovernmental Panel on Climate Change, Geneva
8. Kundzewicz ZW, Kanae S, Seneviratne SI et al (2014) Flood risk and climate change: global and regional perspectives. Hydrol Sci J 59:1–28. https://doi.org/10.1080/02626667.2013.857411

9. Doelle M, Seck S (2020) Loss & damage from climate change: from concept to remedy? Clim Policy 20:669–680. https://doi.org/10.1080/14693062.2019.1630353
10. Reddy V, Reddy PR, Reddy DV (2016) Impact due to sunspot activity on climate change: some salient results article in. Int J Earth Sci Eng
11. Solheim JE, Stordahl K, Humlum O (2012) The long sunspot cycle 23 predicts a significant temperature decrease in cycle 24. J Atmos Solar-Terrestrial Phys 80:267–284. https://doi.org/10.1016/j.jastp.2012.02.008
12. Hady AA (2013) Deep solar minimum and global climate changes. J Adv Res 4:209–214
13. Imbrie J, Imbrie JZ (1980) Modeling the climatic response to orbital variations. Science (80-) 207:943–953. https://doi.org/10.1126/science.207.4434.943
14. Xiao Z-N, Li D-L, Zhou L-M et al (2017) Interdisciplinary studies of solar activity and climate change. Atmos Ocean Sci Lett 10:325–328. https://doi.org/10.1080/16742834.2017.1321951
15. Redman CL (1999) Human impact on ancient environments. University of Arizona Press, Tucson
16. Gelagay HS, Minale AS (2016) Soil loss estimation using GIS and remote sensing techniques: a case of Koga watershed, Northwestern Ethiopia. Int Soil Water Conserv Res 4:126–136. https://doi.org/10.1016/J.ISWCR.2016.01.002
17. Klooster D, Masera O (2000) Community forest management in Mexico: carbon mitigation and biodiversity conservation through rural development. Glob Environ Chang 10:259–272. https://doi.org/10.1016/S0959-3780(00)00033-9
18. Laumonier Y, Uryu Y, Stüwe M et al (2010) Eco-floristic sectors and deforestation threats in Sumatra: Identifying new conservation area network priorities for ecosystem-based land use planning. Biodivers Conserv 19:1153–1174. https://doi.org/10.1007/s10531-010-9784-2
19. Krishnan S, Nandwani D, Smith G, Kankarta V (2016) Sustainable urban agriculture: a growing solution to urban food deserts. In: Nandwani D (ed). Springer International Publishing, Cham, pp 325–340
20. Madaleno I (2000) Urban agriculture in Belem, Brazil. Cities 17:73–77. https://doi.org/10.1016/S0264-2751(99)00053-0
21. Clark KH, Nicholas KA (2013) Introducing urban food forestry: a multifunctional approach to increase food security and provide ecosystem services. Landsc Ecol 28:1649–1669. https://doi.org/10.1007/s10980-013-9903-z
22. FAO (2011) The State of the World's Land and Water Resources for Food and Agriculture (SOLAW): managing systems at risk. Earthscan Publications Ltd., London
23. Koleva-Lizama I (2014) Analysis and assessment of risk and vulnerability in the field of agriculture. In: Analysis and risk and vulnerability assessment of bulgarian economy sectors from climate change, Special part. Ministry of Environment and Water, Sofia, pp 7–27
24. NFSSD (2013) National Framework Strategy on Sustainable Development (NFSSD) for 2012–2024. National Council for Sustainable Development (NFFT), Budapest
25. CSO (2019) The indicators of sustainable development in Hungary. Central Statistical Office, Budapest
26. MFATH (2018) Voluntary national review of hungary on the sustainable development goals of the 2030 agenda. Ministry of Foreign Affairs and Trade of Hungary, Budapest
27. NFSSD (2019) National Framework Strategy on Sustainable Development, Third progress report 2017–2018. National Council for Sustainable Development (NFFT), Budapest
28. OECD (2018) OECD environmental performance reviews: Hungary 2018. OECD Publishing, Paris, OECD Environmental Performance Reviews
29. EEA (2020) Trends and projections in Europe 2020: tracking progress towards Europe's climate and energy targets. European Environmental Agency Reports No. 13/2020, Copenhagen

Sustainable Development Goals, Conflict, and Fragility: Anglophone Crisis in Cameroon

Samuel Mwangi and Giuseppe T. Cirella

Abstract Good governance is critical in creating a conducive business environment and sustained economic growth. However, the weak governance characterized by social inequality has led to instability, fragility, and conflict in many African countries. Part of instability and fragility is deeply rooted in colonialism and dates back to the emergence of modern African states. The Anglophone crisis in Cameroon is such a typical challenge. This chapter presents two key concerns that have not yet been adequately addressed: First, how do the United Nations Sustainable Development Goals (SDGs) and their targets address the challenge of Africa's historical instability and fragility? Second, on the flip side, how does long-standing instability and fragility affect the achievability of the goals? A key lesson from the Anglophone crisis is that, contrary to the universal understanding, SDGs are not just a development framework to pick up from where the United Nations Millennium Development Goals stopped. They are, as well, instruments to reflect on historical and structural foundations that form the root causes of conflict and fragility. SDGs call for international cooperation toward establishing a fundamental change to build back better: replacing the root causes of vulnerability and fragility with pillars of social equity to attain economic growth and sustainable development.

Keywords Human displacement · Sustainable development · Instability · Social equity · Cameroon Anglophone crisis · Africa

1 Introduction

Economic and political instability in any country hinders business operations and sustainable development. The achievement of the United Nations Sustainable Development Goals (SDGs) on the one hand and chronic conflicts and fragility on the

S. Mwangi (✉)
Institute of Political Science, Tübingen University, Tübingen, Germany

G. T. Cirella
Faculty of Economics, University of Gdansk, Sopot, Poland
e-mail: gt.cirella@ug.edu.pl

G. T. Cirella (ed.), *Human Settlements*, Advances in 21st Century Human Settlements,
https://doi.org/10.1007/978-981-16-4031-5_11

other present opposing forces. While SDGs present a guide and an opportunity for inclusivity, collective energy, social justice, and participatory development, political instability manifests a lack of a shared vision at the national level, a force for exclusion, inequality, and insecurity. The achievement of SDGs, in this regard, refers to the implementation of development projects and programs within the framework of the United Nations 2030 vision. Sustainable development is concerned with promoting social equity, economic growth, and environmental concerns to eradicate socioeconomic insecurity and vulnerability of populations [1, 2]. SDGs play an essential role to bring together and quantify development initiatives implemented by the government, the private sector, civil society organizations, and, most importantly, citizens. In other words, SDGs implementation enables the establishment of coherence and consistent coordination between agents, development strategies, and governance structures at the local, national, and international levels.

In Africa, the achievement of the Global Agenda for Sustainable Development and Africa's Agenda 2063 depends on how governments are able to reposition their governance to address the structures of underdevelopment and establish those that enhance social, economic, and political stability. SDGs form a vision that transcends barriers of development, including histo-ethnic challenges. They aim to create a conducive environment for economic activities and enhance livelihoods and human security for present and future generations. SDGs perceive sustainability as both the means and the end of the development process. Hence, it becomes difficult to establish a crystal-clear strategy to implement and achieve them. Africa is an emerging market economy with high prospects for economic growth and sustainable development [3, 4]. But still, in many African countries, socioeconomic and political instability are characterized by a high prevalence of violent conflict. Researchers have widely examined the economic challenges of doing business in Africa, including taxation, cost of production, technology, and access to finance [1, 5, 6]. In this chapter, using the Anglophone crisis in Cameroon, we examine the sociopolitical environment as a factor for the sustenance of business operations. Cameroon is one of many countries still struggling with governance issues related to its colonial legacy. The lack of robust systems for distributive justice is itself a "root cause" that propel inequality, vulnerability, marginalization, and instability. The exclusive development has further been magnified by capitalistic economic systems [7]. A key concern is how the governments and development partners reposition themselves to properly align with the SDGs framework and how that repositioning addresses chronic fragility. Having already had the framework since 2015, one issue that has been inadequately unexplored is how it can synergistically be implemented as a platform to manage conflict, fragility, and vulnerability in different African countries [8, 9]. This is with the understanding that long-standing conflicts and fragility hinder any form of sustainability.

Fragility is characterized by high exposure to risks and weak coping-capacity of a state, an institution, or population. The inability to manage or mitigate those risks causes more fragility in a rent-seeking manner to form what Kodila-Tedika and Simplice [10] refer to as the "fragility trap" or "the vicious cycle of fragility." Situations of fragility and instability are causes and, at the same time, effects of poverty, exclusivity, and violent conflict—forming a vicious cycle. High state fragility increases the prevalence of institution failure, the emergence of conflicts, or other forms of tension. From the institutional political economy, fragility is viewed using risks and the incapacity of formal and informal actors that influence the establishment or impact of security policies in a state [11]. According to the Organization for Economic Cooperation and Development [12], fragility is measured in five perspectives: economic, social, political, security, and environmental weakness. Thus, fragility takes different forms depending on the context and country—utilizing different mechanisms of managing formal or informal situations. This chapter explores two key concerns: (1) how do SDGs and their targets address the challenge of Africa's historical instability and fragility and (2) on the flip side, how does long-standing instability and fragility affect the achievability of the goals? Case research that assesses the Anglophone crisis in Cameroon is used as an example to focus on these questions.

2 Theorization of Change

Achieving SDGs in Africa requires a sustained transformative process. The process of change is explained using models that revolve around the interaction between the structure and the agency [13, 14] or even contingency theory [15]. Any transformation must affect one or both the agency and governance structure. To achieve a changed practice, two forms of change must occur: change in the structure and change in agency. The structuration theory offers a good snapshot for the applicability of the change required to implement development to achieve SDGs in fragile countries. Change in the structure involves a transformation of the institution's structural organization, rules, and regulations. Changing agency involves transforming behavior and attitude of people. This results in a vicious cycle of change where the agency changes the structure, and structure shapes the agency's behavior. The two perspectives of change are co-constitutive, and it is difficult to have a purely one-sided model of change. Antony Giddens [16] refers to this structure-agency cycle as the theory of structuration. Structure refers to the standards for people's behavior and acts as "codes" or templates that guide social actions. They are norms, rules that characterize social systems. On the other hand, agency refers to members of a social system [16]. The interaction between structures and agency within a particular time and space creates a social system. Social systems are the relations of people within specific social settings. Behavior is presumed to continue in a social system within which people are organized [13]. In a social system, structures are not always external to individual members rather they can also be internal and play enabling or constraining roles. It thus becomes necessary to understand how practices (i.e., social behaviors and the knowledge behind those behaviors) are located within societal norms and how they influence social equity and economic activities.

The institutionalization of sustainability is achieved when the agency is willing and able to operate within institutions and when the output and impact of the structure-agency relationship are sustainable. The process is guided by instrumental logic and normative mechanism [17]. A code of behavior is a result of a system of authority, power, and communication of mutual understanding. Fragility and chronic conflict are a malfunction of the sociopolitical order. Ultimately, the transformation between structure and agency should resonate in a harmonious manner that produces the synergy necessary to achieve the desired political and socio-economic environment. Otherwise, when there lacks such resonance, the system will always create a notch, tension, and conflict. SDGs perceive good governance and human development as a means and to an end of sustainable development. In Sub-Saharan Africa, the SDGs' implementation calls for structural functionalist approaches [18, 19] and humanist approaches [20–22].

3 SDG 16: Addressing Conflict, Fragility, and Instability

(All) SDGs form a complex interaction and work in an overarching and multifaceted way to create a holistic development framework. Likewise, the analysis of fragility and instability necessitates examining governance processes in all sectors and at all levels. Good governance is a prerequisite for the achievement of SDGs. The term good governance refers to "an organizational system which is institutionally legitimate for its constituency, efficient in carrying out its purposes characterized by participatory tendencies, consensus-oriented, accountable, transparent, responsive, effective and efficient, equitable and inclusive, and follows the rule of law" [23]. SDG 16 focuses explicitly on good governance practice by "promoting peaceful and inclusive societies for sustainable development, provide access to justice for all, and build effective, accountable, and inclusive institutions at all levels" [24]. It also aims to address governance issues, such as social inclusivity and equity, transparency, accountability, and justice within development processes. As noted by Siakwah et al. [25], justice is not a straightforward concept; it takes different forms, depending on the context. Its fundamental principles include human rights, democracy, fairness and equity, and non-discrimination. It is a governance concept that applies in all social, economic, and political domains of development.

SDG 16 has a particular focus on governance and justice. The first target of SDG 16 (i.e., 16.1) is to "significantly reduce all forms of violence and related death rates everywhere" [24]. The target start at an abstract level without distinguishing the various forms of conflicts. The target prioritizes the need to address violent conflicts to achieve peace and justice within communities. As expected, violence occurs at all levels—from domestic and communal to international conflicts. The second target (i.e., 16.2) focuses on the agency by acknowledging extra vulnerability experienced by children in fragile environments. The vulnerability issue is also highlighted in SDG 1 (on eradicating all forms of poverty) and its targets. Target 16.2 emphasizes "end abuse, exploitation, trafficking, and all forms of violence against and torture children" [24]. This target is closely related to SDG 4, which expresses the need for equitable and quality education for all children. The third target of SDG 16 (i.e., 16.3) aims at attaining peace and stability by strengthening political institutions to "promote the rule of law at the national and international levels and ensure equal access to justice for all" [24].

In economic reams, SDG 16 targets 16.4, 16.5, and 16.6 acknowledge significant sources of financial fragility, including reducing illicit financial and arms flow, reducing corruption and bribery in all its forms, and promoting accountable and transparent institutions at all levels, respectively. The implementation of these targets means building strong economic institutions to track and manage financial fragility. Illicit financial flows are a significant economic challenge connected to economic instability and inequality in Africa [26]. Lack of or unequal access to economic opportunities coupled with capitalistic economic systems are some of the root causes of vulnerability and fragility in developing countries [7, 27]. Indeed, curbing illicit financial flows is a shared goal that requires international cooperation between governments and private sector, and the global North and global South [24, 26, 28]. The existence of illicit financial flow is a manifestation of institutional failure and is associated with corruption, digital divide, and weak governance. Besides, these vices are self-reproducing if left unchecked. They negatively loop back into the system by compelling the agency to engage more in illicit financial activities and corruption—confirming a cyclic corruptive revolving door.

The target 16.7 and 16.8 of SGD 16 reaffirm the inclusive and participatory development process, i.e., sustainability refers to inclusive participation across gender, age, ethnicity, and between regions within a country. Target of 16.7 emphasizes "responsive, inclusive, participatory, and representative decision-making at all levels" [24], urging governments to establish advanced strategies that engage people as active agents in national development. This participation is scaled-up at the international level by target 16.8, which broadens and strengthens participation for developing countries at the global level. The sequencing of these two targets asserts that, for effectiveness, there is the need to bridge the gap and establish resonance between people and institutions. It also indicates an understanding that the ability of institutions in developing countries to participate in the global arena is primarily determined by local community participation in those very institutions.

The process of analysis of individual's participation in the achievement of SDGs begins with capturing civil registration and vital statistics [29]. This explains why SDG 16 target 16.9 is expanded to include provision for the legal identity for all. Target 16.10 enlarges social equity by raising awareness and promoting fundamental freedoms at both national and international levels. Again, the target aims to narrow the divide between national and international processes by increasing participation in developing countries. Overall, achieving sustainability requires participatory decisions from the local level to the international level. The target closely links to SDG 10 and SDG 17, which expound on the need to address inequality within and among countries. Social accountability in governance reflects people's empowerment to engage in the achievement of sustainable development. SDG 16 is summarized by two key targets, i.e., target 16.a and 16.b, which focus on developing countries.

First, national institutions' ability to participate in international affairs is seen as empowerment to address the violent conflict and fragility. Target 16.a establishes a link between national and international institutions as a strategy to build capacity to prevent violence and combat terrorism and crime. In context to developing countries, communities' inequality and marginalization are some of the causes of instability and violent conflict [27, 30]. Secondly, target 16.b centers on promoting and enforcing non-discriminatory policies for equity and sustainable development by envisioning a transition to peaceful, just, and strong institutions.

The concept of governance, within SDGs, is broader and more comprehensive than the government itself. Governance in the SDGs framework is the interaction between agencies and institutions, and citizen participation in all decision-making processes. It is a relationship defined by interdependences, and cooperation starts from local to international levels. In this regard, the government is the critical entity in the wider-scale societal systems of sustainable development. The relationship between government and society is crucial. Non-state actors, such as donor agencies, the private sector, civil society organizations, and the people, are expected to actively engage in the transformation process. Citizenry participation in the development process is achieved through the market and non-market strategies or their hybrid. Overall, governance as an element of sustainable development is characterized by the rule of law, active participation at all levels of decision-making [31].

4 Violent Conflict, Fragility, and Vulnerability as Hindrances of SDGs

Fragility and conflict are severe threats to eradicating extreme poverty and SDGs' achievement among many African countries. Weak state institutions constitute a significant cause of fragility since they create a negative feedback loop that demeans social equity and justice in the development process. That is why many people living under extreme poverty are located within conflict-affected or fragile situations. In Africa, violent conflict and instability appear to be on the rise, especially where political exclusion has widened economic inequality and marginalization. As a result, there has been an increased people's vulnerability, violent extremism, and politics of secession. The number of people in fragile and conflict-torn regions is even expected to rise above 50% by 2030 [32, 33].

Political violence is a form of conflict where real socioeconomic issues are politicized. Violent conflict is transformed via political channels in which political rivalries become the agents of violence. Radicalized groups take political sides, while the ruling class uses the state machinery to suppress non-supporters. In some cases, the difference between government's actions and those of rebels or militia is hardly noticeable. Social violence takes a broader form of interpersonal violence as a means of expressing a grievance within communities. Besides, the political environment is aggravated by fragility and violent conflict and originates from structural weaknesses in governance. They can be traced by examining how systemic socioeconomic and political exclusion produce political instability [34]. Most violent conflicts and instability exist where the rule of law is most absent. Marginalization results in civil wars with minimal chances to grant justice to victimhood.

4.1 The Anglophone Crisis in Cameroon

Cameroon is one of the countries characterized by long-standing fragility and instability referred to as the Anglophone crisis. The crisis started since Cameroon got independence in the early 1960s. Germany was the first country to colonize Cameroon from 1884 to 1914. After World War I, the League of Nations divided Cameroon and reallocated it between Britain and France. Unlike other countries where the jurisdiction of the colonial master defined a country's territory, Cameroon was divided into two parts—with each part identifying its self as a colony but within one country. This gave rise to English-speaking and French-speaking Cameroon. Francophone Cameroon gained independence in 1960, and as a result, the Republic of Cameroon was created. Later in 1961, Anglophone Cameroon followed. Upon independence, British-colonized Cameroon reunited with Francophone Cameroon. The reunification affected the country's governance structure from the initial Republic of Cameroon to the Federal Republic of Cameroon. This form of federation seemed a fair way of acknowledging the different colonial effects, including language. Both English and French were made official national languages. In 1972, Cameroon changed the governance structure from federalism to unitary and, subsequently, the country changed its name to the United Republic of Cameroon. This was an indication of the renewed national cohesion and effort of acknowledging bilingualism in the country. Spatially, Cameroon is about 9% Anglophone (Fig. 1), while demographically, the Anglophone proportion of the country is in constant regression, having decreased from 21% in 1976 to 20% in 1987 and, more recently, to 16% in 2015 [35].

The initial expression of grievances on marginalization by the Anglophone movement took place in the 1970s and was later followed by political mobilizations in the 1980s. The criticisms were placed against the repression from the Francophone-dominated government. The Anglophone crisis again took deeper roots after the 1992 election marked by allegations of massive electoral rigging. Various groups, including the unions of teachers and lawyers, rejuvenated efforts in criticizing the Francophone authorities. Over time, radical pro-Anglophone movements have risen. Most active organizations include the Ambazonia Movement, Cameroon Anglophone Movement, Free West Cameroon Movement, South Cameroon National Council, South Cameroons Youth League, South Cameroon Restoration Movement, and Social Democratic Front. Most of these groups advocate a federal form of governance that would give the Anglophone regions more autonomy over their development process. The Anglophone region took a more radical non-diplomatic position and demanded

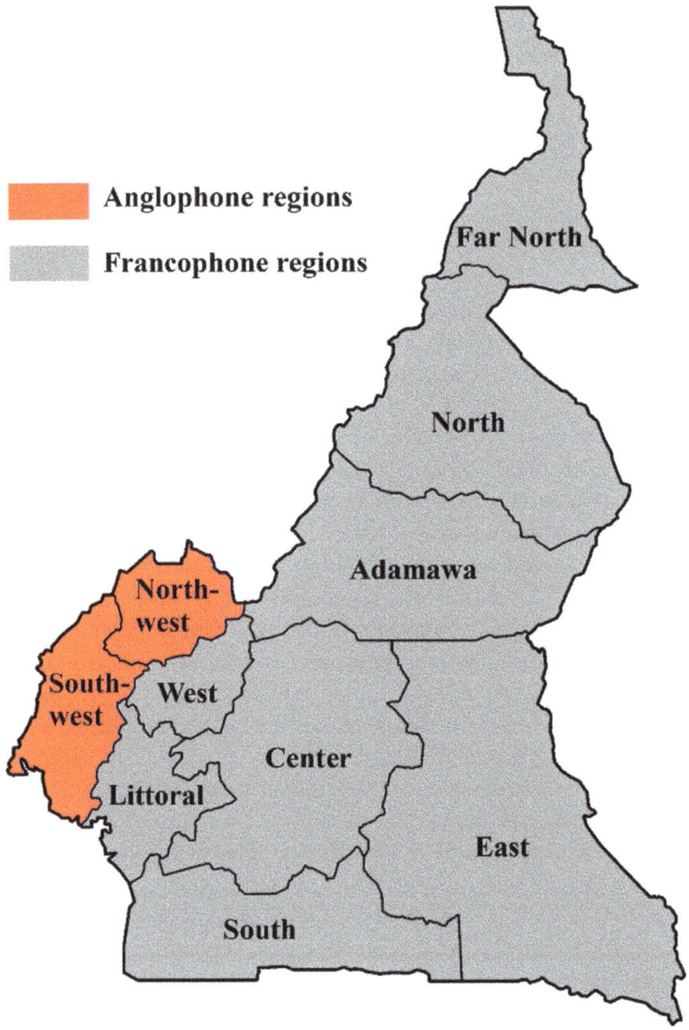

Fig. 1 Map of Cameroon by Anglophone and Francophone regions

full independence through secession bringing the country to the brink of civil war [36, 37].

Since Cameroon obtained independence, the minority rights of Anglophone Cameroonians have overly been subdued [38]. There is consensus within the post-colonial literature that marginalization is a reality in Cameroon. A study by Eyoh [39] found discrimination against English-speakers in public employment. In politics, Anglophones have minimal chances to be appointed to head institutions. The situation is the same in other departments, including the military, judiciary, and the education sector. Marginalization is visible in almost every social, economic, and political domain [38]. The country has since been undergoing a process to promote and streamline bilingualism in schools and public institutions. Nevertheless, analysts have seen bilingualism as problematic and an enormous structural challenge beyond language diversity [40, 41]. In 2016, the English-French bilingualism policy was pronounced as a failure when teachers and lawyers from the Anglophone regions held a demonstration against Francophone speakers' domination. The English-speaking lawyers blamed the government for allocating French-speaking lawyers to court proceedings involving English-speaking citizens in the Anglophone territory. Similar challenges of dominance have been highlighted in the write-up of the judicial documents that indicated conflict between common law and civil law [42]. Some governance analysts like Kiye [43] view the embeddedness of the bilingual judicial challenge as a conflict between and the incompatibility of British common law and French civil law. The Anglophone problem can be seen not just as a policy issue but rather a colonial legacy that manifests through unbalanced economic growth, inadequate social equity, and tension between cultural and ethnic differences. It is a long-standing historical tension that has been worsened by marginalization. Ngeve and Orock [44] explain that any relevant policies need to capture the gist of balanced development and build transparent and inclusive institutions necessary for stability. The historical instability that exists has been reinforced, to some degree, by what Ngeve and Orock [44] term as "cloaked policies" in managing ethnic diversity. In the short run, these policies appeared to achieve national cohesion, but in the long-term, they did not address the root causes related to a colonial-based ethnic identity crisis. They run across nearly all governance areas, including learning institutions, public works, ethnic grouping, and ministerial appointments.

In 2016, the Anglophone problem became pronounced when the Northwest and Southwest regions of the country engaged in strikes and protests. The up rise was led by teachers and lawyers defending the special law provisions and education in the Anglophone territory. During the demonstrations, the government responded with excessive force, using police and military that caused deaths and many injuries. People took pictures and livestreamed the harassment by the police and military and uploaded it via social media. Despite emerging questions concerning the reliability and credibility of information, as noted by Nounkeu [45], the act was a way of externalizing the injustice by the government and seeking intervention from international actors. As a response, the government shut down the internet in the two Anglophone regions for three months [46]. While the separatists conducted violence, government forces reacted through large-scale operations. There were allegations of massive human rights violations in the process, including the killing of suspects and arbitrary arrests [47].

In both the Northwest and Southwest regions, the rate of kidnappings has become excessively high. Teachers and students have been kidnapped for ransom as armed separatist groups such as Ambazonia Defense Forces use terrorist strategies to express their grievances. Pommerolle and Heungoup [37] explain that people from the two Anglophone regions remain repressed by the government. Violent conflict between the armed separatists and military has forced thousands of Cameroonians to flee their homes [48]. Many have settled within Cameroon as internally displaced persons

(IDPs), while others have fled to neighboring countries as refugees. The United Nations High Commission for Refugees (UNHCR) (2019) estimates about half a million IDPs and 50,000 who have sought asylum in Nigeria.

4.2 The Anglophone Crisis Versus the Achievement of SDGs

The Anglophone crisis has destabilized the mobilization of national resources and has adversely affected the implementation of development work. The longer the Anglophone problem remains unsolved, the deeper root it takes. After solving the fundamental challenges of the Anglophone crisis, establishing effective inclusion strategies through disarmament, demobilization, and reintegration (DDR), from the formerly armed groups will be the next future challenge, as witnessed in other countries undergoing post-conflict reconstruction in Africa [49]. Visibly, the United Nations level, the incorporation of DDR compliments the functionality of governance in achieving SDGs [30]. As noted above, SDG 4 ensures inclusive and equitable quality education for all [24]. Any well-functioning system of education, just like other sectors, depends on proper and reliable governance. The Cameroonian government has initiated reforms in the education sector to reflect the people's needs [50]. SDG 16, together with other SDGs, provides a framework for multiagency cooperation and strong linkages between people and institutions, and promotes informational capacity and networking. From the agency perspective, the process starts with information and communication technology-based policy reforms to empower people with the necessary tools [51]. Still, achieving the education goal within a conflict situation is an immense challenge [8]. Despite the government making extensive strides in mainstreaming SDGs in the education sector, in some regions violent conflicts remain a bottleneck. Most kidnappings and abductions have happened in school settings where separatists mainly targeted school buildings and forcefully demand the boycotting of schools [52].

Cameroon has a wide range of biodiversity that forms the backbone of many businesses [4, 53]. However, like in many African countries, ensuring water security in terms of availability, access, and affordability of sufficient and safe water continues to plague most of Sub-Saharan Africa. Water security is one area that SDGs and their targets envision for 2030. There is a direct connection between SDG 1, SGD 2, SGD 3, and SGD 8, and extreme poverty elimination. Other than the effect of changing climatic conditions, a critical concern is poor water sector governance—which has significantly contributed to the water crisis (see Pahl-Wostl [54]). Weak governance has resulted in the inability to manage water insecurity, and constitutes future water crises. Cameroon is one of the African countries with the highest endowment of water resources. Its annual water average availability stands at 21,000 m^3 per capita. This is three times the average global water availability (i.e., 7000 m^3); yet, water scarcity is a significant challenge [55]. In some regions, water scarcity takes both physical and economic forms [56]. The challenge is an inadequate water governance capacity and insufficient investment in the water sector [57]. Cameroon's ability to attain sustainable development depends on its ability to address rural, urban, and peri-urban water crises in all the regions. From the demand side, increasing forced displacement has resulted in a rise in lack of access or affordability to safe water and hygiene challenges [58]. Access to safe water is one of the vital necessities emphasized by UNHCR [59] in conjunction with other humanitarian agencies. Taylor and Sonnenfeld [60] refer to demand-based water challenges as "access crises." The malfunctioning of water demand manifests where people (i.e., often IDPs and refugees) settle in areas whose water supply systems do not satisfy a sudden increase in population density. However, challenges to Cameroon's water resource management system are not limited to just inaccessibility and unaffordability. They also relate to the capability of the water providers to continuously supply water.

The urbanization pattern has mostly been a challenge to urban and peri-urban dwellers [31, 61]. The constellation has drastically been affected by the Anglophone crisis. Focalizing on the Northwest region (i.e., one of the Anglophone regions in Cameroon), Tantoh and Simatele [31] find the weakness in governance as one of the challenges to water resource management. In this regard, fragility exists in the form of weak institutional and policy frameworks that manage water resources. In Cameroon, water is managed by centralized national institutions. The inefficiency in such governance structure has led to the exclusion of some local communities and, commonly, to the Anglophone region. The Northwest's lack of community participation in water management systems is an indicator of failure to address local ecosystem services and human–nature relations [62]. Although the Cameroonian government has indicated the need for devolution, there has been no effort to enhance local communities' capabilities to manage water resources. Other sectors affected by the Anglophone crisis include resources and tourism. First, the two Anglophone regions are endowed with natural resources, including oil reserves. Armed conflict between the separationists and the government has hindered smooth exploitation of resources and significantly hampered SDGs' achievement. Dicklitch [38] argues the Anglophone regions' agenda of secession will always be suppressed by the government due to these resources' value. Second, tourism is a significant revenue source and major economic sectors in achieving SDGs; however, insecurity remains a key challenge [63]. Unfortunately, tourism in Cameroon has been highly negatively impacted by the Anglophone crisis. The Southwest region is one of the tourist hubs for the country, but the high rates of violence, abduction, and kidnapping make it a less competitive tourist destination. Obviously, this has constrained the country's source of revenue to implement the sustainable development agenda [64, 65].

5 Attempts to Address the Anglophone Crisis Within the SDGs Framework

Cameroon is a United Nations member and committed to achieving the 2030 agenda for sustainable development. Against this backdrop, Cameroon has established a policy framework for mainstreaming the SDGs framework in its development processes. The country has recently established an SDGs data portal to monitor the progress in the achievement of the goals. The country has also made an effort to contextualize the SDGs to best fit Cameroon's social and economic setting. This transitional state is based on five cross-cutting and overlapping areas of governance and development: (1) changing relations between the Cameroonian government and people over development issues, (2) the ability of the Cameroonian government to coordinate development initiatives across different sectors in all regions of the country, (3) increased and equal participation of people in the development process, (4) equal distribution of development processes, and (5) governmental ability to deal with conflicts related to the expression of grievances to avoid conflict escalation. These five perspectives are intertwined and can hardly be examined distinctively. However, at the time of writing this chapter, the portal was still new, and SDGs data does not give any significant trend. Two broadly spoken solutions by Anglophone leaders have been established: true federalism and secession. In 1994, the Bamenda Conference on constitutional review led to the Bamenda Proclamation, which provided for the federation system of governance. As such, both Anglophone and Francophone were to be federal units under Cameroon. The concerns of Anglophone leaders were, however, not acknowledged in the constitutional reforms. Since federalism was not an option, Anglophone leaders have been reluctant to dialogue with the government [36, 66]. In 2003, Anglophone Cameroon went to the African Commission on Human and Peoples Rights [67] in an attempt to use legal means to address the crisis. Yet, without hope for either returning to federalism or getting significant intervention from the international community, as Mehler [68] suggests, secession and violent conflict is increasingly becoming the final resort (see also Levintova [69] and Okereke [70]).

Civil society organizations are essential players in the implementation of SDGs. Recently, the Cameroonian government and non-governmental organizations have not had a good working relationship in addressing the Anglophone problem [38]. Due to limited national space for expressing historical marginalization, the Anglophone community has used cyberspace to externalize their grievances. Despite being silenced within Cameroon, different Anglophone-based virtual communities have capitalized on cyberspace to reach out to the international community [66, 71]. This is what Fokwang [71] terms "marginal people moving out of marginal places," i.e., political involvement outside the confinement of formal state structures. One primary form of marginalization in the Anglophone crisis is the lack of opportunity to express social inequality adequately. SDG 16 and its targets encourage governments to strengthen national institutions by building capacities at all levels and guide governments to develop just and non-discriminatory laws. In 1995, Anglophones took the problem (i.e., historical discrimination) to the international community seeking a solution. Southern Cameroon People's Organization had sought legal and diplomatic channels of addressing the crisis by visiting the United Kingdom and the United Nations to present the Anglophone grievances. Unfortunately, this did not yield much due to a lack of internationalization of Cameroon's Anglophone agenda [72]. It also became a challenge to address the issue at the international judicial courts. Specifically, SDG 16 provides a good starting point for international cooperation and a guide for reenergized international community's effort to prevent further fragility originating from the Anglophone crisis. The guerrilla warfare in Anglophone Cameroon is not only due to the government's failure but also by the reluctance of the international community to intervene.

The number of IDPs continually increase in the Anglophone regions and, consequently, a high need for humanitarian assistance. Although the Anglophone crisis appears to restrain the achievement of SDGs, it is not in absolute terms. With the consciousness of SDGs, development actors have increasingly linked the concept of justice to security and sustainable development, i.e., justice–security–inclusive development. Creating a conducive business environment depends on inclusivity that links to economic growth and social equity [73]. This is a development model embedded in the security sector reforms to achieve sustainable development [49, 74]. The conceptualization of security as a precondition for SDGs' achievement is already advancing at the United Nations level [30]. Domesticating SDGs is a critical task to tailor solutions for the local challenges—both institutional and human development issues—and communication for development.

6 Conclusion

On the one hand, SDGs provide a framework for addressing conflict and fragility. On the flip side, violent conflicts are a hindrance to the achievement of SDGs. Implementing the SDGs means changing the structure and the agency. Through human development, human settlements are expected to change from violent to peaceful with lawful approaches of interest articulation. At the same time, strong and legitimate institutions are prerequisites for behavioral change. The fragility in Cameroon takes two major perspectives: the Anglophone crisis and violent conflict by Boko Haram terrorists in the Far North region (this chapter did not address the latter). The Anglophone crisis is a product of weak public institutions marked by a lack of inclusivity—there is a limited space to participate in the development process among Anglophone Cameroonians. SDGs have the potential to convene the international community in resolving the Anglophone crisis. Conventionally, justice, security, and development were seen as three related domains with different actors. The SDGs framework brings them together to establish strong governance and social equity. SDG 16 expands the concept of governance and development as fairness, empowerment, inclusivity, and non-dominance both at the national and international levels. It also calls for national institutions' empowerment through international cooperation to deter violent means of expressing grievances. In implementing SDGs, the conflict between the separationists and the government is a call to prioritize Anglophone crisis management as a development agenda for the country. To resolve the Anglophone crisis, the SDGs themselves urge consolidation between national and international institutions. In this case, SDGs are a means to and the product of addressing the Anglophone crisis.

Contrary to the universal understanding, in the context of countries affected by fragility, SDGs are not just development tools to pick up from where United Nations Millennium Development Goals stopped. They are, as well, instruments to reflect back on human settlements' historical and structural foundations that constitute the root causes of fragility and conflict. SDGs call for international cooperation toward a fundamental change, replacing the root causes of vulnerability and fragility with pillars of justice and inclusivity and human security to attain sustainable development. The effectiveness of enhancing transformation, in turn, depends on how development agents conceptualize and use SGDs in framing sustainable development in countries experiencing similar situations of fragility. This chapter calls for strong "communication for development" in the implementation of SDGs. There is a need for further quantitative research on the extent to which governments and other development actors under the SDGs framework have addressed the root causes of states' fragility and violent conflicts in developing countries. In the case of Cameroon, it is only by tackling the root causes of the Anglophone crisis that the sustainability and future outlook of business operations is guaranteed.

References

1. Asongu SA, Odhiambo NM (2019) Challenges of doing business in africa: a systematic review. J African Bus 20:259–268. https://doi.org/10.1080/15228916.2019.1582294
2. Singh RK, Murty HR, Gupta SK, Dikshit AK (2009) An overview of sustainability assessment methodologies. Ecol Indic 9:189–212. https://doi.org/10.1016/j.ecolind.2008.05.011
3. Burgess SM, Malhotra NK (2020) Marketing in African emerging markets: emerging perspectives. J African Bus 21:433–438. https://doi.org/10.1080/15228916.2020.1785655
4. Ncube M, Soonawalla K, Hausken K (2021) The links between business environment, economic growth and social equity: a study of African Countries. J African Bus 22:61–84. https://doi.org/10.1080/15228916.2019.1695184
5. Asongu S, Folarin O, Biekpe N (2019) The stability of demand for money in the proposed Southern African Monetary Union. Int J Emerg Mark 15:222–244. https://doi.org/10.1108/IJOEM-08-2018-0443
6. Kuada J, Mensah E (2020) Knowledge transfer in the emerging solar energy sector in Ghana. Contemp Soc Sci 15:82–97. https://doi.org/10.1080/21582041.2018.1510132
7. Asongu S, Boateng A (2018) Introduction to special issue: mobile technologies and inclusive development in Africa. J African Bus 19:297–301. https://doi.org/10.1080/15228916.2018.1481307
8. Nygård HM (2017) Achieving the sustainable development agenda: the governance—conflict nexus. Int Area Stud Rev 20:3–18. https://doi.org/10.1177/2233865916683609
9. UNDP (2019) Human development index: United Nations Development Programme. http://www.hdr.undp.org/en/2019-report. Accessed 17 Jul 2019

10. Kodila-Tedika O, Simplice A (2016) State fragility, rent seeking and lobbying: evidence from African data. Int J Soc Econ 43:1016–1030. https://doi.org/10.1108/IJSE-11-2014-0234
11. Bagayoko N, Hutchful E, Luckham R (2016) Hybrid security governance in Africa: rethinking the foundations of security, justice and legitimate public authority. Conflict, Secur Dev 16:1–32. https://doi.org/10.1080/14678802.2016.1136137
12. OECD (2016) States of fragility 2016: understanding violence. Organisation for Economic Co-operation and Development, Paris
13. Busco C (2009) Giddens' structuration theory and its implications for management accounting research. J Manag Gov 13:249–260. https://doi.org/10.1007/s10997-008-9081-6
14. Lamsal M (2012) The structuration approach of Anthony Giddens. Himal J Sociol Anthropol 5:111–122. https://doi.org/10.3126/hjsa.v5i0.7043
15. Burgess SM (2020) Making a global impact the institutions-based view and African contingency theory development. J African Bus 1–23. https://doi.org/10.1080/15228916.2020.1785657
16. Giddens A (1984) The constitution of society. Polity Press, Cambridge
17. Etzion D, Ferraro F (2010) The role of analogy in the institutionalization of sustainability reporting. Organ Sci 21:1092–1107. https://doi.org/10.1287/orsc.1090.0494
18. Rendtorff JD (2020) Corporate citizenship, stakeholder management and Sustainable Development Goals (SDGs) in financial institutions and capital markets. J Cap Mark Stud 4:47–59. https://doi.org/10.1108/jcms-06-2020-0021
19. Scoones I, Stirling A, Abrol D et al (2020) Transformations to sustainability: combining structural, systemic and enabling approaches. Curr Opin Environ Sustain 42:65–75. https://doi.org/10.1016/j.cosust.2019.12.004
20. Goubran S, Cucuzzella C (2019) Integrating the sustainable development goals in building projects. J Sustain Res 1:e190010. https://doi.org/10.20900/jsr20190010
21. Nazar R, Chaudhry IS, Ali S, Faheem M (2018) Role of quality education for sustainable development goals (SDGs). Int J Soc Sci 4:486–501. https://doi.org/10.20319/pijss.2018.42.486501
22. Soudien C (2019) The significance of new humanism for education and development. Prospects 47:309–320. https://doi.org/10.1007/s11125-018-9440-2
23. Lockwood M, Davidson J, Curtis A et al (2010) Governance principles for natural resource management. Soc Nat Resour 23:986–1001. https://doi.org/10.1080/08941920802178214
24. United Nations (2019) Sustainable development goals knowledge platform. https://sustainabledevelopment.un.org/. Accessed 28 May 2019
25. Siakwah P, Musavengane R, Leonard L (2020) Tourism governance and attainment of the sustainable development goals in Africa. Tour Plan Dev 17:355–383. https://doi.org/10.1080/21568316.2019.1600160
26. Schlenther BO (2016) Addressing illicit financial flows in Africa: how broad is the whole of government approach supposed to be? J Financ Crime 23:1074–1091. https://doi.org/10.1108/JFC-01-2016-0002
27. Kolk A, Lenfant F (2015) Cross-sector collaboration, institutional gaps, and fragility: the role of social innovation partnerships in a conflict-affected region. J Public Policy Mark 34:287–303. https://doi.org/10.1509/jppm.14.157
28. OECD (2014) Illicit financial flows from developing countries: measuring OECD responses. Organisation for Economic Co-operation and Development
29. AbouZahr C, De Savigny D, Mikkelsen L et al (2015) Towards universal civil registration and vital statistics systems: the time is now. Lancet 386:1407–1418. https://doi.org/10.1016/S0140-6736(15)60170-2
30. UN Chronicle (2018) Advancing disarmament within the 2030 agenda for sustainable development: volume 2
31. Tantoh HB, Simatele D (2018) Complexity and uncertainty in water resource governance in Northwest Cameroon: reconnoitring the challenges and potential of community-based water resource management. Land Use Policy 75:237–251. https://doi.org/10.1016/j.landusepol.2018.03.044

32. World Bank (2015) The World Bank annual report 2015. World Bank, New York
33. World Bank (2019) Helping Countries navigate a volatile environment. https://www.worldbank.org/en/topic/fragilityconflictviolence/overview. Accessed 2 June 2019
34. Kolk A, Lenfant F (2015) Partnerships for peace and development in fragile states: identifying missing links. Acad Manag Perspect 29:422–437. https://doi.org/10.5465/amp.2013.0122
35. NISC (2015) Population of Cameroon. National Institute of Statistics Cameroon Yaounde, Cameroon
36. Anyefru E (2010) Paradoxes of internationalisation of the Anglophone problem in Cameroon. J Contemp African Stud 28:85–101. https://doi.org/10.1080/02589000903542624
37. Pommerolle M-E, Heungoup HDM (2017) The “Anglophone crisis”: a tale of the Cameroonian postcolony. Afr Aff (Lond) 116:526–538. https://doi.org/10.1093/afraf/adx021
38. Dicklitch S (2011) The Southern Cameroons and minority rights in Cameroon. J Contemp African Stud 29:49–62. https://doi.org/10.1080/02589001.2011.533059
39. Eyoh D (1998) Conflicting narratives of anglophone protest and the politics of identity in Cameroon. J Contemp African Stud 16:249–276. https://doi.org/10.1080/02589009808729630
40. Kouega J-P (2018) The benefits of French-English bilingual competence in Cameroon. Open Access Libr J 5:e4409. https://doi.org/10.4236/oalib.1104409
41. Kouega J-P (2017) The Anglophone problem in Cameroon: focus on education. The Voice 53:6–7. https://doi.org/10.4236/oalib.1104408
42. Willis R (2018) A comparative analysis of widow dispossession in francophone and anglophone Cameroon. J Afr Law 62:147–174. https://doi.org/10.1017/S0021855318000013
43. Kiye ME (2015) The repugnancy and incompatibility tests and customary law in anglophone Cameroon. African Stud Q 15:85–106
44. Eposi Ngeve R, Tabe Egbe Orock R (2012) In the name of “development”: ethnic politics and multicultural public policy in Cameroon. Int J Sociol Soc Policy 32:214–232. https://doi.org/10.1108/01443331211214776
45. Nounkeu CT (2020) Facebook and fake news in the “Anglophone Crisis” in Cameroon. African Journal Stud 41:20–35. https://doi.org/10.1080/23743670.2020.1812102
46. Nganji JT, Cockburn L (2020) Use of Twitter in the Cameroon Anglophone crisis. Behav Inf Technol 39:267–287. https://doi.org/10.1080/0144929X.2019.1620333
47. Ashukem JCN (2021) To give a dog a bad name to kill it–Cameroon’s anti-terrorism law as a strategic framework for human rights’ violations. J Contemp African Stud 39:119–134. https://doi.org/10.1080/02589001.2020.1839633
48. Agbor AA, Njieassam EE (2019) Beyond the contours of normally acceptable political violence: is Cameroon a conflict/transitional society in the offing? Potchefstroom Electron Law J 22:1–32. https://doi.org/10.17159/1727-3781/2019/v22i0a4961
49. Detzner S (2017) Modern post-conflict security sector reform in Africa: patterns of success and failure. African Secur Rev 26:116–142. https://doi.org/10.1080/10246029.2017.1302706
50. Bawack R (2018) Academic libraries in Cameroon: achieving agenda 2030 goals. Int Inf Libr Rev 50:63–66. https://doi.org/10.1080/10572317.2018.1439148
51. Elias GC, Nagy KH (2016) Promoting inclusive information society. In: Elias GC, Nagy KH (eds) Mastering digital transformation (Innovation, technology, and education for growth, Volume). Emerald, London, pp 143–165
52. Robert NB (2020) Uprising and human rights abuses in Southern Cameroon-Ambazonia. In: Nugmanova M, Mikkola H, Rozanov A, Komleva V (eds) Education, human rights and peace in sustainable development. IntechOpen, London
53. Ingenbleek PTM (2019) The endogenous African business: why and how it is different, why it is emerging now and why it matters. J African Bus 20:195–205. https://doi.org/10.1080/15228916.2019.1583979
54. Pahl-Wostl C (2015) Water governance in the face of global change: From understanding to transformation. Springer, New York
55. Ako AA, Eyong GET, Nkeng GE (2010) Water resources management and integrated water resources management (IWRM) in Cameroon. Water Resour Manag 24:871–888. https://doi.org/10.1007/s11269-009-9476-4

56. Tantoh HB, Simatele D (2017) Community-based water resource management in North-west Cameroon: the role of potable water supply in community development. South African Geogr J 99:166–183. https://doi.org/10.1080/03736245.2016.1208589
57. Mafany GT, Fantong WY, Nkeng G (2006) Groundwater quality in Cameroon and its vulnerability to pollution. In: Xu Y, Usher B (eds) Groundwater pollution in Africa. CRC Press, London, pp 61–70
58. Aretouyap Z, Osang Liku E, Bernard Nka L, et al (2017) Water and hygiene quality in the Borgop-Cameroon refugee camp and its potential adverse impacts on environment and public health. J Environ Sci Public Heal 01:139–150. https://doi.org/10.26502/jesph.96120014
59. UNHCR (2018) Update of UNHCR's operations in Africa, Sixty-ninth Session 1–5 October 2018. United Nations High Commissioner for Refugees. https://www.unhcr.org/5b9f537a4.pdf
60. Taylor PL, Sonnenfeld DA (2017) Water crises and institutions: inventing and reinventing governance in an era of uncertainty. Soc Nat Resour 30:395–403. https://doi.org/10.1080/08941920.2017.1274208
61. Fonjong L, Fokum V (2017) Water crisis and options for effective water provision in urban and peri-urban areas in Cameroon. Soc Nat Resour 30:488–505. https://doi.org/10.1080/08941920.2016.1273414
62. Cirella GT, Mwangi SW, Paczoski A, Abebe ST (2020) Chapter 1. Human-nature relations: the unwanted filibuster. In: Cirella GT (ed) Sustainabile human-nature relations: environmental scholarship, economic evaluation, urban strategies. Springer Singapore, Singapore
63. Christie I, Fernandes E, Messerli H, Twining-Ward L (2014) Tourism in Africa: harnessing tourism for growth and improved livelihoods. World Bank, Washington, DC
64. Tichaawa TM, Kimbu AN (2019) Unlocking policy impediments for service delivery in tourism firms: evidence from small and medium sized hotels in Sub-Saharan Africa. Tour Plan Dev 16:179–196. https://doi.org/10.1080/21568316.2018.1556328
65. Unathi Sonwabile K, Tembi MT (2018) Sustainable development goals and Socio-economic development through Tourism in central Africa: myth or reality? Geoj Tour Geosites 23:780. https://doi.org/10.30892/gtg.23314-328
66. Anyefru E (2008) Cyber-nationalism: the imagined Anglophone Cameroon community in cyberspace. Afr Identities 6:253–274. https://doi.org/10.1080/14725840802223572
67. African Commission on Human and Peoples Rights (2003) Dr K. Ngwang Gumne and others, acting on their behalf and on behalf of SCNC/SCAPO and the Southern Cameroons versus La Republique du Cameroun. In: The Banjul Communique 37th session of the African commission on human and peoples rights. Communication no. 266/2003
68. Mehler A (2014) Why federalism did not lead to secession in Cameroon. Ethnopolitics 13:48–66. https://doi.org/10.1080/17449057.2013.844435
69. Levintova E (2017) A tale of two independence movements: how do citizens encounter issues of international law and democratic theory? Natl Pap 45:1167–1188. https://doi.org/10.1080/00905992.2017.1299121
70. Okereke N (2018) Analysing Cameroon's anglophone crisis. Count Terror Trends Anal 10:8–12
71. Fokwang J (2016) Politics at the margins: alternative sites of political involvement among young people in Cameroon. Can J Afr Stud 50:211–228. https://doi.org/10.1080/00083968.2016.1219671
72. Edward H (2018) The Force of argument and the fight for an anglophone identity in Cameroon. SSRN Electron J 2529021. https://doi.org/10.2139/ssrn.2529021
73. Mefoute Badiang A (2018) The meanings of food consumption, consumer effects and the role of "Origin": the case of ethnic food in Cameroon. J African Bus 19:531–549. https://doi.org/10.1080/15228916.2018.1476957
74. Berhe MG, de Waal A (2017) Peace missions and the security sector in Africa. African Secur Rev 26:113–115. https://doi.org/10.1080/10246029.2017.1303074

The Pearl's Social and Environmental Failures: Development Challenges

Michael J. Rosciszewski-Dodgson and Giuseppe T. Cirella

Abstract Identifying the problems in past construction and communicating those lessons in future development will aid the process of successful innovation. This chapter identifies the social and environmental failures of Qatar's Pearl Island, in order to analyze how MIA Park and Hamad Port overcome those challenges. Through the qualitative research process, semi-structured interviews were conducted with key informants and sequences of photographs have been utilized to construct visual investigation. The results demonstrate how Qatar has learnt from past development failures and applied the knowledge in their new ventures. These projects still have several design and construction flaws that require revaluation if Qatar hopes to move forward with their sustainability initiatives. According to the Global Footprint Network, Qatar, alongside other wealthy Gulf states, has the world's largest ecological footprint per person. Lucrative gas and oil revenue, since the 1970s, has rapidly allowed the Middle East to develop and create desert cities with budgets exceeding billions of dollars to create unique landscapes few places in the world could match. Qatar's growth in the past decade has been especially tremendous, with the total population increasing by one million, enabling the country's urban districts to conceptualize artificial islands and mega-structures that have been poorly developed as a result of short-term thinking. This chapter looks at the development of these human settlement projects in light of their immense social and environmental repercussions.

Keywords Artificial Island · Mega-structure · Desert city · Sustainability · MIA Park · Hamad Port · Qatar

M. J. Rosciszewski-Dodgson (✉)
Faculty of Science and Engineering, University of Liverpool, Liverpool, UK

G. T. Cirella
Faculty of Economics, University of Gdansk, Sopot, Poland
e-mail: gt.cirella@ug.edu.pl

G. T. Cirella (ed.), *Human Settlements*, Advances in 21st Century Human Settlements,
https://doi.org/10.1007/978-981-16-4031-5_12

1 Introduction: Projects of Doom

According to the latest Global Footprint Network [1] data, Qatar, alongside other rich Gulf states, has the world's largest ecological footprint per person of any region in the world. As such, lucrative gas and oil revenue, since the 1970s, has rapidly elevated development in certain Middle Eastern countries. Desert cities, described by Di Muzio [2], have sprung up as playgrounds for the rich creating unique, modern landscapes unrivaled anywhere in the world [3]. In the past decade, Qatar's growth has been exceptionally fast with the total population increasing by one million—enabling it to develop a focal mecca of urban projects countrywide [4]. However, these urban districts have been conceptualized to contain destructive developments [2]. Artificial islands from the region, in particular, exemplify the "projects of doom" notion that have carelessly been developed as a result of short-term thinking [2]. These mega-structures were molded in such a way that their unique flattened shapes can be seen from space in order to gain world press and provide prime real estate [3]. Yet, the failure of these projects is widely asserted from a range of academic sources, which conclude that the construction of artificial islands has had immense social and environmental repercussions. Dubai's Palm Jumeriah pioneered artificial island development in the region and encouraged other Arab countries to create their own islands [3, 5–9]. Thus, The Pearl was built in a rushed attempt to keep up with the United Arab Emirates' geopolitical strategy [10]. Plans for The Pearl were first revealed in 2004 and developed using Qatar's old planning laws, which sped up the construction process by undermining environmental mitigation measures. Moreover, since the project's completion, several design and cultural flaws have emerged which have rendered the social life on the island as unpleasant and isolating. As a result, The Pearl and other artificial islands have tarnished the reputation of developments among The Gulf's richest countries, by deeming their expansion as unsustainable.

It would be easy to judge Qatar's unsustainable behavioral patterns and conclude that the country's developments are destructive and ill thought out. However, it is important to acknowledge how Qatar is working to change its poor social and environmental image. In 2011, Qatar's development planning agency launched two key documents under the General Secretariat for Development Planning (GSDP), which set out goals of the country's expansion for a five-year period (i.e., 2011–2016) and their national vision of 2030 [4] that is in-line with the United Nations Sustainable Development Goals (SDGs) [11]. These plans recognize the unsustainability of fast-paced developments and prescribe measures to abolish them [4]. Hence, it would be unrepresentative to criticize Qatar's Pearl Island without identifying how the government has overcome these failures with new projects. This fostering of sustainability in Qatar can be applied to all the other rich Gulf states, who similarly only recently begun implementing measures to improve their future developments [12]. This part of the world needs to expand sustainably in the decades to come, as these countries face the inevitable threat of climate change and the dying of the fossil fuel powered economy (Fig. 1) [2, 13–16]. Therefore, the way these places handle the next phase of developments is crucial for the well-being of people and the environment. As

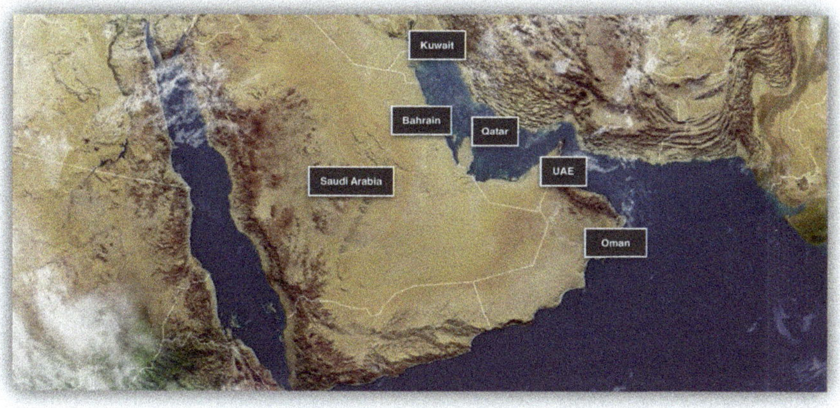

Fig. 1 Satellite image of the Middle East, adapted from Google Earth [17]

such, it is important to evaluate the successes of Qatar's new projects in order to identify how these countries can improve their design and construction processes in the future.

This aim of this chapter is to identify The Pearl Island's social and environmental failures in order to understand how those challenges have been overcome with Qatar's new developments. As a result, this chapter recognizes MIA Park and Hamad Port as post-2011 developments that have utilized the new GSDP legislature to create projects which considered sustainable values. However, Luomi [4] argues that due to their novice understanding of what constitutes as sustainable development, Qatar's recently introduced measures and newly developed projects still have limitations. By including the views of Qatar-based experts as well as people closely linked to the projects, i.e., through the evaluation of past literature, semi-interviews, and photographic evidence, this chapter seeks to address the existing problems and potential pitfalls in Qatar's construction and design process in order to suggest functional solutions. The study is categorized into social and environmental themes which created the following themed questions:

(1) Social theme: Why did The Pearl fail to prevent its social failures? What aspects of MIA Park's design make its socially superior to The Pearl?
(2) Environmental theme: Why did The Pearl fail to mitigate its environmental failures? How did Hamad Port's construction measures reduce its environmental impacts?

A deeper underrating of these questions enables the research to pinpoint the reasons why MIA Park and Hamad Port overcame The Pearl's failures (Fig. 2). In addition, they will identify any social and environmental limitations these projects still have in order to recognize how Qatar's future developments can improve upon them.

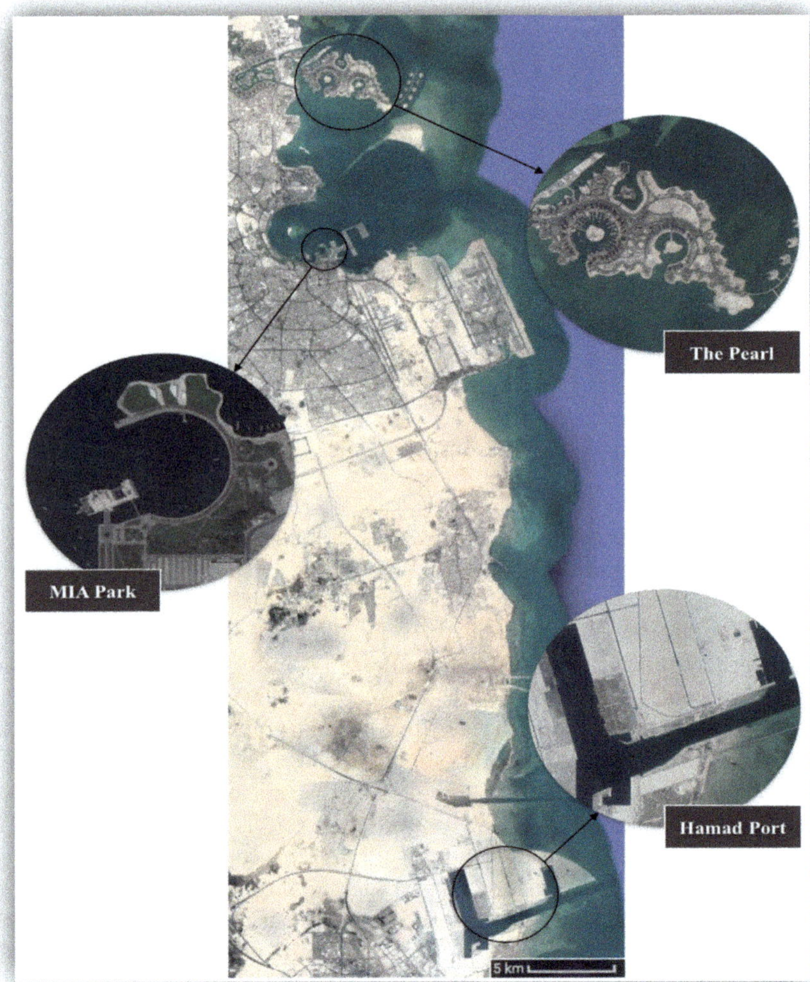

Fig. 2 Satellite view of Qatar's peninsula displaying the locations of the study, adapted from Google Earth [17]

2 Literature Review

The general consensus identified within literature, which approached the environmental and social impacts of artificial islands in the Middle East, deemed their construction a failure. Their failures can be assessed in a number of ways, yet Gibson et al. [18] identified how sustainable development has been adopted as the leading assessment criteria for determining whether projects were successfully built. The

Fig. 3 Aerial photo of The Pearl Islands in Doha, Qatar, photographed by Konevi from Pixabay

National Research Council [19] states that successful projects should generate social benefits that facilitate a responsible use of natural resources in order to minimize impacts. There has been a limited amount of academic studies done on The Pearl. However, Salahuddin [20] emphasized how the construction process of artificial islands in the Middle East has been the same. As such, social and environmental failures of Dubai's Palm Jumeirah can be applied to The Pearl (Fig. 3). Moreover, since all human development entails some risk and design failures are unavoidable [21], there are many lessons to be learnt from failures—so failures should be approached with the benefit of hindsight. With the focus on the social and environmental failures of artificial islands, it can be demonstrated how Qatar's new projects are moving forward with sustainable initiatives.

2.1 Social Failures: Urban Design and Sociology

The scope of social failures identified by academics focused on the artificial islands flawed design. A flawed architectural design according to Campbell [21] is one of the biggest failures any development could have, as it means that the project failed to comply with its requirements. He points out how design flaws are easier to solve at the preliminary design stage; therefore, developers should work closely together to foresee all the complexities of a project [21]. This according to Graham [3] and Muneerudeen et al. [22] expressed how artificial island developers who failed to properly consult pre-development resulted in projects having multiple design flaws. Graham [3] emphasized how Palm Jumeriah's geometrical design may look impressive, but residents have complained that these shapes have created pockets of stagnant water that became breeding grounds for algae and mosquitoes. Muneerudeen et al. [22] who conducted a study on the urban sociology of The Pearl noted how the design has negatively affected the dwellers because it prevents social interaction. They reiterated how sparse shading, lack of foot paths, high traffic, and the absence of green areas have made The Pearl's streetscape hostile. Through their study, it was emphasized that the island needs to convert car lanes into pedestrian-only streets, provide more outdoor seating, plant more vegetation, and supply additional outdoor shading to allow people to stay outside longer. Moreover, it was further suggested that the best urban design for The Pearl would have been mixed development that did not separate commercial and residential zones. Consequently, developers ignored Muneerudeen et al.'s [22] suggestions, and The Pearl's urban design is failing its inhabitants.

Culturally inappropriate design and highly priced dwellings are one of the main reasons why The Pearl has only attracted 33,000 residents out of the 45,000 available dwellings [23]. Graham [3] also discussed how the artificial island's market size was overestimated, since the Westernized design has produced a space that is less attractive except to a select group of people. Apart from the economic setbacks these empty dwellings create, this lack of people has social repercussions whereby The Pearl's residential zones can often feel like a ghost town. Muneerudeen et al. [22] reiterated how the majority of people encountered on the island were tourists; therefore, they argued that The Pearl did not have a stable community because it felt like a tourist destination rather than a residential complex. Jalinoosi and Moradifar [24] explored these community tensions further, and they argued that the development of artificial islands was a mistake because they were built with the intention of being sold to rich Europeans and Americans. They see this as a social failure because such islands create predominately Christian communities in Muslim countries [24]. Their research comes to the conclusion that such non-Muslim minorities destroy the coherent cultural–social structure because they "will lead to a change in the demographic composition regarding religion and ethnicity" [24]. This, however, may not fully reflect the reality of The Pearl's community from a number of in-person accounts from fieldwork over the past few years. Nonetheless, it does show how there is a potential concern over the influence The Pearl's Westernized design and demographic on Qatar's social structure.

2.2 Environmental Failures

To understand the extent of environmental damage caused through the construction of artificial islands, the local ecosystem needs to be understood. This is important because Campbell [21] emphasized that the inadequate information of the local environment causes further depravation to the ecosystem. Salahuddin [20] underscored how island developers underestimated the fragility of the local environment which worsened the extent of failures. Through the use of Himaayat's local research, Salahuddin [20] noted how The Gulf is particularly fragile because the sea is almost completely surrounded by dry land with shallow waters that range in depths of 35–85 m. As such, the environment is more prone to construction damage. It was also brought to attention how summer temperatures climb to 50 °C which can result in evaporation loss of 0.6 m^3 of water daily. In addition, The Gulf experiences strong winds which generate high waves and strong currents. Furthermore, considerable lack of rainfall has increased The Gulf's saltiness ratio to 41.5 g in 1 kg of water. Consequently, these risks have been heightened by artificial island development [20].

The cultural standards under which Palm Jumeirah was built have also been scrutinized by numerous journals. Campbell [21] wrote how a great deal of failures is caused by a contractor's carless approach. Salahuddin [20] emphasized how environmental issues have never been at the forefront of people's consciousness, as a result when artificial islands were being constructed developers dismissed environmental impacts. She highlighted how similarly to The Pearl, developers of Palm Jumeirah did not abide by federal environmental laws, because they were passed after the planning for the island commenced. Ghaffari et al. [10], in particular, used the perspective of international law to evaluate the environmental failures of artificial islands. By citing [20], they explain how Palm Jumeirah's construction was a complete violation of international regulations that would have required further evaluation of the local ecosystem. Other broken environmental laws included the principle of prevention, which would have banned certain construction techniques and the precautionary principle which would have imposed safety measures that avoid possible damage [10]. Salahuddin [20] notes that so many environmental laws were ignored because they were conceptually novel in the Middle East and developers were not skilled at enforcing them. It can be concluded that "if the developers fail to undertake these measures, the marine environment surrounding each artificial island will deteriorate" [20]. Salahuddin's [20] work carries great value because it provided early warning signs on the environmental impacts of artificial islands.

Many articles discussed the destructive methods used in the construction of artificial islands to assess environmental failures. Campbell [21] emphasized how even though the built environment is regulated by rules and procedures, which aim to prevent failures, construction damage is inevitable. With artificial islands, Poole [8] identified how engineering techniques used to build such projects had exceptionally devastating impacts. Using the example of Palm Jumeirah, he explains how the island's foundation was made up of two separate layers of rocks that was placed under geo-textiles and engulfed in 8,500,000 m^3 of sand. Consequently, Dubai lost over 1.2 km^2 of coral reef ecosystems [8]. Gibling's [25] research similarly evaluated post-construction impacts of Palm Jumeirah. By citing Hopkins [26], he wrote how the dredging and land reclamation process deposited and scattered silt in the water which buried coral reefs and sea grass in two inches of sediment [25]. Ultimately, both Gibling [25] and Poole [8] similarly conclude that such techniques result in heavy environmental loss. To estimate the magnitude of Palm Jumeirah's impacts, research was done on the environment after construction had been completed. Hopkin's [26] investigation is particularly valuable because he was instructed by the developers to write a report about the state of the local marine ecosystem. Monitoring was conducted by Scuba divers who used visual inspection and photography to record fish populations. Furthermore, fauna sites were monitored to see if over the period of 25 months their coverage spread. The study found that reefs within six months have demonstrated an ability to recolonize some species [26]. Gibling [25] similarly acknowledged how the breakwater of the island attracted a diverse number of marine animals. By citing Salahuddin [20], he reiterated that polythene and concrete materials used in the construction of islands created artificial reefs. Yet, Hopkins [26] identified several methodology restrictions when the research was conducted. He emphasized how the monitoring schedule was not up to the usual standard of quantitate scientific studies, concluding that the results do not give complete representation of the marine ecosystem. As such, he recommended that follow-up tests should be planned. Later literature such as Poole's [8] research identified how since the construction's completion, every species of fauna that once existed was destroyed and less than half of the 450 local fish species returned. Ghaffari et al. [10] also identified how Palm Jumeirah's construction resulted in environmental threats to the security of migrating birds, destruction of live coral, change in the natural water flow, seabed destruction, and the disposal of waste materials. Thus, even though some literature suggests that Palm Jumeirah's impacts are undetermined, most concluded that its construction had significant environmental repercussions.

Following the 2008 global financial crisis, The Gulf states issued a lengthy hiatus where development of artificial islands slowed down, in that time Qatar completed The Pearl Island but plans to develop another artificial island were abandoned [27]. It became clear that SDGs became the governments drive toward Qatar's national vision of 2030. In the years that followed, the country began developing a manageable system of sustainable development in Qatar's 2011 GSDP policies, which considered social and environmental values [27]. Ibrahim [27] emphasized how the government created an environmental policy for handling a project's direct and indirect impact on the ecosystem. He further reiterated how social values have been enforced thanks to the requirement of creating architecture that is culturally appropriate. Moreover, it underscored how as part of the new monitoring program it has become mandatory for future developments to submit all the appropriate information for pre-government approval. Velegrins and Katodrytis [12] also suggested that with the introduction of these legislatures, Qatar's newly developed projects began to demonstrate the social and environmental vision the country aimed to achieve. They argue that Qatar's new developments have become more vernacular with a focus on livability, green spaces, and outdoor activities. While during the development phase, contractors are paying greater attention to environmental issues [12].

2.3 *Qatar's New Developments*

Some of the developments that overcame The Pearl's social and environmental failures are MIA Park and Hamad Port—both completed in 2016 and 2017, respectively.

To date, there is very limited academic literature written about them, which calls into question the validity of their successes as the content from their official websites is biased; nevertheless, the material displays overwhelmingly positive evidence that shows how Qatar's new projects are more socially and environmentally aware.

MIA Park [28], on the one hand, resolved some of The Pearl's social failures regarding outside activities and cultural appreciation. The park, built as an extension to the Museum of Islamic Art with loads of green space and walkways, highlighted GSDP policies. The development prioritized family play by creating three playgrounds and promoted social activities such as paddle boating, outside fitness and training classes, and bicycle rentals [28]. However, MIA Park's green network system has had certain limitations, including "(1) accessibility to/from the neighboring districts and (2) connectivity to/from the neighboring parks" [29]. Since MIA Park was built with vast open spaces to facilitate annual events such as Bazaar's where the local community can book stalls and sell handcrafts, the park has drawn significant inspiration from Islamic traditions as they highlight how the Bazaar was intended to be a modern version of the Souq Market [28]. In addition, the park's main feature designed by Richard Serra is described as a piece of art that "celebrates the spiritual and scientific significance of the number 7 in Islamic culture" [28].

Hamad Port's official Web site displays a video entitled "Hamad Port Environment" that illustrates how the project overcame some of The Pearl's environmental challenges [30]. The visual media exemplifies how the project's steering committee worked closely with consultants to ensure that a mapping of a detailed environmental strategy is created. It exhibits how the port's location was carefully selected to avoid environmental hotspots and acknowledges that because of the scale of the project, some sensitive habitats could not be avoided [30]. Thus, a lengthy explanation describes how Hamad Port mitigated its environmental impacts by providing stilling basins, independent construction oversights, and relocation programs for sea grass, corals, and mangroves. Detail is provided on how through such measures, i.e., 12,157 corals, 31,656 mangroves, and 14,257 m^2 of sea grass, have been relocated with a mostly successful survival rate. Furthermore, 576 artificial reef balls have been placed on the sea floor and are being continuously monitored [30]. In comparison, information on The Pearl's environmental strategies is unavailable because the project disregarded its local ecosystem. Therefore, even though the validity of these results could be called into question, the fact this information exists shows how Hamad Port acknowledged the importance of conserving its local habitat—indicating important sustainability-oriented development.

3 Case Methodology

This investigation utilized qualitative research analyses that created data using case studies, semi-structured interviews, and organizational photography. The data was collected using thematic analysis and qualitative methodology that focused on enhancing small amounts of information in depth [31]. This method produced in-depth data based on individual experience and perception of Qatar's developments [32]. The case study approach enabled this research to closely examine data within a specific context of three Qatari projects to answer the question regarding how a country learns from past development failures and improves upon them in future projects. Zainal [33] suggests that research which categorizes case studies chronologically can design an effective research method. Table 1 shows how the case studies were grouped and utilized for the methodology.

Overall, the case study method of observing the natural phenomenon of developments created robust in-depth investigations that detailed qualitative accounts for each project, with MIA Park representing social themes, Hamad Port depicting environmental themes, and The Pearl embodying both. The primary data was collected using the qualitative semi-structured interviewing method. This was the most appropriate interviewing method because it allowed for flexibility to probe and expand the interviewee's responses [34]. Interviewees were selected using the in-depth sampling process, whereby five key informants that had knowledge regarding Qatar's projects were contacted (Table 2). In all cases, predetermined social and environmental questions were administered via Skype [34]. Prof. Raffaello Furlan and Dr. Aspassia D. Chatziefthimiou requested to be referenced by name, while other respondents confirmed their anonymity.

In the results section, photographs were used as a means of portraying, describing, and analyzing data [35]. One category of organizational photography allows the respondents to be more involved in the research, by enabling them to supply photographs [35]. This methodology was utilized with the images Hamad Port's construction engineer and the local scuba diver supplied. The respondents were interviewed on the basis of these photographs which showed different construction techniques, relocation procedures, and the local ecosystem. These images served as a predetermined checklist that was expanded upon during the interview [35]. Groups

Table 1 Classification of Qatar's case study projects

Research	Case study	Description
Exploratory	The Pearl Island	The Pearl Island was categorized as a primary exploratory case study that investigated the phenomenon of a development that had substantial environmental and social failures. The artificial island served as a prelude of initial work that pinpointed to a small geographical area, Qatar, which served as a point of interest that shaped the research [33]
Problem solving	MIA Park Hamad Port	Mia Park and Hamad Port were established as secondary problem-solving case studies that tested the theory of how new developments overcame some of The Pearl's social and environmental failures [33]. These projects were singled out by interviewees as Qatar's newly sustainable developments, thus both projects aided in advancing the research and answering the themed questions [33]

Table 2 Position, theme, and purpose of each conducted interview

Position	Theme	Informant
Urban planner	Social	Prof. Raffaello Furlan's interview, co-author of "Urban Revitalization of Public Spaces in the Peal in Qatar" [22] paper, gave insights into the social culture that exists in Qatar. His contribution showed how MIA Park overcame The Pearl's social challenges
Environmental consultant	Environmental	Dr. Aspassia D. Chatziefthimiou as an employer of Qatar's "Richer Environments" consultancy agency revealed the legislative and enforcement setbacks that lead to the environmental failures during The Pearl's construction
Scuba diving instructor	Environmental	The scuba instructor illustrated how the marine environment adjacent to The Pearl's construction was affected
Pearl Island resident	Social	This respondent accounted his personal experiences of what life is like on The Pearl Island
Construction engineer	Environmental	The construction engineer who oversaw Hamad Port's development accounted first-hand experience of how the project mitigated environmental impacts. Furthermore, the respondent discussed changing development patterns in the Middle East and gave information regarding construction site photographs

of photos were categorized into different themes in the results section, and these sequences constructed a visual story that helped impact the investigation's discussion flow.

Thematic analysis was utilized to interpret data from all the gathered semi-structured interviews. As such, the dialog with each responded was fully transcribed and categorized into the overarching social and environmental themes. This was the most appropriate method of analysis because as Braun and Clarke [36] emphasized, thematic analysis offers flexibility to describe data in detail and interpret various aspects of the research topic. The data was then checked against each other and reinforced by previously presented information in the literature review, to pinpoint repeating patterns [36]. Those patterns were grouped into sub-categories, whereby the most relevant quotations and images were extracted from the data and analyzed in a way that would answer the research questions.

4 Piecing Together the Solutions

By reflecting on the in-depth information gathered from the five conducted interviews, analyzing the interviewee's responses, and sequencing the photographs in order to answer the themed questions, conducive findings were formulated. What are The Pearl Island's social and environmental failures and how has Qatar overcome these challenges with their new developments? The first part of the question was extensively analyzed in the literature review which detailed the types of social setbacks and environmental impacts The Pearl's development caused. However, in order to understand how Qatar's new projects overcome those challenges, a further comprehension of those failures has to be acquired. Thus, the previously mentioned themed questions are reviewed.

4.1 Social Theme

4.1.1 Why Did the Pearl Fail to Prevent Its Social Failures?

The scope of social failures identified by academics focused on The Pearl's flawed design. Most notably, Muneerudeen et al.'s [22] research identified a series of design flaws that emphasized how the streetscape lacked adequate greening, shaded areas, and footpaths that made the outside hostile. In the interview with Prof. Furlan (i.e., a co-author of Muneerudeen et al.'s (2016) work), he reiterated how The Pearl lacks green areas and sunshade devices, noting how "in the summer time you cannot even walk outside." When asked why he thought those issues were not addressed at the preliminary design stage, he said, "the real problem in my view, is not practical, its cultural, you can find a lot of green areas between roads, there are green islands at the roundabout but the problem is that these areas are only utilized for decorative purposes." Explaining how, "they've never been exposed to green areas, they don't perceive green areas the way we perceive green areas. This is the difference between for example a place like Italy and this part of the world." Furlan suggests that culturally, Qatari people are not used to undertaking outside activities such as cycling and walking in a park because their natural environment is the desert. They have become accustomed to spending their time indoors, therefore designers felt that The Pearl did not need to have large green spaces. Yet, Furlan emphasized how the problem was not the lack plantation on the island, but the lack of social interaction this caused. For the development of communities, people have to meet and establish relationships. In the Western world, parks cater for such contact by creating a space for social interactions [37–39]. Furlan mentioned how he knows European families living on the island, who find being there "isolating," because even though the outside public area is wide, its inhospitable so people are not socializing. This issue was further emphasized by the local island resident who felt that The Pearl lacks a strong community feeling, reinforcing what Jalinoosi and Moradifar [24] addressed regarding the island's social structure being made up of a rich Western minority. It was also stressed that the Souq Market, organized at MIA Park, has a vibrant lifestyle and culture that The Pearl lacks. As such, it is suggestive that the reason why The Pearl failed to prevent its social failures was because designers overlooked the needs of people.

4.1.2 What Aspects of MIA Park's Design Make Its Socially Superior to the Pearl?

As previously mentioned, the aspects of MIA Park that make it socially superior to The Pearl are its design and culture. Located at the end of Doha's Corniche as an extension of the Museum of Islamic Art, MIA Park has become the main recreation ground for the residents of the city. When asked, Furlan thought MIA park overcame some of The Pearl's social challenges, noting "in MIA park the culture is very different because the green is utilized not just for decorative purposes, they created a usable park that has become the city's biggest green area." Unlike The Pearl, MIA Park has open green spaces, footpaths, and playgrounds for families to come and spend the day. In addition, the benefit of having these large open areas is that they offer flexibility to organize sizeable public events such as bazaars. These gatherings are some of the only free of charge organized events in Doha, which promote understanding, recognition, and appreciation of different cultures [40]. Hence, MIA Park's design and cultural integration have enabled the project to have a socially vibrant community. Furlan also suggested that unlike The Pearl, Doha was in serious need of a large park. Ultimately saying that: "It's really a great intervention and this will definitely allow people to live more outdoor, it's really in my view going to be the life of Doha, this beautiful place, you know, to go for a coffee or to go jogging or meet people. It will become the heart of Doha actually." Description of MIA Park becoming "the heart" of the city draws comparisons to the way people describe New York's Central Park and London's Hyde Park. Yet, upon a closer examination of MIA Park's street view on Google Earth, the park still has several design and maintenance setbacks (Fig. 4). One of the photographs shows people crowding under sparsely planted trees, which suggests that there is still inadequate shading, while the other illustrates how trees have only been planted in certain areas of the field. This hints at the fact that Qatar's designers are still focusing on trying to make the park look aesthetically pleasing rather than practical.

Fig. 4 (left) Streetscape street view of MIA Park depicting people utilizing trees for shading and (right) unmaintained grass, adapted using Google Earth [17], March 15, 2018

4.2 Environmental Theme

4.2.1 Why Did the Pearl Fail to Mitigate Its Environmental Failures?

As shown in the literature review, The Pearl failed to mitigate its environmental impacts at every stage of its design and construction. As Salahuddin [20] underscored, the root of failures can be traced back to the governments inadequate legislation and enforcement mechanisms. In the interview with Dr. Chatziefthimiou, the environmental consultant underscored how Qatar may have introduced new policies to conserve the environment but they have not devised effective strategies to implement them. When asked whether the reason for this is because mitigation strategies have only been introduced in the last few years. She agreed and brought to attention how this is a problem for all the rich Gulf countries. Explaining how, "these are new states, they are being developed in a very fast and short amount of time, and they have the legislature in place, they are signatory to combat conventions, so they mean well but there are some problems within the system." Chatziefthimiou suggests that Qatari people in the past have not been environmentally conscious, and they developed projects that completely disregarded the local ecosystem. With The Pearl, she explained that the reason the project failed to mitigate its environmental impacts was because no prior research of the local habitat was conducted, which underscores what Ghaffari et al. [10] claimed about The Pearl violating federal environmental laws. Consequently, The Pearl's contractors had no awareness of the ecosystem that inhabited the area and could not act accordingly to protect it. When challenged with Campbell's [21] theory which suggests that in construction, ecological damage is inevitable, Chatziefthimiou agreed but felt that projects should do everything possible to reduce their impacts, noting how if The Pearl had a successful mitigation program, the local environment would have been significantly less damaged. In the interview with the local scuba diver, however, he suggested that the reason why The Pearl failed to reduce its environmental damage goes beyond its construction impacts. Highlighting that The Gulf's marine ecosystem is so fragile that mitigation programs such as artificial reef constructions are ineffective. He emphasized Salahuddin's [20] argument by stating how the extreme environment in The Gulf has exceptionally strong currents, shallow seabeds, elevated temperatures, and highly salty water that are, "basically killing everything living." For that reason, only selected species of fish and corals inhabit The Gulf sea, and this smaller diversity makes the ecosystem more prone to construction damage—reiterating Hopkin's [26] and Gibling's [25] belief of ecosystem recovery as showing no signs to date.

4.2.2 How Did Hamad Port's Construction Measures Reduce Its Environmental Impacts?

Hamad Port is a land reclamation project built 37.6 km south of The Pearl on Qatar's coast, with a width of 28.5 km^2 at a depth of 17 m, making it one of the largest ports in the Middle East (Fig. 5) [41]. What makes Hamad Port ideal to compare with to The Pearl is that both projects utilized similar construction techniques and were built around a similar ecosystem. Moreover, the mammoth scale of both endeavors meant their developments should have similarly impacted their environment. Yet, Hamad Port has not been nearly as scrutinized for its environmental damage as The Pearl. In the interview with Hamad Port's construction engineer, he reiterated how the port utilized the same equipment as The Pearl for dredging and rainbowing. Explaining how part of the port required dredging of the channel to allow the shipping to come in. As such, large amounts of sand were picked up from the seabed and transported to form new land. Various photographs provided by the respondents showed the destructive nature of these processes, including long arm excavators digging into the side of the coast and a pipe burst that occurred as a result of high pressure (Fig. 6). He acknowledged that these processes are harmful to the environment because they are

Fig. 5 Hamad Port, adapted from Google Earth [17]

Fig. 6 (left) Pipe burst from dredging; (center) long arm excavators digging in the water; (right) water drained from earth bund, photographed by the respondent in Hamad Port, 2015

invasive, noting how "this type construction will always involve some destruction of the ecosystem."

When asked, the engineer proclaimed Hamad Port reduced its environmental damage by employing an environmental consultancy to oversee the project. As such, there was an entire team of people looking out for environmental risks. He explained, "the environmental consultants we had where touring the project every day and if they found anything that they were concerned about, it would be immediately brought into attention of the contractor. The contractors would then allow a price to help mitigate it." In-line with Chatziefthimiou suggestions, the respondent called to attention how a complete study was carried out in the area, so the developers knew what ecosystem they were impacting—hence supporting Ibrahim's [27] and Velegrins and Katodrytis' [12] findings which argued that Qatar's newly developed projects acknowledged environmental well-being as a major concern. However, regardless of Hamad Port's extensive environmental impact assessment, the rate of successes for many mitigation endeavors was mixed. Photographic evidence provided by the respondents showed six identified species of birds, most notably a large colony of flamingos and osprey. During the construction process, fish would get trapped in the bund and would eventually be picked out by birds as easy prey, this altered the behavior of the birds which would actively wait by the construction site to find more fish. It was also highlighted how no birds were killed by the construction; therefore, their conservation was not a high priority.

In contrast, protection of plants was a high priority for the environmental team. This was done to such an extent that one large tree was fenced off and maintained throughout the project's development. The respondents revealed how the port's location was chosen so that the least number of mangroves would be affected, but large areas of mangoes were still at risk. As such, a great deal of effort was put into collecting and relocating them to a new location. However, contrary to what is presented in Hamad Port's official Web site [29], the interviewees admitted that large pockets of the relocated mangroves died due to not receiving sufficient amounts of water. In that situation, the contractors were penalized for killing the mangroves and suffered financial repercussions. In addition, developers had to rectify their mistakes and plant new mangroves in that location, with the interviewed engineer emphasizing how "nature being what it is will recover over a number of years." This reinforced Hopkin's [26] and Gibling's [25] research which highlighted how wildlife can reintroduce itself around developments. The respondent agreed how just like on The Pearl, Hamad Port's construction had the biggest impact on marine life. To mitigate these consequences, stilling basins were created, which enabled the particles that were being pumped out during the dredging process to settle at the bottom these areas, so that they could be returned to the sea. It was further noted that sea grass and coral reef protection were the project's biggest environmental initiatives—employing more than 20 divers as support. Their job was to replant, monitor, and grow sea grass and artificial reefs even after Hamad Port's construction had been completed.

5 Conclusion

This chapter set out to explore three Qatari projects to answer the research question regarding how a country learns from past development failures and improves upon them in future projects. This investigation utilized thematic analysis to conduct qualitative investigations of case studies. By incorporating the views of Qatar-based experts and people closely linked to the projects, the study evaluated past literature, semi-interviews, and photographic evidence to form conclusions and recommendations. As such, it addressed the existing social and environmental problems of The Pearl Island and identified how MIA Park and Hamad Port overcame those challenges. Furthermore, repeating patterns and overreaching themes revealed potential pitfalls in Qatar's new construction and design process, which helped suggest functional solutions for the country's next phase of developments.

The Pearl's social failures are a consequence of the island's convoluted design and cultural standards. The designers failed to facilitate adequate amounts of car-free zones, shaded areas, and green spaces which made the streetscape sociably unfavorable [22]. Since Qatari people are not used to facilitating infrastructure for outdoor activities, The Pearl's design overlooked the needs of Western people. All the while, in an effort to attract foreign investment, the island lacks traditionally Islamic elements that locals are accustomed to. As underscored by the residential respondent, these flawed designs and sense of community prevent social interactions between the remaining inhabitants who isolate themselves. The Pearl lacks a strong community feeling; thus, its landscape is in-cohesive, unsociable, and unable to attract an adequate number of both foreign and local investors. MIA Park overcame The Pearl's social failures by complying with Qatar's new GSDP legislatures that encouraged vernacular developments with a focus on livability, green spaces, and outdoor activities. Furlan argued that, this was achieved by utilizing green spaces not just for decorative purposes, but with a functional usage in mind. MIA Park's large green fields create a space to socialize and offer the flexibility to organize sizeable public events that promote local culture. Furthermore, Doha lacked a sizeable park, as such, once the project was completed, people actively began to use it. MIA Park was a necessary development for the country's sustainability vision of 2030 that has become an oasis in Qatar's harsh desert climate.

The root of The Pearl's environmental failures can be traced back to the issues ignored at the preliminary design stage. As Chatziefthimiou emphasized, when the island was developed, Qatari people had not been environmentally conscious, so no prior research of the habitat was conducted. Moreover, the local scuba diver underscored that constructional impacts were heightened by the extreme natural conditions of The Gulf. Consequently, environmental damage was inevitable [21]. Nonetheless, The Pearl could have mitigated its repercussions through various relocation programs. Yet, contractors ignored environmental issues in an effort to speed up the development process [20]. Ultimately, The Pearl's development process was careless and outdated, which created an unsustainable project that degraded its local environment. Hamad Port overcame The Pearl's environmental failures by complying with Qatar's new GSDP legislatures that enforced sustainable development. The project's engineer highlighted how a study of the local habitat was carried out, enabling Hamad Port to be positioned in way that would least affect the environment. However, some valuable habitats could not be avoided, so contractors were required to employ a greater number of environmental consultants whose job was to identify ecological concerns. Furthermore, the project had an increased budget for mitigation programs. Hamad Port employed large groups of people to relocate sea grass and mangroves as well as install stilling basins and artificial reef balls. Thus, Hamad Port's operations that prioritized environmental well-being reflected SDGs initiatives the Qatari government aims to develop for their 2030 vision [11].

Qatar has sustainably developed more refined methodologies that enabled MIA Park and Hamad Port to overcome The Pearl's social and environmental failures. As such, GSDP legislatures achieved their five-year plan. However, these projects still have fundamental design and construction flaws that would need to be addressed, in order for Qatar to successfully achieve their national vision of 2030. The following recommendations are identified as suggested measures that could enhance sustainability initiatives in Qatar's next phase of developments: (1) emphasis on functionality, i.e., developing projects that are necessary and practically utilized, (2) adequate maintenance and upkeep, i.e., regardless of Qatar's harsh desert climate, significant funding for maintenance needs to be applied to every future development to insure their long-lasting success, (3) sharing of data, i.e., lack of data availability can cause miscommunication between the different construction departments because valuable information is not shared and, consequently, unnecessary ecological damage can occur, (4) increased penalties, i.e., environmental failures may require increased penalties to mitigate impacts, (5) superior protecting and policing, i.e., mitigations measures by government should police and protect those newly replanted ecosystems too insure that other factors are not destroying the habitat afterward, and (6) continuous revision of legislatures, i.e., policies may change, as such, contractors should progressively review legislatures to gain newly required skills that will help them mitigate environmental impacts in future development. The overarching theme of identifying problems in past constructions and communicating those lessons in future developments is universal. The solutions and recommendations presented can be used to evaluate all developments among The Gulf's richest countries, because the culture, ecosystem, and construction process in the entire region has been similar. Thus, this chapter contributes to changing The Gulf's poor social and environmental image by suggesting functional human settlement solutions for the region's next phase of developments.

References

1. Global Footprint Network (2021) Open data platform. https://data.footprintnetwork.org/. Accessed 4 Apr 2021
2. Di Muzio T (2010) The world's affluent playground: Dubai's architecture of doom and the future of globalized social reproduction. Standing Gr Int Relations 9:18–23
3. Graham S (2016) Vertical: the city from satellites to bunkers. Verso, London
4. Luomi M (2012) Qatar's natural sustainability: plans, perceptions, and pitfalls. CIRS Occas Pap 11:6–69. https://doi.org/10.2139/ssrn.2825934
5. United Development Company (2011) The Pearl-Qatar redefines luxury. Port Technology. https://www.porttechnology.org/technical-papers/the_pearl_qatar_redefines_luxury/. Accessed 4 Apr 2021
6. Furlan R (2015) Livability and social capital in West Bay, The New Business Precinct of Doha. Arts Soc Sci J 6. https://doi.org/10.4172/2151-6200.1000116
7. Shin YJ (2009) Understanding spatial differentiation of social interaction: suggesting a conceptual framework for spatial mediation. Commun Theory 19:423–444. https://doi.org/10.1111/j.1468-2885.2009.01350.x
8. Poole E (2009) The Dubai palms: construction and environmental consequences. In: Proceedings of World environmental and water resources congress 2009—World Environmental and Water Resources Congress 2009: Great Rivers. American Society of Civil Engineers, pp 5277–5282
9. Cavalcante GH, Kjerfve B, Feary DA (2012) Examination of residence time and its relevance to water quality within a coastal mega-structure: The Palm Jumeirah Lagoon. J Hydrol 468–469:111–119. https://doi.org/10.1016/j.jhydrol.2012.08.027
10. Ghaffari K, Habibzadeh T, Asfad MN, Mousazadeh R (2017) Construction of artificial Islands in Southern Coast of the Persian Gulf from the viewpoint of international environmental law. J Polit Law 10:271. https://doi.org/10.5539/jpl.v10n2p264
11. United Nations (2015) Sustainable development goals. United Nations Resolution. A/RES/70/1 25 Sept. 2015. http://www.un.org/ga/search/view_doc.asp?symbol=A/RES/70/1&Lang=E. Accessed 3 May 2019
12. Velegrinis S, Katodrytis G (2015) Drawing on sand: cities in the making. Archit Des 85:72–79. https://doi.org/10.1002/ad.1856
13. Bayomi N, Fernandez JE (2019) Towards sustainable energy trends in the Middle East: a study of four major emitters. Energies 12:1615. https://doi.org/10.3390/en12091615
14. Natural Resource Governance Institute (2020) Clean energy and fossil fuels in the Middle East: A Virtuous Cycle? Natural Resource Government Institute. https://resourcegovernance.org/blog/clean-energy-fossil-fuels-middle-east-virtuous-cycle. Accessed 4 Apr 2021
15. Springborg R (2014) The energy revolution's impacts on the Arab World. Middle East Institute. https://www.mei.edu/publications/energy-revolutions-impacts-arab-world. Accessed 4 Apr 2021
16. Gordon PH (2020) How curbing reliance on fossil fuels will change the World | Wilson Center. West. Cent. https://www.wilsoncenter.org/article/how-curbing-reliance-fossil-fuels-will-change-world. Accessed 4 Apr 2021
17. Google Earth (2021) Google Earth. https://www.google.com/earth/. Accessed 4 Apr 2021
18. Gibson B, Hassan S, Tansey J (2013) Sustainability assessment: criteria and processes. Routledge, London
19. National Research Council (2011) Sustainability and the U.S. EPA. National Academies Press, Washington, DC
20. Salahuddin B (2006) The Marine environmental impacts of artificial island construction Dubai, UAE. Nicholas School of the Environment and Earth Sciences of Duke University, Durham, NC
21. Campbell P (2001) Learning from construction failures: applied forensic engineering: Campbell, Peter: 9780470399491: Amazon.com: Books. Whittlers Publishing, Latheronwheel

22. Muneerudeen A, Al F, Furlan R (2016) Urban revitalization of public spaces in the Pearl in Qatar. Am J Sociol Res 6:1–9
23. The Pearl (2021) The Pearl, Qatar. The Pearl Qatar. https://www.thepearlqatar.com/The-Island/The-Pearl-Qatar. Accessed 4 Apr 2021
24. Jalinoosi A, Moradifar S (2017) Analyzing the consequences of the UAE creating artificial Islands in The Persian Gulf (Considering the Copenhagen School). J World Sociopolitical Stud 1:281–313. https://doi.org/10.22059/wsps.2017.242117.1030
25. Gibling C (2013) Construction process and post-construction impacts of the Palm Jumeirah in Dubai, United Arab Emirates. J Undergrad Eng Res Scholarsh 1:1–8
26. Hopkins C (2007) Dubai pilot project: test report for reef systems at Palm Jumeirah. AquaMarine Advis 1–6
27. Ibrahim HG (2012) Ecological indicators for landscape assessment-moving forward with sustainability initiatives for Qatar. J Ecosyst Ecography 02:119. https://doi.org/10.4172/2157-7625.1000119
28. MIA (2020) MIA Park: a wonderful green, open space in the city. http://www.mia.org.qa/en/mia-park. Accessed 4 Apr 2021
29. Furlan R, Sinclair BR (2021) Planning for a neighborhood and city-scale green network system in Qatar: the case of MIA Park. Environ Dev Sustain. https://doi.org/10.1007/s10668-021-01280-9
30. New Port Project Qatar (2018) Hamad port environment 2016 English (video). https://www.youtube.com/watch?v=wXJtpAEjoz0. Accessed 4 Apr 2021
31. Neuman WL (2009) Social research methods: qualitative and quantitative approaches, 7th edn. Pearson, Boston
32. Lester S (1999) An introduction to phenomenological research. Stan Lester Developments, Taunton
33. Zainal Z (2002) An investigation into the effects of discipline-specific knowledge proficiency and genre on reading comprehension and strategies of Malaysia ESP Students. University of Reading, Reading
34. Alshenqeeti H (2014) Interviewing as a data collection method: a critical review. English Linguist Res 3:39. https://doi.org/10.5430/elr.v3n1p39
35. Petersen NJ (2003) Organisational photography as a research method: what, how and why. Copenhagen Business School, Copenhagen
36. Braun V, Clarke V (2006) Using thematic analysis in psychology. Qual Res Psychol 3:77–101. https://doi.org/10.1191/1478088706qp063oa
37. Pauleit S, Jones N, Nyhuus S et al (2005) Urban forest resources in European Cities. In: Konijnendijk C, Nilsson K, Randrup T, Schipperijn J (eds) Urban forests and trees. Springer, Berlin Heidelberg, pp 49–80
38. Russo A, Cirella GT (2018) Modern compact cities: how much greenery do we need? Int J Environ Res Public Health 15:2180. https://doi.org/10.3390/ijerph15102180
39. Konijnendijk C, Annerstedt M, Nielsen A, Maruthaveeran S (2013) Benefits of urban parks. IFPRA, Copenhagen
40. Marhaba Qatar (2017) International cultures celebrated at MIA Park. Marhaba Qatar Destination Guide. https://www.marhaba.qa/international-cultures-celebrated-at-mia-park/. Accessed 4 Apr 2021
41. Oman Observer (2017) Al Futaisi at opening of Hamad Port in Doha. Oman Daily Observer. https://www.omanobserver.om/al-futaisi-opening-hamad-port-doha/. Accessed 4 Apr 2021

Changing Dynamics with COVID-19: Future Outlook

Cengiz Kahraman, Christian Orobello, and Giuseppe T. Cirella

Abstract Today, human activities are developing at a drastic speed with little limitation. Excessive development can be problematic to human health as well as reduce the sustainability of natural resources. At present, the COVID-19 pandemic has affected the whole world and adversely has impacted a broad range of human activities, including the economy, agriculture, tourism, and transportation. The excessive increase from globalization, unplanned urbanization, and the destruction of nature (i.e., due to ongoing human activity) have been partially to blame for the current state of the planet. The human population has tripled over the last 70 years. In addition, existing agricultural areas of the world yield lower rates—especially in underdeveloped or developing countries. Moreover, climate change has affected agricultural areas and caused a shrinkage in production patterns. As a result, an increasing population has decreased access to their cultural foods, leaving them with few options, and these unhealthy dietary options frame detrimental long-term concerns. As a result, the human–nature relationship must be considered and decisions made. In sensitive areas, especially in mountainous regions, tropical forests, deserts, and glaciers, human habitation and human and economic activity should be limited. In this chapter, natural and human dynamics are examined by analyzing current changes and future population trends in light of the COVID-19 pandemic. A critical question that should be asked is what can be done regarding environmental sustainability during the crisis and what lessons will be learnt to best apply them in a post-COVID-19 world?

Keywords COVID-19 pandemic · World population · Demographic projection · Sustainable agriculture · Healthy environment

---

C. Kahraman (✉)
Faculty of Engineering, Istanbul University-Cerrahpasa, Istanbul, Turkey
e-mail: cengiz.kahraman@istanbul.edu.tr

C. Orobello · G. T. Cirella
Faculty of Economics, University of Gdansk, Sopot, Poland
e-mail: christian.orobello@ug.edu.pl

G. T. Cirella
e-mail: gt.cirella@ug.edu.pl

G. T. Cirella (ed.), *Human Settlements*, Advances in 21st Century Human Settlements,
https://doi.org/10.1007/978-981-16-4031-5_13

1 Introduction

Humanity's struggle with nature is partly due to its success in altering the environment, human ingenuity, and a strong survival instinct. With the assistance of technological developments, human beings have been able to produce solutions in the face of natural disasters such as earthquakes, volcanic activity, floods, and forest fires. Today, humanity is faced with the global COVID-19 pandemic, which should consider be considered in the context of the lessons learnt from previous pandemics. COVID-19, named by the combination of the initials "CO" corona, "VI" virus, and "D" disease, is a disease caused by a new coronavirus defined as SARS-CoV-2 [1]. According to current widespread acceptance, the virus-borne disease, which first appeared in Wuhan, China, spread worldwide in a short period of time. The World Health Organization (WHO) announced that it first learnt of this new virus on December 31, 2019, after a series of reports of viral pneumonia cases in Wuhan—declaring it a pandemic on March 11, 2020 [2]. Although the severity and prevalence of the disease caused by the virus fluctuate according to various parameters such as seasonal conditions, restrictions, and personal precautions, infections have increased to date. 36 COVID-19 cases diagnosed in researchers at the Chilean General Bernardo O'Higgins Riquelme research base in Antarctica in December 2020 show that there is no continent that has not been infected with this virus [3]. According to WHO data [4], dated March 28, 2021, 126,359,540 confirmed COVID-19 cases and 2,769,473 related deaths were reported worldwide [4]. This data is updated daily alongside a number of other outlets, notably John Hopkins University's Center for Systems Science and Engineering [5]. Historically speaking, the effects of the Spanish flu in 1918 (i.e., caused by influenza A (H1N1) virus), the Asian flu (i.e., caused by influenza A (H2N2) virus) in 1957, the Hong Kong flu (i.e., caused by influenza A (H3N2) virus in 1968), and the swine flu in 2009 were all pandemics from the last two centuries pre-COVID-19; however, all were somewhat limited in comparison to the current situation [6]. The COVID-19 pandemic, which affects social life, health, and human mobility, has caused damage to the economies of countries on a global scale that will take many years, perhaps decades before a full recovery is reached. COVID-19 has become a truly global, systemic, and gradual emergency that poses a real threat to global sustainable development [7].

The enormous socioeconomic changes in Great Britain in the mid-eighteenth and early nineteenth centuries (i.e., 1760–1840) are known as the First Industrial Revolution [8]. The industrialization movement evolved into the Second Industrial Revolution that spread to Europe and the USA, giving positive momentum and the speedy influx of people toward cities (i.e., the increasing of urbanization) and population growth—the world over. In terms of urbanization and industrialization, magnified throughout the world since the 1950s, there has been a dramatic increase and pressure to utilize natural and rural areas, including coastal zones, glaciers, seas, underground and surface water resources, agricultural areas, and forests. Uncontrolled growth since the first two industrial revolutions, i.e., a very short period of time in the whole of human history, forms the basis of today's leading problems such as deforestation, desertification, an increase in greenhouse gases, a decrease in biodiversity, and water–air–soil pollution—to mention a few. The widespread environmental problems on earth endanger the sustainability of the environment regardless of political boundaries and pose a threat to human safety, survival of other living species, food safety, and water resources. Correspondingly, these problems affect agricultural lands and access to food as agricultural production activities are improperly and poorly managed. It is a known fact that as a result of pressure from human beings, other species habitation space, and pathogens and microorganisms that had previously existed in certain areas are released from animal hosts and moved toward new hosts, including humans. There is no doubt that human beings' excessive interference with the balance of nature, e.g., illicit wildlife trade [9–26], animal laboratory experimentation, and genetic engineering [27–30], has aided in producing this result. The COVID-19 pandemic has once again demonstrated how important it is for humanity to adopt a standard that is compatible with nature—developing a civilizational direction that promotes health and well-being [31] for the 7.85 billion people of the world [32]. The enormous increase in population, i.e., approximately three times higher than in 1950, has caused vertical growth in cities and triggered their development toward rural areas by continually expanding and encroaching upon natural habitats. The gradual deterioration of the natural environment in conjunction with economic concerns has piggybacked off each other to cause a great migratory influx from rural to urban areas [33–45].

Inadequate health services in many countries, accompanied by deaths, indicate the population is highly susceptible to unpredictable shocks, such as the current COVID-19 pandemic. The current pandemic has not only affected the physical health of millions of people but has spread across the globe, drastically affecting the economy—with shocks reaching socioeconomics and medical services [31, 46]. Moreover, due to restrictions—e.g., quarantines on a global scale—purchasing power has decreased, unemployment has increased, and problems in accessing healthy food have been amplified—particularly in underdeveloped or developing countries. At large, declaring a pandemic is considered a historical event as it is not just a health crisis; it encompasses economic, political, and social dimensions—globally [6].

In many areas around the world, sustainability has become impossible with the negative impacts of the pandemic. It is necessary to rethink sustainability as an inter-sectorial concept of the economy, the environment, and society, over time. It is essential to consider human health as one of the basic components of sustainable development [46]. The common disease caused by the SARS-CoV-2 has clearly manifested itself widely, from accessing healthy food and water to the socioeconomic imbalances all around the world. In order to prevent such pandemics in the future, the critical underlying factors that should necessitate the limitation, improvement, and sustainability of human life are shown in Fig. 1.

The prevalence of the COVID-19 pandemic is seen as one of the adverse effects of globalization—a process that has gained momentum in the last half-century. Additional scientific studies are needed to redefine the concept of globalization, considering the impact of the pandemic. Increasing global tourism, trade-driven human

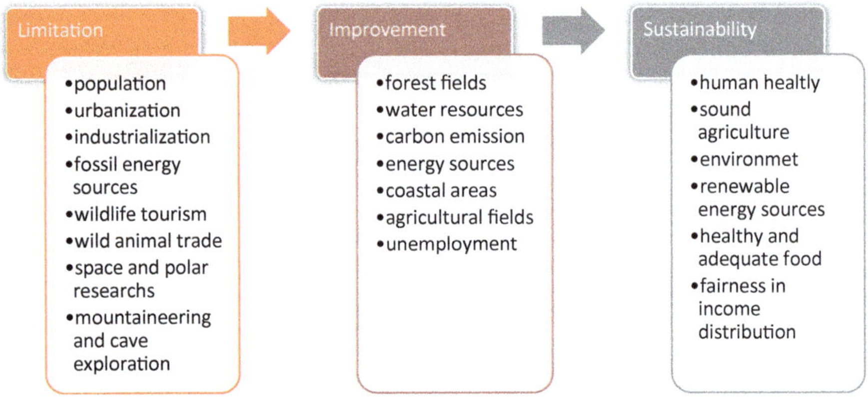

Fig. 1 Sustainability of human life

movement, goods mobility, and the development of the airline industry contributed to the spread of the disease. Human activity combined with globalization and an increase in international mobility alongside weakened public health systems has created the most favorable conditions for the emergence and rapid spread of SARS-CoV-2 [47]. On the other hand, reduced biodiversity, deforestation, and ecosystem instability increase the likelihood of outbreaks and instability [47–49]. To date, the global economy has been heavily affected by the COVID-19 pandemic, as it is highly integrated and interdependent through global supply chains. This chapter considers much of the discussion regarding human–nature relations and sustainable development theory. In particular, the economic crisis created by the pandemic, which has exposed farmers to multiple difficulties such as uncertainty in production, marketing, transportation, and unstable income, puts the food supply at risk. The pandemic, coupled with unpredictable climate conditions, has created unfavorable conditions that make it difficult for large masses of people to access healthy food while increasing the cost of food supply. In a sense, the effects of the pandemic necessitate permanent restrictions on human and economic life. This chapter, specifically, will also explore population in conjunction with sustainable agriculture as two important dynamics for a strong response against such shocks in the hope that they may be better dealt with in the future.

2 COVID-19: Demographic Outlook to De-Urbanize Mega-Cities

Population, an expression of the number of inhabitants living in any geographical unit, is a dynamic concept that constantly varies. Human settlements are at the center of sustainable development, and demographic data contains important implications for the economy, social development, and environmental sustainability [50]. One of the most important reasons for the global prevalence of the COVID-19 pandemic is the mass movement of people. With the increase in urbanization, the creation of metropolises and megalopolises, where millions of people live in multi-story buildings, has emerged. These mega-cities are not limited to the geographical areas where they are established but tend to continually grow and impede upon the natural areas that feed them, including development and growth toward surrounding regions. The COVID-19 pandemic offers an important opportunity for humanity to question the classical understanding of urbanization and better understand that many urban populations have reached an undesirable tipping point. As such, masses of people exiting cities before a lockdown order is enforced have been reported all over the world. Table 1 illustrates the world's population between 1950–2020 and population projections for 2020–2100 by region. The world population in 1960 was 2.54 billion, and in 2020 7.79 billion, i.e., a threefold population increase or 307% in 70 years. As a result, urban centers must be optimized for maximum health and well-being. Undoubtedly, the decrease in infant mortality and the increase in health services together with longer life spans have contributed to the growth. Future projections from Table 1 (i.e., according to the medium forecast of the United Nations) are that the world's population will reach 9.73 billion in 2050 and 10.88 billion in 2100.

These future population projections should be viewed as very optimistic estimates unless necessary measures are taken. It seems likely that the world may encounter a population of even 12–13 billion in the year 2100. It is clear that such a population will dangerously affect the sustainability of natural resources—especially for city dwellers. Indeed, even the current situation imposes pressure on healthcare systems and the food supply. It is fair to point out the COVID-19 outbreak has acted as a catalyst to show potential demographic dangers. In terms of the extent of the pandemic, seen in many countries at certain times, including developed ones, healthcare systems have almost collapsed, with many people dying due to inadequate intensive care support [51–65]. It is a well-known fact that the economies of developing and underdeveloped countries have been adversely affected by the restrictions imposed globally, with nutritional problems becoming more apparent [66–68].

Population growth varies according to geographic and economic region. When it is evaluated in terms of the United Nations development groups, it is evident that the population increase in underdeveloped regions is higher. While the total population in these regions was 1.72 billion in 1950, it increased by 378% in 2020, reaching 6.52 billion. In comparison, this rate has been around 56% in developed countries. The most important reason for the increase in population in underdeveloped regions is the high fertility rate. Although the average life expectancy in these regions is relatively short, the rate of having five or more children on average increases the rate of population growth. When the population of continental-based sub-regions is examined (i.e., between 1950 and 2020), in accordance with the degree of development, the highest increase is in the African continent with 588%. It is understood that the least increase is on the European continent with 136%. Looking at future population projections via global sub-regions, it can be projected that, unlike other continents, the African continent will be subjected to a large population growth wave. When

Table 1 Population in the world and sub-regions with medium projection, 1950–2100

Location	1950	1975	2000	2020	2050	2075	2100
World	2,536,431	4,079,480	6,143,494	7,794,799	9,735,034	10,577,288	10,875,394
United Nations development groups							
More developed regions	814,819	1,048,416	1,188,359	1,273,304	1,279,913	1,249,365	1,244,296
Less developed regions	1,721,612	3,031,064	4,955,134	6,521,494	8,455,121	9,327,923	9,631,098
Geographic regions							
Africa	227,794	414,675	810,984	1,340,598	2,489,275	3,498,757	4,280,127
Asia	1,404,909	2,401,171	3,741,263	4,641,055	5,290,263	5,142,761	4,719,907
Europe	549,329	676,895	725,558	747,636	710,486	657,283	629,563
Latin America and the Caribbean	168,821	322,777	521,836	653,962	762,432	749,876	679,993
North America	172,603	242,251	312,427	368,870	425,200	461,329	490,889
Oceania	12,976	21,710	31,425	42,678	57,376	67,282	74,916

examined further, it is calculated that there will be a significant decrease in the population growth in other continents, and even an absolute reduction in the population in Asia until 2100 (Fig. 2). However, the rise in the African continent is thought to be associated with a high birth rate. Despite the decreasing trend in the birth rate in this region, it is still 4.5 children per woman [50]. Controlling the population in the African continent, which has been associated with a number of pandemic-like events, will reduce potential outbreaks. As such, this is also important for the health of other parts of the world.

The world's population from 1950 (i.e., including the projections for the year 2100) and the growth rate of the population covering the same period show a remarkable point in which the growth rates in the population tend to decrease. It is predicted that this ratio, which is around 1% today, will decrease until 2100 and reach 0% by the same date. However, this situation is far from achieving an absolute reduction in the population. To manage unpredictable threats, such as pandemics, it is essential that the excessive increases in population reach sustainable levels—especially in urban areas for livability and access to services. A post-COVID-19 world will need to develop an innovative approach to social, political, and economic lifestyles, which may make the future of mega-cities, i.e., with populations of over 30 million, questionable. As crowded cities, which started to trend after the First and Second Industrial Revolution, lose their popularity, it is thought that cleanliness, health, and hygiene in social life will be a priority with the desire to live in smaller, more community-oriented places [6]. Today, overpopulated cities such as Tokyo, Delhi,

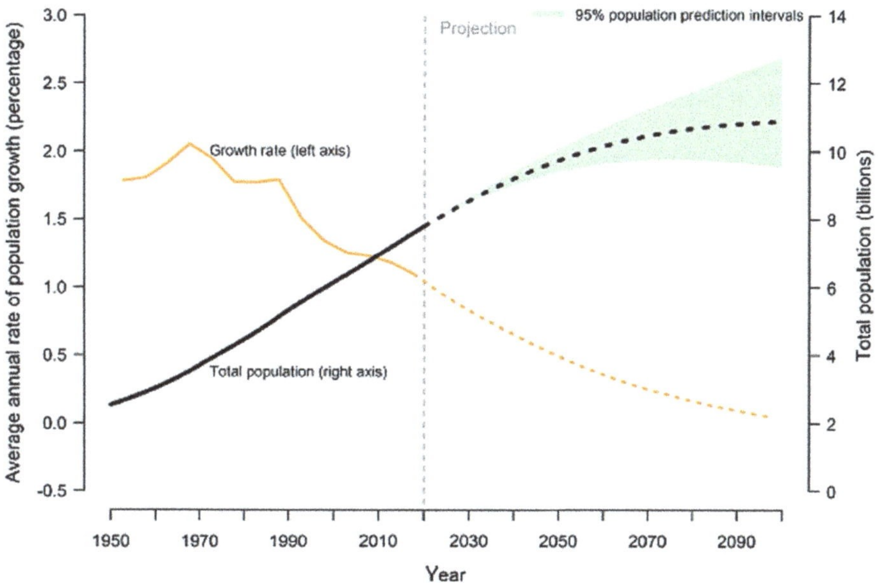

Fig. 2 United Nations population projections of the world population and population growth rates, 1950–2100 [50]

Shanghai, Sao Paulo, and Mexico City are candidate cities that may be the source of future concern. The unlimited human activity necessary to meet the needs of large population groups causes irreversible damage to nature—harming agricultural lands, food stocks and creates a high urban-food distribution dependency. The worldwide population growth rate between 2000 and 2025 indicates that the average rate is 11.5%, i.e., the value is 0.7% for developed regions and 20.1% for less developed ones. It is estimated that the population growth rate of less developed countries will gradually decrease, reaching 1.5% relative from 2075 to 2100. Again, the most important conclusions drawn from these statistical projections are that more than half of the global population growth is expected to occur in Africa between today and 2050—specifically Sub-Saharan Africa [50]. The situation in developed countries, on the other hand, reveals a completely opposite projection. Future modeling for these countries predicts the total population will decrease due to negative population growth rates.

One of the critical parameters used in demographic evaluations is the ratio of the number of people living in rural and urban areas. Table 2 shows the urban and rural population numbers and the proportion of those living in cities, according to the United Nations [69] and Ritchie and Roser [70]. At present, the population living in cities worldwide is 55.3% [70, 71]. This ratio has been reversed from the rural population being predominate, in the recent past, in favor of cities and urban life—consequential of the rapid urbanization in the last 70 years. The urban population ratio, which was 29.6% in 1950, exceeded the rural population worldwide in 2007. Considering these development levels, the rate of urban dwellers in developed regions currently stands at 78.7% and underdeveloped regions at 50.6%. The continent with the highest urban population is North America with 82.2%, while the continent with the least is Africa with 42.5%. Increases in the urban population worldwide carry a risk in terms of pandemic-like shocks.

Table 2 Population of urban and rural areas at mid-year (thousands) and percentage urban, 2018 [70]

Region, subregion, country, or area	Urban	Rural	Total	Urban %
World	4,219,817	3,413,002	7,632,819	55.3
United Nations development groups				
More developed regions	993,837	269,363	1,263,200	78.7
Less developed regions	3,225,980	3,143,639	6,369,620	50.6
Geographic regions				
Africa	547,602	740,318	1,287,921	42.5
Asia	2,266,131	2,279,003	4,545,133	49.9
Europe	552,911	189,737	742,648	74.5
Latin America and the Caribbean	526,057	125,955	652,012	80.7
North America	298,987	64,857	363,844	82.2
Oceania	28,129	13,132	41,261	68.2

Life in multi-story and adjacent buildings in big cities, whose populations are expressed in the millions, creates favorable conditions for shock events. For this reason, considering the balance between urban and rural populations, it is imperative to oversee any fluctuation in agricultural output—including proper management and sector sustainability. The fact that the world population is encroaching upon 10 billion increases the number of people per km^2 in older settlements. This rate, according to 2020 data, is 49.9 people per km^2. It is estimated that this value will reach 83.6 people in 2100. At the end of this century, the United Nations projects that the number of people per km^2 will reach 114 in Africa and 152 in Asia [50]. This increase has a direct relationship with agricultural areas due to the opening up, i.e., land development, of new suburbs and settlements.

The boundary between human beings and wildlife is gradually being blurred due to human encroachment on nature, e.g., deforestation, which causes the gradual distancing of agricultural lands. (However, it is worth noting that edible green infrastructure is a partial solution [72–75] but is still not developed to support mega-city populations.) The fact that a number of pandemics occurring in the recent past originated in China is highly characteristic of this demographic, i.e., spatiotemporal linkage [76–79]. Likewise, the health conditions and standards in India, another country with a high population, are obvious risk factors [80, 81]. It is also noteworthy that population density is extremely compatible, i.e., they overlay very closely with the current geography of the COVID-19 pandemic. Hence, urban population control in such countries, reducing the population density, proper management of agricultural lands, and making the natural habitats of wildlife uninterrupted will be critical factors in determining a brighter future—generationally.

3 Post-COVID-19: Steps Forward

The negative effect of rapid urban population growth is evident throughout the world. Natural environments and agricultural areas must come first. Reflecting on the last century, dramatic population growth increased human and economic activity around these areas by expanding residential zoning almost exponentially. Although the destruction of forests in a previously untouched region seems to be only for research nowadays, the delusion of many countries to grab space from the polar mainland, the construction of technological cities in desert areas, the efforts to reclaim tropical forests for agriculture and settlement purposes, oil exploration activities in the open seas, and other human activities cause irreversible damage to the natural environment. Moreover, the misapplication of human activity, especially mega-human settlements and industrialization, where, because of development, contact between the soil and the sky is impossible, excessive pollution is created, and hydrographic elements are overly built, creating unhealthy environments that pose a new challenge to human beings [82]. This increases the possibility of pathogen exposure and transmission. Due to degraded ecosystems living species that have never contacted each other—introduce novel cross-habitation of species. As a result of the unlimited use and destruction of the natural environment, nature does seem to have a way of responding to human overreach. With the COVID-19 pandemic, regardless of its origin, societies around the world have had to rethink their human–nature relationship and (re-)evaluate custodial priorities. With the absence of another planet, the paradoxical destruction of nature by human beings is alarming, to say the least. The COVID-19 pandemic should be considered a milestone for environmental management and human–nature relations. The sustainability of human interaction with nature post-pandemic is one of the topics discussed both by governments and scientists alike. Figure 3 illustrates the main environmental problems in terms of geographic parameters and source causes. A post-COVID-19 reality will need to consider more planning to eliminate these shock events and associated issues. These considerations should be underscored as extremely important for future generations and the sustainability of human health and well-being.

More than three-quarters of the terrestrial biosphere and geosphere on earth have been reshaped by humans, namely the anthroposphere [83]. Today, most of the net total of the terrestrial biosphere is used by humans for human activities. Only 20% of the earth represents semi-natural environments and less than 25% wild environments. The symbiotic system of the earth, i.e., its self-regulating system, has been damaged by increasing globalization, urbanization, industrial animal husbandry, unsustainable agricultural intensification, and exploitation of natural resources [83]. In order to minimize the risks of COVID-19 and any derivative-like pandemic, it is necessary to reconstruct the boundary between human beings and nature. To stop contamination, it is essential to identify the populations and geographic areas where infectious risk is highest. Reducing the risks of infection in the most affected populations ultimately serves the entire population [84]. Hence, the proposed human depopulation of certain

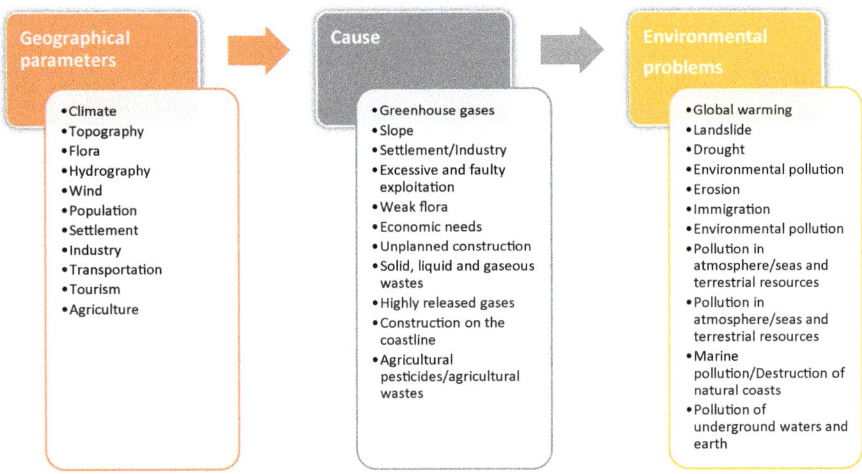

Fig. 3 Environmental problems associated with geographical parameters and source causes, adapted from Kahraman [82]

sensitive areas should be considered, including glaciers (i.e., the polar regions), caves with wildlife, deserts, tropical forests (i.e., especially the remaining virgin forests), high mountain ranges after a certain altitude, and the Sub-Saharan African savannahs. With the exception of scientific research and tourism, these areas should be carefully managed and considered (Fig. 4).

As a result of technological and industrial developments in the last century, a rapid negative transformation has occurred in the atmosphere, lithosphere, hydrosphere, and biosphere: climate change. The negative impact of climatic change has also become more visible in agriculture in recent years. Climate change causes a narrowing of the product pattern and a decrease in agriculture production. According to the World Meteorological Organization, there has been a decrease in the rate of CO_2 concentrations in the atmosphere due to restrictions and lockdowns resulting from COVID-19—with reports of replenishment of air quality in major cities around the world [85]. It is thought that a reduction in emission levels may gain momentum with possible post-COVID-19 normalization. Polluting emissions from industrial activities to tourism need to be brought under control and lowered rapidly to prevent widespread climate change and agricultural concerns. Apart from threatening human life, the COVID-19 pandemic has shockingly revealed the importance of a sustainable environment. Future generations' use of natural resources will be subject to sustainability standards, i.e., in-line with the United Nations Sustainable Development Goals [86], as well as the human–nature relationship that embodies a holistic economy aligned socially and environmentally [82, 87]. To date, climate change, destruction of fertile agricultural lands, pollution, overuse of natural resources, deforestation, poor

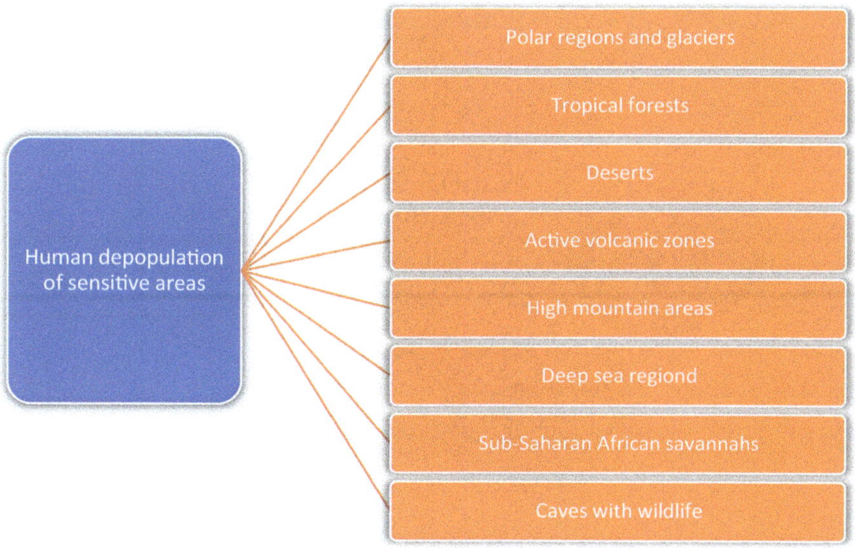

Fig. 4 Human depopulation of sensitive areas

waste management, reduction of biodiversity, ecological degradation, etc., are environmental problems that affect all countries. This situation causes gradual depletion and a decrease of underground and aboveground resources. The continued burning of fossil fuels causes rapid pollution of atmospheric and hydrographic resources in an unprecedented manner. If this situation continues, it is clear that future generations will face significant problems in meeting their vital needs—especially food.

First, a paradigm shift is needed. The perception that technological developments are beneficial to humanity in every situation is incorrect. Although the unlimited use of industry and technology facilitates human life in the short-term, it has the potential to make life impossible in the long-term. In this respect, the question of how to find the balance between the natural environment and technology post-COVID-19 will be one of the questions the scientific community will need to face. In order to achieve this balance, urgent action plans at the global level covering the next few decades are required. Second, there should be an aim to achieve a significant decrease in urban population and, consequently, a reduction in urban pollution and emissions. This reduction will reduce environmental degradation and contribute to improving many human components, especially agriculture and health. Figure 5 illustrates some steps to be taken in the short- and middle-term for sustainability post-COVID-19. It is getting harder (and harder) to ensure sustainable opportunities and

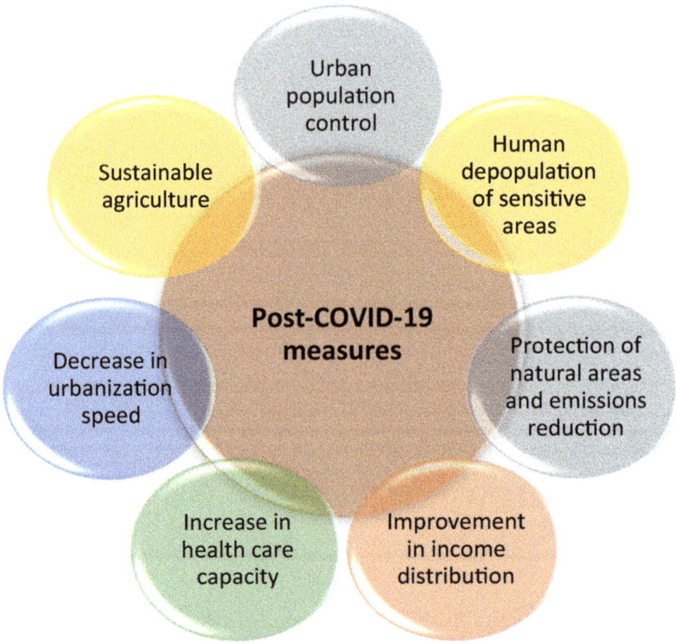

Fig. 5 Post-COVID-19 measures for environmental sustainability

nature-based services remain uninterrupted. Unpredictable threats like COVID-19 have made humanity understand this type of shock-fact. Planning should commence in terms of the environment and sustainability post-pandemic—cooperatively and globally.

4 Conclusion

Human beings have been shaping the earth with human activities ever since early human history. The fact that people engaged in agriculture and spatial activities is a significant turning point in human history. The most important consequence of the transition from gatherer-hunter-nomad life to settled life is the increase in population and the increase in the assets of people. As a natural result of this, underground and aboveground resources have started to be used at an alarming rate due to the development of industry with increased urbanization. The world's population has increased dramatically since the industrial revolution. It has increased approximately eightfold in the last two centuries—rapidly consuming natural resources by establishing technological and crowded cities as well as destroying the ecological order—and exponentially since the early twentieth century. Moreover, while modernity and technology claim to make human life easier, the disruptive natural dynamics they are causing cannot be adequately predicted. Unwilling to come to grips with the natural order, a need for maintenance and renewal is evident post-COVID-19. The classical concept and definition of sustainability, which is widely covered in academic dialogue, is understood to be dysfunctional [88, 89]. It has become necessary to redefine the concept of sustainability, taking into account the COVID-19 crisis and future outbreaks. A shift needs to take place within the paradigm of sustainability, and we should focus on sustaining the earth and its inhabitant rather than sustaining our urban and industrial activities. It has also become evident that human health is the most crucial parameter in this definition. In order to prevent environmental problems and protect human health, the improvement stages proposed in this chapter are a strong starting point that primarily depend on making binding and deep-seated decisions by countries—the world over. The effect of technological developments and their compatibility with nature must be considered and investigated [90–92]. The COVID-19 pandemic has resulted in shutdowns that human beings have not experienced (i.e., on such a large scale), macro-scale disruptions in economies, and a decline in agricultural production. Above all, millions of people have lost their lives, reason enough to reverse the drastic global scenario and start taking steps toward a humanity post-COVID-19 in harmony with nature.

References

1. UNICEF, WHO, IFRC (2019) Key messages and actions for COVID-19 prevention and control in schools. UNICEF, New York
2. WHO (2019) Coronavirus disease: COVID-19. World Health Organization. https://www.who.int/emergencies/diseases/novel-coronavirus-2019/question-and-answers-hub/q-a-detail/coronavirus-disease-covid-19. Accessed 13 Mar 2021
3. BBC (2020) Koronavirüs: Antarktika'da ilk Covid vakaları tespit edildi. BBC News Turkey. https://www.bbc.com/turkce/haberler-dunya-55411819. Accessed 13 Mar 2021
4. WHO (2021) WHO Coronavirus (COVID-19) dashboard. World Health Organization. https://covid19.who.int/. Accessed 13 Mar 2021
5. CSSE (2021) COVID-19 Dashboard. In: Johns Hopkins Univ. https://coronavirus.jhu.edu/map.html. Accessed 14 Mar 2021
6. Açikgöz Ö, Günay A (2020) The early impact of the Covid-19 pandemic on the global and Turkish economy. Turkish J Med Sci 50:520–526. https://doi.org/10.3906/sag-2004-6
7. Li X, Zhao Z, Liu F (2020) Big data assimilation to improve the predictability of COVID-19. Geogr Sustain 1:317–320. https://doi.org/10.1016/j.geosus.2020.11.005
8. Mohajan H (2019) The first industrial revolution: creation of a new global human era. J Soc Sci Humanit 5:317–320
9. Borzée A, McNeely J, Magellan K et al (2020) COVID-19 highlights the need for more effective wildlife trade legislation. Trends Ecol Evol 35:1052–1055. https://doi.org/10.1016/j.tree.2020.10.001
10. Lam WK, Zhong NS, Tan WC (2003) Overview on SARS in Asia and the World. Respirology 8:S2–S5. https://doi.org/10.1046/j.1440-1843.2003.00516.x
11. Gao F, Balles E, Robertson DL et al (1999) Origin of HIV-1 in the chimpanzee Pan troglodytes troglodytes. Nature 397:436–441. https://doi.org/10.1038/17130
12. Leroy EM, Rouquet P, Formenty P et al (2004) Multiple ebola virus transmission events and rapid decline of Central African Wildlife. Science (80-) 303:387–390. https://doi.org/10.1126/science.1092528
13. Van Borm S, Thomas I, Hanquet G et al (2005) Highly pathogenic H5N1 influenza virus in smuggled Thai eagles, Belgium. Emerg Infect Dis 11:702–705. https://doi.org/10.3201/eid1105.050211
14. Taylor LH, Latham SM, Woolhouse MEJ (2001) Risk factors for human disease emergence. Philos Trans R Soc B Biol Sci 356:983–989. https://doi.org/10.1098/rstb.2001.0888
15. Karesh WB, Cook RA, Bennett EL, Newcomb J (2005) Wildlife trade and global disease emergence. Emerg Infect Dis 11:1000–1002. https://doi.org/10.3201/eid1107.050194
16. Walsh JF, Molyneux DH, Birley MH (1993) Deforestation effects on vector-borne disease. Parasitology 106:S55–S75. https://doi.org/10.1017/S0031182000086121
17. Lilley B, Lammie P, Dickerson J, Eberhard M (1997) An increase in hookworm infection temporally associated with ecologic change. Emerg Infect Dis 3:391–393. https://doi.org/10.3201/eid0303.970321
18. Bell D, Roberton S, Hunter PR (2004) Animal origins of SARS coronavirus: possible links with the international trade in small carnivores. Philos Trans R Soc B Biol Sci 359:1107–1114. https://doi.org/10.1098/rstb.2004.1492
19. Tu C, Crameri G, Kong X et al (2004) Antibodies to SARS coronavirus in civets. Emerg Infect Dis 10:2244–2248. https://doi.org/10.3201/eid1012.040520
20. Greatorex ZF, Olson SH, Singhalath S et al (2016) Wildlife trade and human health in Lao PDR: an assessment of the zoonotic disease risk in markets. PLoS ONE 11:e0150666. https://doi.org/10.1371/journal.pone.0150666
21. Jones KE, Patel NG, Levy MA et al (2008) Global trends in emerging infectious diseases. Nature 451:990–993. https://doi.org/10.1038/nature06536
22. Hai W, Zhao Z, Wang J, Hou Z-G (2004) The short-term impact of SARS on the Chinese economy. Asian Econ Pap 3:57–61. https://doi.org/10.1162/1535351041747905

23. Aguirre AA, Catherina R, Frye H, Shelley L (2020) Illicit wildlife trade, wet markets, and COVID-19: preventing future pandemics. World Med Heal Policy 12:256–265. https://doi.org/10.1002/wmh3.348
24. Patz JA, Graczyk TK, Geller N, Vittor AY (2000) Effects of environmental change on emerging parasitic diseases. Int J Parasitol 30:1395–1405. https://doi.org/10.1016/S0020-7519(00)00141-7
25. Kung NY, Guan Y, Perkins NR et al (2003) The impact of a monthly rest day on avian influenza virus isolation rates in retail live poultry markets in Hong Kong. Avian Dis 47:1037–1041. https://doi.org/10.1637/0005-2086-47.s3.1037
26. Dezső Z, Barabási AL (2002) Halting viruses in scale-free networks. Phys Rev E Stat Physics, Plasmas, Fluids, Relat Interdiscip Top 65:4. https://doi.org/10.1103/PhysRevE.65.055103
27. Gilsdorf JR, Zilinskas RA (2005) New considerations in infectious disease outbreaks: the threat of genetically modified microbes. Clin Infect Dis 40:1160–1165. https://doi.org/10.1086/428843
28. Andersen KG, Rambaut A, Lipkin WI et al (2020) The proximal origin of SARS-CoV-2. Nat Med 26:450–452. https://doi.org/10.1038/s41591-020-0820-9
29. Segreto R, Deigin Y (2021) The genetic structure of SARS-CoV-2 does not rule out a laboratory origin. BioEssays 43:2000240. https://doi.org/10.1002/bies.202000240
30. van Aken J, Hammond E (2003) Genetic engineering and biological weapons. New technologies, desires and threats from biological research. EMBO Rep 4:S57. https://doi.org/10.1038/sj.embor.embor860
31. Simon M (2020) The coronavirus lockdown is a threat for many animals, not a blessing. WIRED. https://www.wired.com/story/coronavirus-lockdown-conservation/. Accessed 13 Mar 2021
32. Worldometer (2021) World population clock: 7.9 billion people in 2021. https://www.worldometers.info/world-population/. Accessed 13 Mar 2021
33. Qiu P, Yang Y, Zhang J, Ma X (2011) Rural-to-urban migration and its implication for new cooperative medical scheme coverage and utilization in China. BMC Public Health 11:520. https://doi.org/10.1186/1471-2458-11-520
34. Østby G (2016) Rural–urban migration, inequality and urban social disorder: evidence from African and Asian cities. Confl Manag Peace Sci 33:491–515. https://doi.org/10.1177/0738894215581315
35. Conning AM (1972) Rural urban destinations of migrants and community differentiation in a rural region of Chile. Int Migr Rev 6:148–157. https://doi.org/10.1177/019791837200600203
36. Rhoda R (1983) Rural development and urban migration: can we keep them down on the farm? Int Migr Rev 17:34–64. https://doi.org/10.1177/019791838301700102
37. Byerlee D (1974) Rural-urban migration in Africa: theory, policy and research implications. Int Migr Rev 8:543–566. https://doi.org/10.1177/019791837400800404
38. Marta J, Fauzi A, Juanda B, Rustiadi E (2020) Understanding migration motives and its impact on household welfare: evidence from rural–urban migration in Indonesia. Reg Stud Reg Sci 7:118–132. https://doi.org/10.1080/21681376.2020.1746194
39. Lee ES (1966) A theory of migration. Demography 3:47–57. https://doi.org/10.2307/2060063
40. Penninx R (1982) A critical review of theory and practice: the case of Turkey. Int Migr Rev 16:781–818. https://doi.org/10.1177/019791838201600404
41. Ajaero CK, Onokala PC (2013) The effects of rural-urban migration on rural communities of Southeastern Nigeria. Int J Popul Res 2013:1–10. https://doi.org/10.1155/2013/610193
42. Carlos MRD (2002) On the determinants of international migration in the Philippines: an empirical analysis. Int Migr Rev 36:81–102. https://doi.org/10.1111/j.1747-7379.2002.tb00072.x
43. Fuller G, Chapman M (1974) On the role of mental maps in migration research. Int Migr Rev 8:491–506. https://doi.org/10.1177/019791837400800401
44. de Haas H (2021) A theory of migration: the aspirations-capabilities framework. Comp Migr Stud 9:1–35. https://doi.org/10.1186/s40878-020-00210-4

45. Rollwagen JR (1971) Region of origin and rural-urban migration in Mexico: some general comments and a case study of entrepreneurial migration from the West. Int Migr Rev 5:325–338. https://doi.org/10.1177/019791837100500305
46. Hakovirta M, Denuwara N (2020) How COVID-19 redefines the concept of sustainability. Sustainability 12:3727. https://doi.org/10.3390/su12093727
47. Barouki R, Kogevinas M, Audouze K et al (2021) The COVID-19 pandemic and global environmental change: emerging research needs. Environ Int 146:1–5. https://doi.org/10.1016/j.envint.2020.106272
48. Cardinale BJ, Duffy JE, Gonzalez A et al (2012) Biodiversity loss and its impact on humanity. Nature 486:59–67. https://doi.org/10.1038/nature11148
49. Morand S, Lajaunie C (2018) Loss of biological diversity and emergence of infectious diseases. In: Gaill F (ed) Biodiversity and health: linking life, ecosystems and societies. Elsevier, London, pp 29–47
50. United Nations (2019) World population prospects 2019: highlights. United Nations, Department of Economic and Social Affairs, Population Division, New York
51. Blumenthal D, Fowler EJ, Abrams M, Collins SR (2020) Covid-19—implications for the health care system. N Engl J Med 383:1483–1488. https://doi.org/10.1056/nejmsb2021088
52. Kamel MI (2020) A view of the health services after COVID-19: an Egyptian perspective. Alexandria J Med 56:118–129. https://doi.org/10.1080/20905068.2020.1789391
53. Dzinamarira T, Dzobo M, Chitungo I (2020) COVID-19: a perspective on Africa's capacity and response. J Med Virol 92:2465–2472. https://doi.org/10.1002/jmv.26159
54. EJ Ridley AF-SKH (2020) Surge capacity for critical care specialised allied health professionals in Australia during COVID-19. Aust Crit Care 12:90
55. Bourgeault IL, Maier CB, Dieleman M et al (2020) The COVID-19 pandemic presents an opportunity to develop more sustainable health workforces. Hum Resour Health 18:83. https://doi.org/10.1186/s12960-020-00529-0
56. Chen C, Pittman P, Westergaard S et al (2020) Emerging health workforce strategies to address COVID-19. J Heal Aff. https://doi.org/10.1377/hblog20200414.197056/full/
57. Fernández-Portillo LA, Sianes A, Santos-Carrillo F (2020) How will COVID-19 impact on the governance of global health in the 2030 agenda framework? The opinion of experts. Healthcare 8:356. https://doi.org/10.3390/healthcare8040356
58. Shadmi E, Chen Y, Dourado I et al (2020) Health equity and COVID-19: Global perspectives. Int J Equity Health 19:1–16. https://doi.org/10.1186/s12939-020-01218-z
59. Armocida B, Formenti B, Ussai S et al (2020) The Italian health system and the COVID-19 challenge. Lancet Public Heal 5:e253. https://doi.org/10.1016/S2468-2667(20)30074-8
60. Adini B, Goldberg A, Laor D et al (2010) Do standard operating procedures for pandemic influenza impact on emergency preparedness? Prehosp Disaster Med 25:S25–S25. https://doi.org/10.1017/s1049023x00022329
61. Khalid A, Ali S (2020) COVID-19 and its challenges for the healthcare system in Pakistan. Asian Bioeth Rev 12:551–564. https://doi.org/10.1007/s41649-020-00139-x
62. Narain JP, Dawa N, Bhatia R (2020) Health system response to COVID-19 and future pandemics. J Health Manag 22:138–145. https://doi.org/10.1177/0972063420935538
63. Lal A, Ashworth HC, Dada S et al (2020) Optimizing pandemic preparedness and response through health information systems: lessons learned from ebola to COVID-19. Disaster Med Public Health Prep, pp 1–8. https://doi.org/10.1017/dmp.2020.361
64. Sturmberg JP, Tsasis P, Hoemeke L (2020) COVID-19–an opportunity to redesign health policy thinking. Int J Heal Policy Manag. https://doi.org/10.34172/ijhpm.2020.132
65. Lal A, Erondu NA, Heymann DL et al (2021) Fragmented health systems in COVID-19: rectifying the misalignment between global health security and universal health coverage. Lancet 397:61–67. https://doi.org/10.1016/S0140-6736(20)32228-5
66. Headey D, Heidkamp R, Osendarp S et al (2020) Impacts of COVID-19 on childhood malnutrition and nutrition-related mortality. Lancet 396:519–521. https://doi.org/10.1016/S0140-6736(20)31647-0

67. Knorr D, Khoo C-SH (2020) COVID-19 and food: challenges and research needs. Front Nutr 7:598913. https://doi.org/10.3389/fnut.2020.598913
68. Naja F, Hamadeh R (2020) Nutrition amid the COVID-19 pandemic: a multi-level framework for action. Eur J Clin Nutr 74:1117–1121. https://doi.org/10.1038/s41430-020-0634-3
69. United Nations (2021) World population prospects: data query. Population Division of the Department of Economic and Socials Affairs United Nations Secretary. https://population.un.org/wpp/DataQuery/. Accessed 13 Mar 2021
70. Ritchie H, Roser M (2018) Urbanization. In: Our world data. https://ourworldindata.org/urbanization. Accessed 13 Mar 2021
71. World Bank (2021) Urban population (% of total population). https://data.worldbank.org/indicator/SP.URB.TOTL.IN.ZS. Accessed 14 Mar 2021
72. Russo A, Cirella GT (2020) Edible green infrastructure for urban regeneration and food security: case studies from the Campania Region. Agriculture 10:358. https://doi.org/10.3390/agriculture10080358
73. Russo A, Cirella GT (2018) Edible green infrastructure for urban regeneration: case studies from the Campania Region, Italy. In: 4th International symposium on infrastructure development, 12–14 Oct 2018. Hasanuddin University and Manado State Polytechnic, Manado, Indonesia
74. Russo A, Cirella GT (2019) Edible urbanism 5.0. Palgrave Commun 5:1–9. https://doi.org/10.1057/s41599-019-0377-8
75. Russo A, Escobedo FJ, Cirella GT, Zerbe S (2017) Edible green infrastructure: an approach and review of provisioning ecosystem services and disservices in urban environments. Agric Ecosyst Environ 242:53–66. https://doi.org/10.1016/j.agee.2017.03.026
76. Cao C, Chen W, Zheng S et al (2016) Analysis of spatiotemporal characteristics of pandemic SARS spread in Mainland China. Biomed Res Int 2016. https://doi.org/10.1155/2016/7247983
77. Meyers LA, Pourbohloul B, Newman MEJ et al (2005) Network theory and SARS: predicting outbreak diversity. J Theor Biol 232:71–81. https://doi.org/10.1016/j.jtbi.2004.07.026
78. Hsieh YH, Chen CWS, Hsu SB (2004) SARS outbreak, Taiwan, 2003. Emerg Infect Dis 10:201–206. https://doi.org/10.3201/eid1002.030515
79. Jiang J, Luo L (2020) Influence of population mobility on the novel coronavirus disease (COVID-19) epidemic: based on panel data from Hubei China. Glob Heal Res Policy 5:30. https://doi.org/10.1186/s41256-020-00151-6
80. John J, Kang G (2020) Public health during the pandemic in India. Science (80-) 370:663–664. https://doi.org/10.1126/science.abe9707
81. Institut Montaigne (2020) Public health in India: lessons of a pandemic. https://www.institutmontaigne.org/en/blog/public-health-india-lessons-pandemic. Accessed 14 Mar 2021
82. Kahraman C (2016) Role of geography in environmental education. Int J Humanit Arts Soc Sci 2:121–125. https://doi.org/10.20469/ijhss.2.20001-4
83. Ellis EC, Klein Goldewijk K, Siebert S et al (2010) Anthropogenic transformation of the biomes, 1700 to 2000. Glob Ecol Biogeogr 19:589–606. https://doi.org/10.1111/j.1466-8238.2010.00540.x
84. von Seidlein L, Alabaster G, Deen J, Knudsen J (2021) Crowding has consequences: Prevention and management of COVID-19 in informal urban settlements. Build Environ 188:107472. https://doi.org/10.1016/j.buildenv.2020.107472
85. WMO (2021) WMO publications and resources. World Meteorological Organization. https://library.wmo.int/index.php?lvl=infopages&lang=en_UK&pagesid=1. YEAicWgzbIU. Accessed 13 Mar 2021
86. United Nations (2015) Sustainable development goals. United Nations Resolution A/RES/70/1 25 Sept. 2015. http://www.un.org/ga/search/view_doc.asp?symbol=A/RES/70/1&Lang=E. Accessed 3 May 2019
87. Cirella GT, Mwangi SW, Paczoski A, Abebe ST (2020) Human-nature relations: the unwanted filibuster. In: Cirella GT (ed) Sustainabile human-nature relations: Environmental scholarship, economic evaluation, urban strategies. Springer Singapore, Singapore
88. Cirella GT, Tao L (2008) Measuring sustainability: an application using the index of sustainable functionality in South East Queensland, Australia. Int J Interdiscip Soc Sci 3:231–240

89. Cirella GT, Zerbe S (2014) Quizzical societies: a closer look at sustainability and principles of unlocking its measurability. Int J Sci Soc 5:29–45
90. Diffenbaugh NS, Field CB, Appel EA et al (2020) The COVID-19 lockdowns: a window into the Earth System. Nat Rev Earth Environ 1:470–481. https://doi.org/10.1038/s43017-020-0079-1
91. Bouman T, Steg L, Dietz T (2020) Insights from early COVID-19 responses about promoting sustainable action. Nat Sustain 4:194–200. https://doi.org/10.1038/s41893-020-00626-x
92. Leal Filho W, Azul AM, Wall T et al (2021) COVID-19: the impact of a global crisis on sustainable development research. Sustain Sci 16:85–99. https://doi.org/10.1007/s11625-020-00866-y

Glossary

Air pollution The presence of toxic chemicals or compounds, including those of biological origin, in the air, at levels that pose a health risk. Broadly speaking, air pollution means the presence of chemicals or compounds in the air which are usually not present and which lower the quality of the air or cause detrimental changes to the quality of life [1].

Atmosphere A layer of gases surrounding the planet made up of 99% nitrogen and oxygen plus other dry air gases such as argon, carbon dioxide, helium, neon, and other gases [2].

Climate action Stepped-up efforts to reduce greenhouse gas emissions and strengthen resilience and adaptive capacity to climate-induced impacts, including: climate-related hazards in all countries; integrating climate change measures into national policies, strategies and planning; and improving education, awareness-raising and human and institutional capacity with respect to climate change mitigation, adaptation, impact reduction and early warning [3].

Climate change Any change in the state of the climate that can be identified by changes in the mean and variability of its properties, and that persists for an extended period, typically decades or longer [4].

Commodity A raw material or primary agricultural product that can be traded, bought and sold, such as oil or pineapple [5].

Complexity theory Provides an understanding of how systems, such as the economy and global corporations, grow, adapt, and evolve. It explains how the relationships between members of these systems give rise to the collective behavior and sheds light on how a system interacts with its environment [6].

Decentralization The transfer of authority and responsibility for public functions from the central government to intermediate and local governments or quasi-independent government organizations or the private sector [7].

Development The economic, social, and institutional evolution of national states [8]. Can be interpreted to mean the process of becoming larger, more mature, and better organized—often in terms of economic criteria [9].

G. T. Cirella (ed.), *Human Settlements*, Advances in 21st Century Human Settlements,
https://doi.org/10.1007/978-981-16-4031-5

Disaster management A systematic process of using the body of policy, administrative decisions, organization, and operational skills and capacities to reduce the risk of natural or man-made disasters and their adverse effects. Disaster management can be realized at various levels—sub-national (i.e., local, regional), national, and international, and at various times—before, during, and after a disaster [6].

Disaster risk reduction and management The policy objective of anticipating and reducing risk is called disaster risk reduction (DRR). Although often used interchangeably with DRR, disaster risk reduction and management (DRRM) can be thought of as the implementation of DRR, since it describes the actions that aim to achieve the objective of reducing risk [10].

Ecology The study of relationships between organisms and their environments [9].

Ecosystem services The direct and indirect contributions of ecosystems to human well-being. There are four types of ecosystem services: provisioning, regulating, supporting, and cultural [11].

Emissions trading Sometimes referred to as cap-and-trade or allowance trading, itis an approach to reducing pollution that has been used successfully to protect human health and the environment. Emissions trading programs have two key components: a limit (or cap) on pollution, and tradable allowances equal to the limit that authorize allowance holders to emit a specific quantity (e.g., one ton) of the pollutant. This limit ensures that the environmental goal is met and the tradable allowances provide flexibility for individual emissions sources to set their own compliance path. Because allowances can be bought and sold in an allowance market, these programs are often referred to as market-based [12].

Environmental education It is a process that allows individuals to explore environmental issues, engage in problem solving, and take action to improve the environment. As a result, individuals develop a deeper understanding of environmental issues and have the skills to make informed and responsible decisions. Key components of environmental education include awareness and sensitivity to, knowledge and understanding of, attitudes of concern for, and skills to identify and participation in the environment and environmental challenges [13].

Environmental pollution Defined as the contamination of the physical and biological components of the earth or atmosphere system to such an extent that normal environmental processes are adversely affected [14].

Food security Defined as all people, at all times, have physical, social, and economic access to sufficient, safe, and nutritious food that meets their food preferences and dietary needs for an active and healthy life [15].

Global North Represent the economically developed societies of Europe, North America, Australia, Israel, South Africa, Japan, South Korea, amongst others. Global North countries are wealthy, technologically advanced, politically stable, and aging as their societies tend towards zero population growth [16].

Global South Represent the economically backward countries of Africa, Central and South America, the Indian subcontinent, the Middle East, amongst others. The Global South countries are agrarian based, dependent economically and politically on the Global North, the Global North has continued to dominate and direct the Global South in international trade and politics [16].

Green city A loose association of a city focused on sustainability, striving to lessen its environmental impact by reducing waste, expanding recycling, lowering emissions, increasing housing density while expanding open blue and green spaces, and encouraging the development of sustainable local businesses [17].

Green energy Green energy is any energy type that is generated from natural resources, such as sunlight, wind, or water. It often comes from renewable energy sources (see definition below) although there are some differences between renewable and green energy. The key with these energy resources are that they do not harm the environment through factors such as releasing greenhouse gases into the atmosphere [18].

Green infrastructure A strategically planned network of high-quality natural and semi-natural areas with other environmental features, which is designed and managed to deliver a wide range of ecosystem services and protect biodiversity in both rural and urban settings [19].

Green tax A tax that is imposed on the environment polluting goods or activities, to discourage people from anti-ecological behavior and make them sensitive towards it [20]. They are based on a simple principle—those who pollute, pay—and are designed to halting environmental pollution [21].

Habitation The state of living somewhere. When an area has no human habitation, it means that no people live there [22].

Human ecology Concerned with the organizational aspects of human populations that arise from their sustenance-producing activities [23]. It interfaces with economic progress that meets all of our needs without leaving future generations with fewer resources than those we enjoy, hence, a way of living from nature's income rather than mining its capital account [24].

Human settlement A form of human habitation which ranges from a single dwelling to large urban center. It is a process of opening up and settling of a previously uninhabited area by people. The study of human settlement is basic to human geography as it examines the forms of settlement in any particular region reflective of the human relationship with the environment [25].

Informal settlement An area where groups of housing units have been constructed on land that the occupants have no legal claim to, or occupy illegally. Unplanned settlements and areas where housing is not in compliance with current planning and building regulations (i.e., unauthorized housing) [26].

Internally displaced persons Someone who is forced to leave their home but who remains within their country's borders [27]. Note, they are often thought of as refugees, although they do not fall within the legal definition [28].

Knowledge management The process of creating, sharing, using, and managing the knowledge and information of an organization—i.e., referring to a multidisciplinary approach to achieving organizational objectives by making the best use of knowledge [29, 30].

Landscape ecology The science and art of studying and improving the relationship between spatial pattern and ecological processes on a multitude of scales and organizational levels. In a broad sense, landscape ecology represents both a field of study and a scientific perspective or paradigm. As a highly interdisciplinary and transdisciplinary enterprise, landscape ecology integrates biophysical and analytical approaches with humanistic and holistic perspectives across natural and social sciences [31].

Legislation The preparing and enacting of laws by local, state, or national legislatures. In other contexts, it is sometimes used to apply to municipal ordinances and to the rules and regulations of administrative agencies passed in the exercise of delegated legislative functions [32].

Methodology Is a theory of producing knowledge through research and provides a rationale for the way a researcher proceeds. It is the philosophical underpinning of a given research practice. Methodology is the discussion of the theory upon which research is based and indicates how the experimentation is performed [33].

Natural disasters Catastrophic events with atmospheric, geological, and hydrological origins (e.g., droughts, earthquakes, floods, hurricanes, landslides) that can cause fatalities, property damage, and social environmental disruption [34].

Policy A law, regulation, procedure, administrative action, incentive, or voluntary practice of governments and other institutions [35].

Renewable energy Energy that is derived from sources or processes that are constantly replenished. These sources of energy include solar energy, wind energy, geothermal energy, and hydroelectric power [18].

Shelterbelts Belts of trees, planted in a grid-like format pattern within areas of farmland, to act as windbreaks [36, 37].

Supply chain Defined as the entire process of making and selling commercial goods, including every stage from the supply of materials and the manufacture of the goods through to their distribution and sale [38].

Survey A research method used for collecting data from a predefined group of respondents to gain information and insights into various topics of interest [39].

Sustainability Development that meets the needs of the present without compromising the ability of future generations to meet their own needs [40].

Sustainable agriculture Ensuring the continuing availability to the consumer of adequate supplies of, wholesome, varied, and reasonably priced food, produced within accordance with generally accepted environmental and social standards. Maintaining a flexible and competitive industry which contributes to an economically viable rural society. Ensuring effective protection of the environment and prudent use of natural resources. Conserving and enhancing the landscape, wildlife, cultural, and archaeological value of agricultural land. Respecting a high level of animal welfare [41].

Systematic review Defined as a review of the evidence on a clearly formulated question that uses systematic and explicit methods to identify, select, and critically appraise relevant primary research, and to extract and analyze data from the studies that are included in the review. The methods used must be reproducible and transparent [42].

Trade regulations The removal or reduction of trade barriers in accordance with the former General Agreement on Tariffs and Trade and most recently the World Trade Organization presents opportunities for businesses. Likewise, multinational agreements that remove trade barriers to create largely duty-free and tariff-free trading zones allow for freer flow of goods and services between specific countries. These agreements create tremendous opportunities for businesses because they lower the costs associated with importing and exporting, which is a primary consideration for many companies [43].

Urban morphology Refers to the study of urban form that focuses on the formation and transformation of urban forms of cities, towns, and villages over time; their spatial patterns at different scales; and physical characteristics to inform appropriate urban interventions to promote sustainable urban development [44].

Vulnerability Within a migration context, it is the limited capacity to avoid, resist, cope with, or recover from harm. This limited capacity is the result of the unique interaction of individual, household, community, and structural characteristics and conditions [45].

Well-being The degree to which the needs and wants of a society are satisfied [9]. An individual's perception of their position in life in the context of the culture and value systems in which they live and in relation to their goals, expectations, standards, and concerns. It is a broad ranging concept affected in a complex way by the person's physical health, psychological state, personal beliefs, social relationships, and their relationship to salient features of their environment [46].

References

1. Environmental Pollution Centers (2021) What is air pollution. Environment Pollution Centers. https://www.environmentalpollutioncenters.org/air/. Accessed 24 Mar 2021
2. National Geographic (2021) Atmosphere. National Geography Society. https://www.nationalgeographic.org/encyclopedia/atmosphere/. Accessed 24 Mar 2021
3. UNDP (2015) Goal 13: climate action. In: United Nations sustainable development goals. https://www.sdfinance.undp.org/content/sdfinance/en/home/sdg/goal-13--climate-action.html. Accessed 24 Mar 2021
4. IPCC (2007) Climate change 2007: the physical science basis. https://www.ipcc.ch/report/ar4/wg1/. Accessed 30 Oct 2019
5. Cambridge Dictionary (2021) Commodity meaning. Cambridge English Dictionary. https://dictionary.cambridge.org/dictionary/english/commodity. Accessed 24 Mar 2021
6. Sammut-Bonnici T (2015) Complexity theory. Wiley Encyclopedia of Management. Wiley, Chichester, pp 1–2
7. World Bank (2001) What, why, and where. World Bank. http://www1.worldbank.org/publicsector/decentralization/what.htm. Accessed 30 Oct 2019
8. De Blij HJ, Murphy AB (2003) Human geography: culture, society, and space, 7th edn. Wiley, Hoboken, NJ
9. Norton W (1998) Human geography. Oxford University Press
10. UNDRR (2021) Disaster risk reduction & disaster risk management. PreventionWeb.net. https://www.preventionweb.net/disaster-risk/concepts/drr-drm/. Accessed 24 Mar 2021

11. TEEB (2019) Ecosystem services. The Economics of Ecosystems and Biodiversity. http://www.teebweb.org/resources/ecosystem-services/. Accessed 30 Oct 2019
12. US EPA O (2021) What is emissions trading? United States Environmental Protection Agency
13. US EPA O (2021) What is environmental education? In United States Environmental Protection Agency. https://www.epa.gov/education/what-environmental-education. Accessed 24 Mar 2021
14. Muralikrishna IV, Manickam V (2017) Introduction. In: Environmental management. Elsevier, pp 1–4
15. IFPRI (2021) Food security. International Food Policy Research Institute, United Nations' Communications World Food Security. https://www.ifpri.org/topic/food-security. Accessed 24 Mar 2021
16. Odeh L (2012) A comparative analysis of global north and global south economies. Int J Curr Res Humanit 12:67–82
17. E/The Environmental Magazine (2009) What are "green cities"? Phys.org. https://phys.org/news/2009-03-green-cities.html. Accessed 24 Mar 2021
18. TWI (2021) What is green energy? https://www.twi-global.com/technical-knowledge/faqs/what-is-green-energy. Accessed 24 Mar 2021
19. Directorate-General for Environment (2013) Building a green infrastructure for Europe. European Commission. https://op.europa.eu/en/publication-detail/-/publication/738d80bb-7d10-47bc-b131-ba8110e7c2d6/language-en. Accessed 30 Oct 2019
20. Aditya Birla Capital (2020) What is green tax?—definition and meaning. In: ABC Money. https://www.adityabirlacapital.com/abc-of-money/what-is-green-tax. Accessed 24 Mar 2021
21. Iberdrola (2021) What are the green and environmental taxes? https://www.iberdrola.com/environment/green-and-environmental-taxes. Accessed 24 Mar 2021
22. Vocabulary.com (2019) Habitation. Vocabulary.com. https://www.vocabulary.com/dictionary/habitation. Accessed 30 Oct 2019
23. Poston DL (2015) Human ecology. In: International encyclopedia of the social & behavioral sciences: Second Edition. Elsevier Inc., pp 283–288
24. Steiner F (2008) Human ecology: overview. In: Encyclopedia of ecology, Five-Volume Set. Elsevier Inc., pp 1898–1906
25. Josh J (2015) Human settlement. https://www.jagranjosh.com/general-knowledge/human-settlement-1448446636-1. Accessed 30 Oct 2019
26. OECD (1997) OECD glossary of statistical terms—informal settlements definition. Glossary of Environment Statistics, Studies in Methods, United Nations. https://stats.oecd.org/glossary/detail.asp?ID=1351. Accessed 24 Mar 2021
27. UNHCR (2021) Internally displaced people. United Nations High Communications Refugee. https://www.unhcr.org/en-us/internally-displaced-people.html. Accessed 28 Mar 2021
28. Afolabi AO, Oluwatayo A, Oyeyipo O et al (2018) Assessment of designers' perception of post conflict housing schemes for internally displaced persons. Constr Econ Build 18:27–47. https://doi.org/10.5130/AJCEB.v18i1.5780
29. University of North Carolina at Chapel Hill (2008) Introduction to knowledge management. www.unc.edu. Accessed 24 Mar 2021
30. Girard J, Girard J (2015) Online journal of applied knowledge management defining knowledge management: toward an applied compendium. Online J Appl Knowl Manag 3:1–20
31. Wu J (2008) Landscape ecology. In: Encyclopedia of ecology, Five-Volume Set. Elsevier Inc., pp 2103–2108
32. Britannica (2021) Legislation | Definition. British Dictionary. https://www.britannica.com/topic/legislation-politics. Accessed 24 Mar 2021
33. Zeegers M, Barron D, Zeegers M, Barron D (2015) Milestone 5: methodology. Milestone moments get your Ph.D. Qual Res 61–74. https://doi.org/10.1016/B978-0-08-100231-5.00005-5
34. Xu J, Wang Z, Shen F et al (2016) Natural disasters and social conflict: a systematic literature review. Int J Disaster Risk Reduct 17:38–48. https://doi.org/10.1016/j.ijdrr.2016.04.001

35. CDC (2021) Definition of policy. Centers for Disease Control and Prevention. https://www.cdc.gov/policy/analysis/process/definition.html. Accessed 24 Mar 2021
36. FAO (2007) Shelterbelts. Food and Agriculture Organization of the United Nations, Rome
37. Ritchie KA (1988) 24. Shelterbelt plantings in semi-arid areas. Agric Ecosyst Environ 22–23:425–440. https://doi.org/10.1016/0167-8809(88)90037-0
38. Supply Chain (2021) What is supply chain? A definitive guide. Supply chain digital. https://www.supplychaindigital.com/supply-chain-2/what-supply-chain-definitive-guide. Accessed 24 Mar 2021
39. QuestionPro (2020) What is a survey | Definition, characteristics, examples of a good survey. https://www.questionpro.com/blog/surveys/. Accessed 24 Mar 2021
40. Brundtland GH (1987) Our common future: the World Commission on environment and development. Oxford University Press, Oxford
41. DEFRA (2020) Sustainable agriculture. Department for Environment, Food & Rural Affairs. https://www.gov.uk/government/organisations/department-for-environment-food-rural-affairs. Accessed 24 Mar 2021
42. Centre for Reviews and Dissemination (2009) Systematic reviews: CRD's guidance for undertaking reviews in health care. York Publishing, York
43. Saylor Academy (2012) Trade regulations, contracts, and prohibited activities in international business. Legal Ethical Environmental Business. https://saylordotorg.github.io/text_the-legal-and-ethical-environment-of-business/s16-02-trade-regulations-contracts-an.html. Accessed 24 Mar 2021
44. Chen F (2014) Urban morphology and citizens' Life. Encyclopedia of quality of life and well-being research. Springer, Netherlands, pp 6850–6855
45. IOM (2019) Key migration terms. International Organization Migration. https://www.iom.int/key-migration-terms. Accessed 30 Oct 2019
46. WHO (2014) WHOQOL: measuring quality of life. World Health Organization. https://www.who.int/healthinfo/survey/whoqol-qualityoflife/en/. Accessed 30 Oct 2019

Index

G. T. Cirella (ed.), *Human Settlements*, Advances in 21st Century Human Settlements, https://doi.org/10.1007/978-981-16-4031-5

F

G

W

www.ingramcontent.com/pod-product-compliance
Ingram Content Group UK Ltd.
Pitfield, Milton Keynes, MK11 3LW, UK
UKHW021315130926
15127UKWH00002B/7
* 9 7 8 9 8 1 1 6 4 0 3 0 8 *